# KEY ENVIRONMENTS
# ANTARCTICA

*Edited by*

## W. N. BONNER and D. W. H. WALTON

*British Antarctic Survey, Cambridge, UK*

*Foreword by*

## HRH THE DUKE OF EDINBURGH

Published in collaboration with the

INTERNATIONAL UNION FOR CONSERVATION OF
NATURE AND NATURAL RESOURCES

by

## PERGAMON PRESS

OXFORD · NEW YORK · TORONTO · SYDNEY · PARIS · FRANKFURT

| U.K. | Pergamon Press Ltd., Headington Hill Hall, Oxford OX3 0BW, England. |
| U.S.A. | Pergamon Press Inc., Maxwell House, Fairview Park, Elmsford, New York 10523, U.S.A. |
| CANADA | Pergamon Press Canada Ltd., Suite 104, 150 Consumers Rd., Willowdale, Ontario M2J 1P9, Canada |
| AUSTRALIA | Pergamon Press (Aust.) Pty. Ltd., P.O. Box 544, Potts Point, N.S.W. 2011, Australia |
| FRANCE | Pergamon Press SARL, 24 rue des Ecoles, 75240 Paris, Cedex 05, France |
| FEDERAL REPUBLIC OF GERMANY | Pergamon Press GmbH, Hammerweg 6, D-6242 Kronberg-Taunus, Federal Republic of Germany |

First edition 1985

**Library of Congress Cataloging in Publication Data**
Main entry under title:
Key environments — Antarctica
(Key environments)
Includes index.
1. Ecology — Antarctic Regions. I. Bonner, W. Nigel (William Nigel) II. Walton, D. W. H. III. Title: Antarctica. IV. Series.
QH84.2.K49     1985     574.5′0998′9     84-10973

**British Library Cataloguing in Publication Data**

Antarctica — (Key environments)
1. Ecology — Antarctic Regions
I. Bonner, W. Nigel     II. Walton, D. W. H.
III. Series
574.5′0998′9     QH84.2

ISBN 0-08-028881-2

*Printed in Great Britain by A. Wheaton & Co. Ltd., Exeter*

The general problems of conservation are understood
by most people who take an intelligent interest in the state
of the natural environment.  But if adequate measures are to
be taken, there is an urgent need for the problems to be
spelled out in accurate detail.

    This series of volumes on "Key Environments" concentrates
attention on those areas of the world of nature that are under
the most severe threat of disturbance and destruction.  The
authors expose the stark reality of the situation without
rhetoric or prejudice.

    The value of this project is that it provides specialists,
as well as those who have an interest in the conservation of
nature as a whole, with the essential facts without which it is
quite impossible to develop any practical and effective
conservation action.

1984

# General Preface

The increasing rates of exploitation and pollution are producing unprecedented environmental changes in all parts of the world. In many cases it is not possible to predict the ultimate consequences of such changes, while in some, environmental destruction has already resulted in ecological disasters.

A major obstacle, which hinders the formulation of rational strategies of conservation and management, is the difficulty in obtaining reliable information. At the present time the results of scientific research in many threatened environments are scattered in various specialist journals, in the reports of expeditions and scientific commissions and in a variety of conference proceedings. It is, thus, frequently difficult even for professional biologists to locate important information. There is consequently an urgent need for scientifically accurate, concise and well-illustrated accounts of major environments which are now or soon will be, under threat. It is this need which these volumes attempt to meet.

The series is produced in collaboration with the International Union for the Conservation of Nature. It aims to identify environments of international ecological importance, to summarize the present knowledge of the flora and fauna, to relate this to recent environmental changes and to suggest where possible, effective management and conservation strategies for the future. The selected environments will be re-examined in subsequent editions to indicate the extent and characteristics of significant changes.

The volume editors and authors are all acknowledged experts who have contributed significantly to the knowledge of their particular environments.

The volumes are aimed at a wide readership, including: academic biologists, environmentalists, conservationists, professional ecologists, some geographers as well as graduate students and informed lay people.

John Treherne

# Contents

Contents

# CHAPTER 1

# History and Exploration in Antarctic Biology

## D. W. H. WALTON and W. N. BONNER

British Antarctic Survey, Natural Environment Research Council, High Cross, Madingley Road, Cambridge CB3 0ET, U.K.

## CONTENTS

## 1.1. EARLY DAYS

Biology, or perhaps more appropriately natural history, has a surprisingly early start in the records of Antarctic science. True, these first attempts were often piecemeal, or ancillary to the main occupation of the collector, and at times decidedly partisan in the selection of interesting organisms. Nevertheless, there is no doubting the enthusiasm of these eighteenth and nineteenth-century natural historians in a new and exciting ecosystem.

In that earlier age of sailing ships and grand national expeditions to explore the world, the biologist was often employed as the ship's surgeon, as, for instance, was J. D. Hooker on the James Clark Ross expedition. In a few cases professional scientists were employed specifically as natural historians. It was in this capacity that Captain Cook embarked the famous German naturalists J. R. and G. Forster.

Natural history, even in those early days, was recognized as a desirable pursuit on any official expedition, but only rarely were official funds available to pay for professional biologists. Their remit was wide. Not for them the limitations of arcane specialisms, but a consuming interest in the whole natural world, from ethnology to meteorology, geology to ornithology.

The outstanding scientists amongst these early Antarctic explorers made and documented collections of major significance in many fields of biology. J. D. Hooker, Titian Ramsay Peale and the Forsters were unusual in that they all analysed and wrote up substantial parts of their own collections. Later

Fig. 1.1. Portraits of some early Antarctic scientists.
a. James Weddell (Scott Polar Research Institute).

expeditions generally turned to specialists for the analysis of their biological collections and in many of these cases the name of the expedition's biologist is now largely forgotten.

Considering the working conditions on the small wooden sailing ships and the limited storage facilities, it is truly amazing not only that so much material returned to the safety of museums but that much of it was still in a usable condition. Taxonomically this material is of great importance, since the early naturalists provided the first descriptions and type specimens for many of the most significant Antarctic species in all groups.

The achievements of these early naturalists varied considerably. For the Forsters Antarctic collecting was a fairly short episode in a comparatively impoverished area during a circumnavigation of the globe. Titian Ramsay Peale must have had a rather similar experience on the Wilkes Expedition. The Antarctic material from both these expeditions was overshadowed scientifically by the collections made in South America, the Pacific, and Australasia. Who can blame them if they placed less significance on the depauperate floras of the subantarctic islands than on the dazzling riches of Australia?

Yet not all were blind to the significance of the Antarctic material. J. D. Hooker was able both to collect in the field and classify all his own collections later in England. His careful analysis allowed him to recognize the relationships between the South American and Australasian floras and to connect both the Antarctic and its islands with fundamental patterns in the Southern Hemisphere biogeography. Hooker's *Flora Antarctica* (1847) stands as a major milestone in plant biogeography and illustrates clearly some of the distribution patterns that later tectonic theory has only recently begun to explain.

Fig. 1.1. Portraits of some early Antarctic scientists.
b. Joseph Dalton Hooker (Scott Polar Research Institute).

An earlier discovery, unknown to Hooker, would have added considerable weight to his biogeographical theory. James Eights, who visited the Antarctic Peninsula (*c.* 1829-31) with Captains Pendleton and Palmer, was the first scientist to discover fossil wood in the Antarctic. His reports on this and on the natural history of the South Shetland Islands (Eights, 1833) apparently remained unknown to Antarctic scientists until the twentieth century. He collected thirteen cases of natural history specimens which included rocks, lichens and marine animals, and his discovery of the ten-legged sea spider *Decolopoda australis*, published in 1837, preceded that of its next collector, W. S. Bruce, by 66 years. Being alone in the field of Antarctic research and publishing his findings just at the time when Congress

Fig. 1.1. Portraits of some early Antarctic scientists.
c. William S. Bruce (Scott Polar Research Institute).

was agreeing to send Wilkes off to circumnavigate the world obviously did not help Eights find fame. Yet it is difficult to understand how his important collections and published papers remained unknown for so long. That fate was not one to trouble the naturalists accompanying Charles Wilkes on his explorations. As befitted a nationally sponsored expedition, the U.S. Exploring Expedition had many natural historians aboard — J. D. Dana as geologist, W. Rich as botanist, W. D. Brackenridge as horticulturist, J. P. Couthony as conchologist, H. E. Hale as philologist, together with two naturalists, Dr. C. Pickering and T. R. Peale, and two artists. Of all this assembled talent only Peale accompanied the Antarctic cruises. He concentrated on collecting birds and some marine specimens and writing his informative journal, but the later loss of many of his specimens in a shipwreck and the problems surrounding the publishing of the expedition results effectively obscured any important contribution to Antarctic science.

Another naturalist who made a particular contribution in this early period was Gaudichaud (1826) who reported on marine algae, whilst both Dumont d'Urville's expedition and Bouganville's expedition made collections of the Antarctic marine invertebrate fauna. The only Antarctic fishes from this period were those collected from subantarctic Kerguelen by Ross's expedition and described by Richardson and Gray (1844 — 5), and those of the earlier Wilkes Expedition from near Wilkes Land, the Antarctic Peninsula and the South Shetland Islands. Unfortunately the manuscript of the latter, prepared by Louis Agassiz, was never completed and was finally lost along with many of the illustrations. This important collection of material was allowed to fall into a disastrous condition and when finally examined by Henry Fowler in 1920 all the Antarctic material had been lost — one of the many scientific catastrophes which marked the Wilkes Expedition.

One other early major expedition to Antarctic waters remains to be examined, that of the *Challenger* led by Charles Wyville Thomson, Professor of Natural History at the University of Edinburgh. In

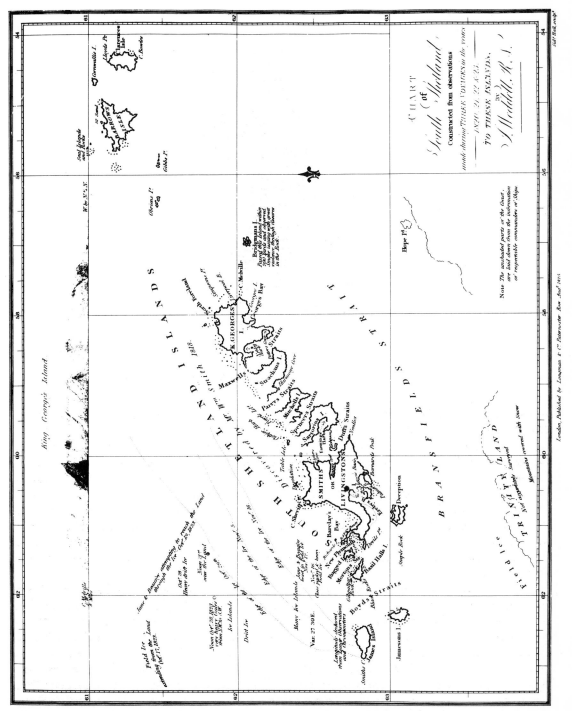

Fig. 1.2. Early map of the South Shetland Islands by James Weddell (1825).

1870 – 1 he persuaded the Royal Society to ask the British Government to lend him a ship to explore the depths of the world's oceans in a way never previously attempted. The cruise eventually covered nearly

69,000 nautical miles with 362 observing stations. *Challenger*, a three-masted square rigged wooden ship of around 2300 tons, set sail in 1872 on a voyage lasting three and a half years. It resulted, as Wyville Thomson was justly able to claim, in 'the first general survey of deep sea collections undertaken with a knowledge of the circumstances under which the specimens were procured' (Wyville Thomson, 1877).

The Antarctic leg of the voyage was between Cape Town and Melbourne. Marion Island, Heard Island and Iles Kerguelen were all visited and collections were made ashore. Iles Crozet were seen but the weather was too poor for a party to land. Throughout this part of the voyage trawling and dredging were carried out, resulting in the capture of innumerable new species and genera of marine animals unknown to science. Beyond Kerguelen the ship's track took her to 65°S, along the edge of the pack ice.

The contributions to Antarctic science by the *Challenger* scientists were many and varied. The detailed oceanographic data, the terrestrial collections from the subantarctic islands and above all the marine organisms and bottom sediment samples provided much of the fundamental framework used by later expeditions. The fifty large volumes comprising the complete expedition reports stand as a monument to the original inspiration of Wyville Thomson, the enthusiasm of the scientific team and the industry of John Murray — a major achievement in the history of biology.

## 1.2. SCIENCE AND NATIONALISM

Towards the end of the nineteenth century new forces began to appear on the Antarctic stage. The report by Ross of right whales in the Antarctic was sufficient to instigate exploratory whaling voyages to Antarctic waters in 1892 by Scottish whalers and in 1893 by Norwegians and Germans. No right whales were found but these voyages provided the impetus for further important developments.

Fig. 1.3. Killing Elephant Seals at Kerguelen in the early nineteenth century (from Goodridge, 1851).

Fig. 1.4. Naturalists workroom aboard H.M.S. Challenger (from W.J.J. Spry, 1876).

The voyage in 1893 was led by Captain C. A. Larsen whose observations on the numbers of whales present led in 1904 to the establishment of the Antarctic whaling industry, initially at Grytviken on South Georgia. Thus commerce had returned to the Antarctic and with it a major change in international interest in the region (Tønnesson and Johnson, 1982).

On the earlier voyage in 1892 W. S. Bruce had travelled as the naturalist. So impressed was he by the Continent and its islands that he determined to return with a scientific expedition — the Scottish National Antarctic Expedition. This latter expedition was one of several, born out of a ferment of scientific discussion and support for Antarctic investigations that developed in Europe in the late nineteenth century. In Germany Baron von Humbolt, Professor Gauss and Dr. von Neumayer had all urged the case for magnetic research in high southern latitudes, the Belgian Government gave grudging support to Lieutenant de Gerlache's expedition and in England first Sir John Murray and then Sir Clement Markham provided powerful arguments in favour of British involvement. In a lecture to the Royal Geographical Society in 1893 Murray, after reviewing the Antarctic results of the *Challenger* cruise, put forward forceful scientific reasons for landbased Antarctic expeditions. He demonstrated how the existence of an Antarctic continent could be deduced from the rocks dredged up by *Challenger* but then pointed out that proof of this had yet to be obtained by land survey.

Markham, enthused with the romantic idea of heroic exploration, had seen in the Antarctic new lands still to be conquered by British explorers. He set about recruiting assistance with all his formidable energy and political skill. Despite getting the International Geographical Congress in 1895 to pass a unanimous resolution recommending new Antarctic endeavours, Markham had no success in persuading the British Government to part with money. In 1897 he persuaded the Royal Geographical Society Council to put up £5000 in support of a public appeal for the purpose but all to no avail. Instead, with the support of the magazine proprietor Sir George Newnes and despite Markham's determined opposition, the Norwegian C. E. Borchgrevink sailed south in the *Southern Cross* leading the first expedition to winter on land in the Antarctic. With W. Colbeck as magnetician and surveyor, Louis Bernacchi as meteorologist and Nicolai Hanson as biologist the expedition was well staffed scientifically.

Unfortunately Hanson died whilst on the expedition and all that remains of his work are his detailed observations recorded in a daily diary, and a variety of specimens which were worked on by other scientists. Many of his seal specimens were useless as they were unlabelled. About 200 fish specimens yielded sixteen species of which eight were new. Many of the marine invertebrates were poorly preserved but it still proved possible to name fifteen polychaete worms (three new species), a few species of sponges, hydrozoans, anthozoans, tunicates, molluscs and echinoderms. Dr. Klovstad, the surgeon, collected the first springtails from the Continent and the second species of mite. Very few botanical specimens were obtained. Many of the rock and bird specimens lacked labels and field notes, making them of little value. How much more might have been achieved if Hanson had lived!

Several other major expeditions can be considered as belonging to this period and having their roots in the Geographical Congress resolutions. The Swedish South Polar Expedition, led by Otto Nordenskjöld and supported entirely by private funds, arose out of Nordenskjöld's earlier work on the geology of Tierra del Fuego. Despite the loss of their ship the *Antarctic*, and with it many specimens, the expedition was a great success. Scientifically this expedition was also well staffed. Nordenskjöld himself took charge of the geology, whilst Carl Skottsberg was the botanist, K. A. Anderson was in charge of zoology, and Erick Ekelöf was both doctor and bacteriologist. Two further scientists, S. A. Duse and G. Bodman, looked after the mapping and hydrographical work. The scientific results of the expedition were considerable. Skottsberg prepared the first detailed descriptive accounts of the floras of the Falkland Islands and Tierra del Fuego, and his account of the South Georgian flora greatly supplemented that published by Will (German Transit of Venus expedition 1882 — 1883) in 1890. The geological work on Snow Hill Island was of considerable importance. Ekelöf's work, along with that of Pirie, with the *Scotia* expedition in the South Orkney Islands, and Gazert, with the *Gauss* expedition, marked the start of scientific microbiology in the Antarctic.

The Scottish expedition to the South Orkney Islands did have, in some ways, a nationalistic flavour. One of Bruce's aims in organizing the Scottish National Antarctic Expedition was to show that a small country was still capable of making a significant scientific contribution and in this he succeeded admirably. Wintering on Laurie Island, South Orkney Islands, and with summer cruises on *Scotia* as far south as 72°S in the Weddell Sea his able team of biologists collected large numbers of specimens. Their work considerably extended the collections made by T. V. Hodgson on *Discovery* in the Ross Sea and constituted the major Antarctic marine investigations of that period. *Scotia* trawled, with a variety of nets, 18 times in water over 1000 fathoms and four times in water deeper than 2500 fathoms south of 60°S. One hundred and fifty dredge specimens were obtained as well as large numbers of fish from traps and baited lines. Special attention was paid to plankton sampling with vertical nets being taken as far south as 71° 50'S.

On land R. N. R. Brown collected lichens, mosses and freshwater algae in the South Orkneys together with flowering plants from Gough Island and Ascension Island on the voyage down. J. H. H. Pirie carried out pioneer work on the microbiology of the intestines of Antarctic animals, as well as examining sea water and sediments for bacteria. Collections of tardigrades and nematodes together with springtails greatly increased the knowledge of the land fauna, whilst the marine invertebrate collections were both diverse and novel. In almost every group new species were found — for example hydroids (seven new species and a new genus), sea-spiders (five new species), brachiopods (three new species), holothurians (ten new species), etc.

The fish collection resulted in seven new genera and twenty-one new species. Bird collections and observations were made throughout the voyage by Brown and Bruce together with R. Davidson, the second mate. A complete series of penguin embryos was obtained as well as egg and plumage specimens from many other Antarctic birds. Seal work concentrated particularly on the morphology and anatomy of the Weddell seal, adding considerably to the limited data then available from the few museum specimens.

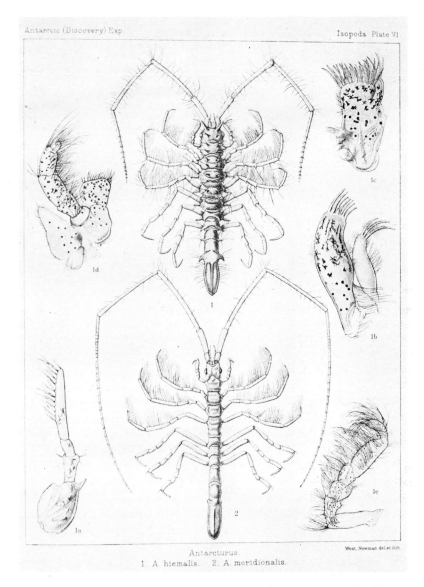

Fig. 1.5. Drawings of marine isopods collected on the British National Antarctic Expedition.

## 1.3. THE QUEST FOR THE POLE

The early part of the twentieth century has often been described, perhaps aptly, as the Heroic Age of Antarctic exploration. Within this period four British and one Norwegian expeditions strove to attain the all-important pinnacle of southern geographical exploration — the South Pole. All the expeditions were nationalist in concept and all were led and planned by non-scientists.

Although Scott's *Discovery* expedition was originally conceived as being principally scientific, its major objectives were changed after a violent dispute between Professor J. W. Gregory and Sir Clement

Markham. Gregory, originally appointed as leader of the scientific staff and of the land parties, resigned when Markham insisted that Scott should remain in command of all the parties. Despite this the scientists recruited were able men, adequately trained and equipped and their discoveries were duly published in a splendid set of official expedition reports. H. T. Ferrar's geological account of the McMurdo Sound area was a major scientific achievement in Antarctic geology, whilst the bird studies by Edward Wilson set new standards in Antarctic ornithology. Regneld Koettlitz as doctor-*cum*-botanist and Thomas Hodgson as the marine biologist completed the scientific team.

Scott was indeed lucky to again attract a powerful scientific team for his second expedition in *Terra Nova*. Wilson rejoined him as senior zoologist with Apsley Cherry-Garrard as his assistant, whilst Raymond Priestley, Frank Debenham and Griffith Taylor made a formidable group of geologists. Edward Nelson as biologist with Murray Levick and Edward Atkinson as doctors completed the natural history team on shore. Nelson concentrated on sampling both the freshwater lakes and the marine life below the sea ice. He collected sea water samples for later analysis and compiled detailed tide records. Levick, Wilson and Cherry-Garrard concentrated on the birds and seals. Atkinson concerned himself with bacteriology and the parasitology of the penguins, seals and fish. On board the ship D. G. Lillie was collecting dredge samples, and plankton from both tow nets and vertical hauls. Good collections of

Fig. 1.6. Sketch of adult skuas at a nest on Hut Point, Ross Island by Edward Wilson (Scott Polar Research Institute).

sea anemones, worms, urchins, starfish, crustacea, sea-spiders, molluscs and fishes were also obtained. Meanwhile the geologists were examining the Royal Society Range, the Dry Valleys, Mount Erebus and the area around Granite Harbour as well as the coast west of Cape Adare. The plant fossils collected by Wilson and Bowers from the Beardmore Glacier and found with their bodies, were of particular significance. Although the expedition failed to reach the Pole first and ended in tragedy its scientific achievements were more than sufficient to justify it. Amundsen won the race but his efforts provided virtually nothing in terms of scientific information.

Much of this geological investigation was planned on the then unpublished data collected by Edgeworth David, Raymond Priestley and Douglas Mawson, the geologists with Shackleton's 1907 — 9 expedition. They had observed an eruption on Mount Erebus, discovered an organic deposit in one of the lakes, reported on mirabilite deposits around old lakes and made considerable observations on physiography, geomorphology and petrology. James Murray, the biologist, concentrated on the behaviour of the Adélie penguin but managed also to collect rotifers and tardigrades from the pools, and marine organisms by dredging.

The aim of the first Australian Antarctic expedition (1911 — 14) led by Douglas Mawson was most definitely scientific investigation and geographical discovery of the sort originally advocated by Sir John Murray back in 1893. By splitting the land party into three groups, including one on Macquarie Island, his expedition achieved far more than the previous single base expeditions. On the Continent J. G. Hunter, C. F. Laseron and A. C. McLean looked after the biology, with F. C. Stillwell, C. A. Hoadley, A. D. Watson and Mawson himself as geologists. On Macquarie Island L. R. Black and H. Hamilton were geologist and biologist respectively. In organization and scientific achievements this expedition set new standards.

Mawson returned again in 1929 — 31 with the British — Australian — New Zealand Antarctic Research Expedition to further extend scientific knowledge by exploring the last major stretch of unknown Antarctic coast. With J. W. S. Marr as oceanographer and plankton specialist, R. A. Falla as ornithologist and three other biologists (T. H. Johnston, W. W. Ingram, H. O. Fletcher) the biological output of this expedition was considerable.

It was during this period that the United States began to take a fresh interest in Antarctica. The years 1928 — 30 marked the start of Admiral Richard Byrd's lifetime interest in the scientific exploration of the Continent and his expeditions were remarkable both for their complexity and their costs. Although he was principally interested in geographical exploration Byrd also actively promoted the more academic scientific disciplines.

The first expedition achieved relatively little scientifically, although useful geological observations were made in the Queen Maud Mountains and some zoological specimens were collected. The second expedition, 1933 — 35, was much more productive. This time there were four over-wintering biologists (E. B. Perkins, P. A. Siple, A. A. Lindsay and J. M. Sterrett), and two geologists (F. A. Wade and D. Stewart). A geological reconnaissance was made of the Rockefeller Mountains. Life history studies were carried out on Weddell and crabeater seals, bird distribution data were collected, mosses and lichens were collected from 215 localities, plankton hauls were made and the microbiology of snow, mud and water was investigated.

Between the World Wars several other expeditions, mostly funded privately, made a variety of worthwhile scientific advances. The Norwegian, Lars Christensen, supported some work from his whaling factory fleets. France began work on its subantarctic islands, whilst the British Grahamland Expedition (BGLE) led by John Rymill, investigated the area around Alexander Island. BGLE was probably the most scientifically productive of all of these smaller expeditions, with a comprehensive programme in geology, glaciology, meteorology and biology (Fleming *et al.*, 1948). W. L. S. Fleming, the geologist, collected specimens mainly from coastal outcrops but managed to provide a very sound survey of a complex and mainly unknown area.

Fig. 1.7. Naturalist Apsley Cherry-Garrard skinning an Emperor Penguin during R.F. Scott's "Terra Nova" expedition (Scott Polar Research Institute).

The biologist G. C. L. Bertram found the area around the Argentine Islands to be exceptionally floristically rich and made extensive collections of plants and invertebrates in the moss mat. Copepods and crustacea were found in freshwater ponds. Bertram devoted most of his attention, however, to seals.

B. B. Roberts, the expedition ornithologist, had a very wide-ranging programme. He concentrated mainly on the behaviour and ecology of a broad spectrum of birds, from penguins to petrels, as well as collecting specimens for improving their classification and descriptions. He mapped penguin rookeries, investigated the relationships between their behaviour and sexual development and produced the first comprehensive bibliography for Antarctic ornithology.

This rapid survey of the biological contributions of the early Antarctic expeditions is by no means complete. Little or no mention has been made of the science carried out by the German expeditions led by Erich von Drygalski (1901 – 3) and Wilhelm Filchner (1911 – 12) nor of the two expeditions to the

Antarctic Peninsula by Jean Charcot (1903 – 5 and 1908 – 10). The German Deep Sea Expedition aboard the *Valdivia* (1898 – 9) investigated Antarctic waters and visited Iles Kerguelen, adding to the information collected by the German Transit of Venus Expedition to the islands in 1874 – 6.

Despite Ernest Shackleton's death at South Georgia *en route* for the Antarctic a third time, the scientists on board the *Quest* collected useful material during the expedition although much of it was

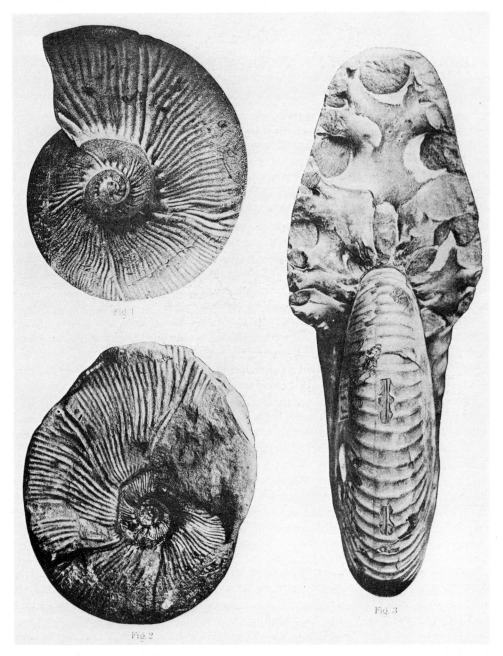

Fig. 1.8. Fossils (*Maorites* and *Pachydiscus*) collected on Seymour Island during the Swedish South Polar Expedition (from Kilian and Reboul, 1909).

from the South Atlantic islands rather than from the Antarctic. Assistance from whalers allowed Thomas Bagshawe to carry out his detailed study of Adélie penguins at Waterboat Point, and it was also passage on a whaling ship that allowed Ola Olstad and Olaf Holtedahl to visit South Georgia, Port Lockroy and the South Shetland Islands in 1927 − 8. The use of whaling transports became an important feature of many of the smaller Antarctic expeditions between 1920 and 1950, an interesting alliance between science and commerce.

## 1.4. EXPLOITATION IN THE ANTARCTIC AND THE GROWTH OF KNOWLEDGE

The many voyages made by the sealers round the fringes of the Antarctic as they searched for fur and elephant seals in the course of the nineteenth century could have afforded great opportunities for biological discovery. With few exceptions, however, these opportunities were not taken. In general, it was not in the interests of the sealers to record the places they visited and what they found there. The discovery of a new group of rocks was of little interest unless there were seals on them, and if there were, then this was kept as secret as possible until the harvest had been reaped. Perhaps in any case rather few of the sealing captains were capable of writing more than the briefest log book entry.

One of those who was capable of fluent writing and genuinely scientific observation was the Scottish sealing captain James Weddell. His account *A Voyage Towards the South Pole Performed in the Years 1822 − 1824* (Weddell, 1825) contains much of interest to the natural historian. In 1823 Weddell sailed through open waters to reach latitude 74°15′S in the sea that now bears his name, a record that was to stand for the rest of that century and well into the present one. Weddell described many novel forms of life, and a seal new to science and brought back by him was named in his honour by Lesson in 1826. Though *Leptonychotes weddellii* is now universally known as the Weddell seal it was known to its discoverer as the leopard or leopardine seal (on account of its spotting), a fact which has caused some confusion with the species currently called the leopard seal, *Hydrurga leptonyx*.

James Eights, an American naturalist who sailed with the New England sealers Palmer and Pendleton, and who gave his name to a number of marine invertebrates from the Antarctic, has already been mentioned.

Another sealer who left voluminous accounts of his travels was Benjamin Morrell, but his writings are a good deal less factual than Weddell's. Some of the events described by Morrell and the animals he claimed to have seen owed more to his imagination than to his observation.

An outstanding contribution to science was made right at the end of the era of the old-time sealers and whalemen when Robert Cushman Murphy signed on as 'assistant navigator' on the whaling brig *Daisy* of New Bedford. Murphy, travelling on behalf of the American Museum of Natural History, used the brig as a base for a study of the oceanic sea birds of the South Atlantic in 1912 − 13, which was eventually to form the basis for a major work on the subject (Murphy, 1936). Murphy made many other observations on natural history and the contemporary scene around him which are entertainingly recounted in *Logbook for Grace* (Murphy, 1948).

The second phase of exploitation that followed the sealing was the Antarctic steam whaling industry. This started in South Georgia in 1904 and in the South Shetland Islands in the following year. Whaling was to have a much greater role than sealing in adding to our knowledge of the structure and functioning of the Antarctic marine ecosystem. Unlike the sealers, who were essentially itinerant opportunists without the need or the opportunity to establish permanent bases, the whalers in the early years of this century were obliged by technological limitations to operate from established harbours. The small steam-powered catching boats brought their whales back to a factory for processing. This could either be built ashore, as at South Georgia, or be brought down intact in an old steamer converted to a 'floating factory', as at the South Shetland Islands. In either case a sheltered anchorage and an abundant supply of fresh water were needed.

At the beginning of the twentieth century the great majority of whaling took place in the Scotia Sea and around the Antarctic Peninsula. The land in this area constituted the Falkland Islands Dependencies and operations there were controlled by the British Colonial Office, acting through the administration in the Falkland Islands. The Falklands Islands government had instituted a licensing system for the whalers, mainly as a means of raising revenue, but also as a means of controlling the industry and substantiating British claims to the Dependencies.

Whaling in the Falkland Islands Dependencies developed very quickly. By the second decade of the century production from the region was greater than that from all the rest of the world. However, there was already an awareness that the stocks of whales were not inexhaustible. The earlier collapse of the Greenland whaling industry was evidence enough of that, whilst nearer to hand there were clear indications that the stocks of humpback whales in the Antarctic had already been greatly diminished.

In 1917 the British Colonial Office set up an interdepartmental committee on research and development in the Dependencies of the Falkland Islands. The main concern of the committee was the whaling industry and it showed a very enlightened attitude to the investigation of the biological background to this. The committee examined what was known of the breeding and migration of whales and their food in Antarctic waters. It considered also the role of seals, penguins, fish and other animals, though it is fair to say that its interest in these was mainly exploitative.

The interdepartmental committee reported in 1920 (Colonial Office, 1920) and amongst its recommendations was a proposal to set up an intensive scientific research programme which would provide data for the rational management of the whaling industry and additionally investigate other aspects of the biology of the region. It stressed that a complete hydrographical survey of the area was needed. The whole operation was to be funded by an increase in the duty paid by the various whaling companies operating in the Falkland Islands Dependencies.

This was the origin of the *Discovery* Investigations, which were active in the Antarctic from 1925 up until the onset of the Second World War. For a time it seemed possible that the whaling industry might be reformed by the adoption of management practices soundly based on a proper knowledge of the biology of the whales. This hope was dashed by the development of the stern-slip factory ship (Chapter 5b) which enabled the whalers to conduct all their business on the high seas wholly outside national jurisdiction. The stern-slip factory ship was developed as a means of avoiding the payment of duty and also, and perhaps more significantly, the limitations on catching effort imposed by the Falkland Islands government. It was ironic that the first of these vessels, the Norwegian factory ship *Lancing* first came to the Antarctic in the same season that the *Discovery* Investigations started work.

*Discovery* Investigations took its name from Captain Scott's old ship from his 1901 – 3 expedition, which had been bought back from the Hudson Bay Company. She was refitted as an oceanographic research vessel, though of a very different kind from those we know today. She arrived at Grytviken in South Georgia on 20 February 1925, where a laboratory had already been built, and immediately started work. Later she was joined by a smaller vessel, the *William Scoresby*, purpose-built on the lines of a whale catcher to undertake whale marking.

The director of research and expedition leader was Dr. Stanley Kemp, who was succeeded in 1936 by Dr. N. A. Mackintosh. A fascinating account of the work of *Discovery* has been given by Sir Alistair Hardy in his book *Great Waters* (1967). Hardy was the first of the expedition's scientific staff. His lucid prose and fine drawings and water-colour sketches provide a vivid picture of the life of the expedition and the contribution it made to science.

By any standards *Discovery* Investigations made great progress in oceanography and must be rated one of the greatest oceanographical expeditions. The last Antarctic cruise (in a new ship, *Discovery II*, which replaced the original in 1929) was in 1950 — 1 and the identity of the original organization was finally lost in the National Institute of Oceanography (now the Institute of Oceanographical Sciences) yet the impressive series of *Discovery* Reports continues to be published; they are now in their thirty-seventh volume.

Most of the papers in these volumes are monographic treatments of species or natural groups. James Marr's great volume on the natural history and geography of Antarctic krill (Marr, 1962) is one of the more recent contributions. As important in a real understanding of the animals and plants in the sea are the detailed accounts of the physical and chemical oceanography of the Antarctic. Sir George Deacon's account (1937) of the hydrology of the Southern Ocean, in which that highly significant feature, the Antarctic Convergence, was first described and explained, was an outstanding example in this vein.

## 1.5. TERRITORIAL CLAIMS AND THE INTERNATIONAL GEOPHYSICAL YEAR

Activity in the Antarctic between the wars was not exclusively scientific, even when the intense commercial activity of the whaling industry has been taken into account. Following Great Britain's claim to the Falkland Islands Dependencies made in 1917, other nations made formal territorial claims; New Zealand in 1923, France in 1924, Argentina in 1925, Australia in 1931 and Norway in 1939. These claims were based on various grounds, most (but not all) of which included an element of discovery. The United States of America, through the activities of nineteenth century explorers and more recently through Richard Byrd's expeditions of 1928 — 30 and 1933 — 5 and Lincoln Ellsworth's trans-Antarctic flight in 1925, had made substantial discoveries in Antarctica and several of their explorers had made territorial claims on behalf of their government. These, however, were never taken up, though the U.S. Government has retained the right to do so (Auburn, 1982).

Formal territorial claims created potential political difficulties. These were most acute in the Antarctic Peninsula area where the claims of Great Britain and Argentine (and after 1950, Chile also) overlapped substantially. Partly as a response to Argentinian claims in what was regarded by Great Britain as the Falkland Islands Dependencies, the British Government in 1943 sent a naval party under the code-name Operation Tabarin to the Antarctic. Early in 1944 bases were established at Deception Island in the South Shetland Islands and at Port Lockroy on Wiencke Island at the eastern end of the Bismark Strait. When the war ended in 1945 Operation Tabarin was reconstituted as the Falkland Islands Dependencies Survey (FIDS) with a permanent programme of scientific research and a further seventeen research stations were established in the Antarctic. In 1962 FIDS was renamed the British Antarctic Survey (BAS) which continues today as the United Kingdom's main contribution to Antarctic research.

In 1947 both Argentina and Chile established permanently occupied meteorological stations in the Peninsula region. Argentina, it may be noted, had operated a meteorological station inherited from the *Scotia* expedition at Laurie Island since 1904, the longest permanently-manned research station in the Antarctic by far. Also in 1947 the Australian Government set up the Australian National Antarctic Expeditions (ANARE) to maintain a series of scientific stations in the Australian claimed sector of the Antarctic. France, after sending a number of expeditions south, established a permanent station in Terre Adélie in 1950. Admiral Byrd brought the American presence back to the Antarctic in 1946 — 7 when he returned to launch Operation High Jump. This largely political and military expedition employed 13 ships and 4000 men and was the biggest venture of its kind ever seen in the Antarctic (King, 1969).

The next main development in Antarctica, and one which was to draw together all the nations engaged in Antarctic research, was one associated with geophysics. Many geophysical phenomena are best studied at high latitudes and this had led to co-operative 'Polar Years' in 1882 — 3 and 1932 — 3. In

1950 discussions took place between influential geophysicists which resulted in 1957 — 8 being chosen for a third Polar Year as it was to be a period of maximum sunspot activity. Under the auspices of the International Council of Scientific Unions (ICSU) the project was formalized as the International Geophysical Year (IGY).

Despite the possible, and indeed probable, political clashes the various nations involved (some of which had no official contact other than that involved with planning the IGY) succeeded in working out a programme which avoided duplication of effort and used the resources of the countries involved to the best advantage. The U.S.A. proposed to establish a station at the South Pole itself, while the U.S.S.R. planned stations at the South Geomagnetic Pole and the Pole of Inaccessibility.

The American Operation Deep Freeze I in 1955 — 6 established two stations, one on Ross Island at McMurdo Sound, the other near the site of Byrd's Little America. In December 1955 the first of what was to become a continuing series of large cargo planes operated by the U.S. Navy arrived at McMurdo from Christchurch, New Zealand, and established a vital speedy logistic link between Antarctica and the rest of the world. The following season Operation Deep Freeze II established Byrd Station in Marie Byrd Land and Amundsen-Scott station at the Pole. When Admiral Dufek, in charge of Deep Freeze, stepped out of his aircraft at the Pole on 31 October 1956 he was the first man to set foot there since Scott's party in 1912.

The Russians, lacking the facility of cargo-carrying planes, had a more difficult task, but from Mirny station on the Knox Coast in the Australian sector, the Geomagnetic Pole station, Vostok, was set up in the 1957 — 8 season. The Pole of Inaccessibility was reached the following year from the meteorological station Sovetskaya.


## 1.6. SCAR AND THE ANTARCTIC TREATY


During the IGY there were 47 research stations operating in the Antarctic south of latitude 60°S and a further eight stations on islands north of this. Nations co-operated freely and cordial relations existed between scientists in the field. To consolidate the advances made in international science during the IGY, ICSU established in September 1957 the Special (later, the Scientific) Committee on Antarctic Research (SCAR). This was charged with the task of bringing together scientists from the various countries active in research in the Antarctic so as to promote the development of Antarctic science and provide a means of co-ordinating activities in different disciplines.

SCAR is made up of a member for each of the countries with a major research commitment in the Antarctic, together with members for those countries which were active at the time of the formation of SCAR, but are no longer so today. Besides these 15 countries,* there are representatives of certain member unions of ICSU, such as the International Council of Geodesy and Geophysics, and the International Union of Biological Sciences. The World Meteorological Organization is also a member.

SCAR has been highly successful in its task and Antarctic science continues to flourish. One SCAR initiative of particular interest is the BIOMASS programme, which will be referred to later (Chapter 4a).

The achievements in scientific co-operation and the diplomatic advantages accruing from this during the IGY prompted the U.S.A. in 1958 to urge representatives of the nations concerned to confer together 'to seek an effective joint means of keeping Antarctica open to all nations to conduct scientific

*Argentina, Australia, Belgium, Chile, France, German Democratic Republic, Germany (Federal Republic), Japan, New Zealand, Norway, Poland, South Africa, United Kingdom, U.S.A., U.S.S.R.

Fig. 1.9. Marine zoologist diving for winter specimens under sea ice at Signy Island, South Orkney Islands (M.G. White/British Antarctic Survey).

or peaceful activities there'. The conference resulting from this drew up the Antarctic Treaty which was signed in December 1959 and came into force in 1961.

The Antarctic Treaty ensured that, by freezing all territorial claims south of latitude 60°S, there would be freedom of scientific investigation and co-operation. Other important provisions of the Treaty concerned the exchange of scientific information, the opening of all research stations and equipment to international inspection, and the prohibition of military operations (except in support of the science). Nuclear testing and the disposal of radioactive waste were also prohibited under the Treaty. From the treaty arose the Agreed Measures for the Conservation of Antarctic Fauna and Flora, and associated legislation on the conservation of Antarctic seals and Antarctic marine living resources. These are considered in more detail in Chapter 8.

The modern phase of research in the Antarctic has shifted from what could be done in the course of one or two seasons' work from an isolated expedition, to longer term programmes supported by centrally organized institutions with sophisticated logistical support. Some nations, such as the U.S.A. and New Zealand, develop their Antarctic programmes from a series of projects put forward by universities. Others, such as the U.K. and the U.S.S.R., have a research institute devoted solely to organizing and carrying out Antarctic research. The regular biennial meetings of SCAR help to provide a forum for discussion and an opportunity for co-ordination.

Environmentally, the most important joint programme of SCAR has been the BIOMASS (Biological Investigations of Marine Antarctic Systems and Stocks) programme. This study was initiated in 1972 when SCAR charged a sub-committee of its Working Group on Biology to examine the current state of knowledge of the Antarctic marine ecosystem. The report of this group drew attention to the key role played by Antarctic krill, *Euphausia superba*, in the system and to our insufficient knowledge of the biology of krill. Collaboration with another ICSU body, the Scientific Committee on Oceanic Research

(SCOR), resulted in a programme designed to gain a deeper understanding of the structure and dynamic functioning of the Antarctic marine ecosystem as a basis for the future management of potential living resources, of which krill would certainly be a major one.

Such a programme needed to be international in character since it impinged directly on multi-national economic interests (Smith, 1978). The information needed could only come through a multi-ship operation and SCAR provided the means for organizing this. In the austral summer of 1980-1 the First International BIOMASS Experiment (FIBEX) was held and proved to be the largest ever biological oceanographic operation of its kind.

BIOMASS continues to add to our knowledge of the Antarctic environment and may be a powerful influence in ensuring that the commercial exploitation of krill does not have the same devastating consequences as did the exploitation of whales half a century earlier (Chapter 5b).

Today the era of privately financed expeditions has ended. Many governments have a strong interest in the Antarctic, but their programmes demand specialists rather than naturalists. Indeed, in this respect the change had come earlier, growing out of the experiences of the Scott and Shackleton expeditions. Casual observation and collecting across the whole field of biology has gradually given way to carefully planned studies on behaviour, distribution, physiology and later biochemistry. Our knowledge of the distribution of Antarctic organisms is still imperfect. Collecting is still required, especially in those groups with an inadequate taxonomic structure, but the sum of Antarctic biological knowledge is now far beyond one man to encompass. In Antarctica the transition from naturalist to scientist was later than elsewhere in the world but more rapid when it did occur.

Antarctica, the frozen and for so long unknown continent, now occupies a special place in the world order. Its treaty is a model of peaceful agreement between sovereign nations, its population a shifting

Fig. 1.10. Modern ornithological work in the sub Antarctic. The Greyhead Albatross chick is sitting on an artificial nest which weighs the bird automatically every 10 minutes to provide data on feed size and growth rate (P.A. Prince/British Antarctic Survey).

mix of nationalities dedicated to the furtherance of science. It all seems a long way from those early explorers in their small wooden ships sailing to push back the limits of the known world, but perhaps the motivation of many of the present Antarctic scientists is similar. There is still adventure in the Antarctic and it remains a dangerous place for the novice. Our biological progress, reviewed in the remainder of this volume, is only a part of the scientific story of the Antarctic and advances within other disciplines have been equally exciting. Much remains to be learnt and it is to be hoped that further work will be able to take place within the present framework of international collaboration and agreement.

## REFERENCES

Auburn, F. M. (1982) *Antarctic Law and Politics*. C. Hurst, London.

Colonial Office (1920) *Report of the Interdepartmental Committee on Research and Development in the Dependencies of the Falkland Islands*. Command Paper No. 657. HMSO, London.

Deacon, G. E .R. (1937) The hydrology of the Southern Ocean. *Discovery Rep.* 15, 1 – 214.

Eights, J. (1833) Description of a new crustaceous animal found on the shores of the South Shetland Islands, with remarks on their natural history. *Trans. Albany Inst. II*, 1, 53 – 69.

Fleming, W. L. S., Stephenson, A., Roberts, B. B. and Bertram, G. C. L. (1938) Notes on the scientific work of the British Grahamland Expedition, 1934 – 37. *Geog. J.* 91, 508 – 32.

Hardy, A. (1967) *Great Waters: a voyage of natural history to study whales, plankton and the waters of the Southern Ocean in the old Royal Research Ship Discovery*. Collins, London.

Hooker, J. D. (1847) *The botany of the Antarctic voyage of H.M. Discovery ships Erebus and Terror in the years 1839 – 1843. I. Flora Antarctica*. Reeve Brothers, London.

Gaudichaud, C. (1826) Botanique. In Freycinet, L. (ed.) *Voyage autour de la monde sur l'Uranie et la Physicienne, pendant 1817 – 1820*. Paris.

King, H. G. R. (1969) *The Antarctic*. Blandford Press, London.

Lesson, R.-P. (1826) Sur le phoque léopard de mer (sea leopard) des Orcades australes; par James Weddell. *Bull. Sci. Nat. Géol.* 7, 437 – 8.

Marr, J. S. W. (1962) The natural history and geography of the Antarctic krill *Euphasia superba* Dana. *Discovery Rep.* 32, 37 – 464.

Murphy, R. C. (1936) *Oceanic Birds of South America*. Macmillan, New York.

Murphy, R. C. (1948) *Logbook for Grace*. Robert Hale, London.

Richardson, J. and Gray, J. E. (1844 – 45) *The zoology of the voyage of H.M.S. Erebus and Terror under the command of Capt. Sir James Clark Ross, 1839, 1840, 1841, 1842 and 1843*. Longmans, London.

Smith, E. J. (1978). The BIOMASS Report: a review. *Polar Record* 19, 204 – 206.

Thomson, C. W. (1877) *The voyage of the "Challenger". The Atlantic. A preliminary account of the exploring voyage of H.M.S. "Challenger" during the years 1873 – 1876*. Macmillan, London.

Tønnesson, J. N. and Johnson, A. O. (1982) *The History of Modern Whaling*. C. Hurst, London.

Weddell, J. (1825) *A voyage towards the South Pole performed in the years 1822 – 1824 containing an examination of the Antarctic Sea to the 74th degree of latitude and a visit to Tierra del Fuego with a particular account of the inhabitants*. Longmans, Hurst et al., London.

Will, H. (1890) Vegetationsverhaltnisse Süd-Georgiens. Die Internationale Polarforschung 1882 – 1883. *Die Deutsch Expeditionen und ihre Ergebnisse*, 2, 172 – 94.

# CHAPTER 2

# Physical Geography — Introduction

## D. W. H. WALTON and W. N. BONNER

British Antarctic Survey, Natural Environment Research Council, High Cross,
Madingley Road, Cambridge CB3 0ET, U.K.

Although in 1984 Antarctica can hardly be regarded any longer as Terra Incognita by the geographer, our knowledge of many aspects of the Continent is still relatively rudimentary. Cartographically the maps of this area owe a great deal to satellite photography and it is through the fortunate application of certain major advances in twentieth-century technology that so much progress has been made in less than 100 years. At the turn of the century, by adventurous sledge journeys and daring feats in small ships, much of the Antarctic coastline was initially surveyed. The advent of aircraft with BGLE, Byrd's expeditions, BANZARE and those of Lars Christensen allowed rapid progress to be made in photo-mapping the interior. The great land journeys through the interior supplemented those observations and now satellite pictures have completed virtually all of the primary mapping.

The availability of aircraft and the development of airborne radar and magnetometer systems are doing much to change our perception of the icesheet and its underlying rocks. A substantial proportion of Antarctica now has accurate under-ice maps, clearly showing the division of the landmass into two major areas, Greater and Lesser Antarctica.

Magnetic studies, combined with data from geology and palaeontology, have substantiated the Gondwana hypothesis and put the historical biogeography of the Southern Hemisphere into a new perspective. This collaboration between geologists, glaciologists and pedologists, reviewed in detail by David Elliot, is producing a detailed timescale for the historical changes in the Antarctic icesheet, whilst much of the present efforts in glaciology are concentrated on developing theories to explain modern icesheet dynamics.

There is clearly a close relationship between historical glaciology and historical climates. Exciting developments in analytical techniques have led to the use of oxygen isotopes to establish historical temperatures, analysis of the gas content of snow cores for evidence of previous atmospheric composition and the use of microparticle layers and heavy metal concentrations to date sections of snow and ice cores. Antarctica, as the cleanest area of the world, can also supply baseline data for comparative contamination studies.

An understanding of climate, as clearly pointed out by Henry Phillpot, needs both detailed and long term records. We must wait for the latter but present meteorological activities are now providing much of the former. The development of unattended automatic weather stations, the proliferation of manned stations reporting synoptic weather data, and the use of satellites and large computers has greatly improved the forecast potential for much of the Southern Hemisphere. Detailed macro- and micro-climatic studies are also under way at various locations in an effort to understand particular unique features of the weather system.

Soil development, reviewed here by Graham Claridge and Ian Campbell, has been studied under both the present climate and those of the past four million years. The saline soils of the Dry Valleys present very special characteristics to the pedologist, akin only to other desert areas of the world. Yet in the maritime Antarctic and in the subantarctic islands are to be found tundra soils similar in many ways to those from Arctic regions.

Thus in some respects the outlook for this area of science is promising. The importance, on a global scale, of the Antarctic climate is now fully appreciated at both the scientific and political levels. The recent interest in $CO_2$ and climatic trends has focused even more attention on the role of Antarctica. Scientifically, the development of the many facets of the Gondwana hypothesis has stimulated renewed academic interest in all areas of Antarctic earth sciences. A corollary of this is a governmental interest in the assessment of mineral and hydrocarbon potential for the area, which could have unfortunate consequences if any development is not strictly controlled.

In this chapter we try to draw together some of the most important facets of the present knowledge in earth sciences and climatology. Perhaps the most exciting development of all is that scientists from such diverse disciplines are working together effectively to make major advances in our understanding of the history of the Continent.

# CHAPTER 2a

# Physical Geography — Climate

## H. R. PHILLPOT

Department of Meteorology, University of Melbourne, Parkville, Victoria 3052, Australia

## CONTENTS

## 2a.1. ANTARCTICA'S PRESENT CLIMATE

Meteorological stations in the Antarctic are far apart (Fig. 2a.1), sites having often been chosen for ease of access from the sea, rather than for meteorological requirements. Few have records extending over more than about 20 years, hence the data available for analysis are much more limited than for the Arctic regions. Despite these limitations we can see quite clearly that the climate of Antarctica today is dominated by three features — the waters of the Southern Ocean, the variable sea-ice cover, and the continental ice sheet. In the Southern Ocean the two main surface water currents (Fig. 2a.2) are the West Wind Drift, and the water current located just off the Antarctic continent which moves toward the west. Both are driven by the surface wind at speeds of up to 0.5 ms$^{-1}$. The major oceanic feature of biological importance is the Antarctic Convergence (also shown on Fig. 2a.2), one of the fundamental boundary zones of the world's oceans. The Convergence marks the area in which Antarctic upper water sinks below and mixes with warmer sub-Antarctic surface water, and in practice is generally identified by sub-surface temperature and salinity behaviour. From the meteorological point of view the significance of the Convergence is hard to assess, because it has been found that in the ocean sector south of Australia the zone of steepest surface temperature transition appears to be located not across, but some hundreds of kilometres north of the mean position of the Convergence.

    The mean positions of the sea-ice edge, between 1973 – 80, are shown in Fig. 2a.3 for February – March (minimum cover) and for September (maximum cover) with the range within which the maximum was found. The area covered varies from about $3 \times 10^6$km$^2$ in summer to $19 \times 10^6$km$^2$ in winter, but the cover is not uniform; in September the concentration has been found to be between 15 and 85% over half the total area, and more than 85% over the other half (Zwally et al., 1979). Sea-ice thickness also varies; over the ocean in winter it is still largely unknown, but it is known to vary from about 0.3 m in latitudes $60-65°$S in early winter, to between 2 and 3 m in latitude 75°S in the late spring – early summer period (Budd, 1981a).

The Antarctic continent, some $14 \times 10^6$km$^2$ in area, ranges in height from sea-level to more than 4 km, with a mean altitude exceeding 2 km. Although there is much variety in continental features — including exposed mountain chains, 'dry' valleys, and coastal 'oases' — more than 95% of the total area is covered by snow and ice. The continent is very cold, windy and over much of its area, very dry; it has a profound effect on Southern Hemisphere weather systems, but as yet our understanding of the climate and its complex short-term and long-term variations is poor.

The relative uniformity of the continental plateau has allowed realistic climatic descriptions of a large part of the interior, despite the paucity of stations, but the steep coastal-slope zone presents much more complex problems. The 'oases', representing areas exposed by the recession of the ice cap, are too small

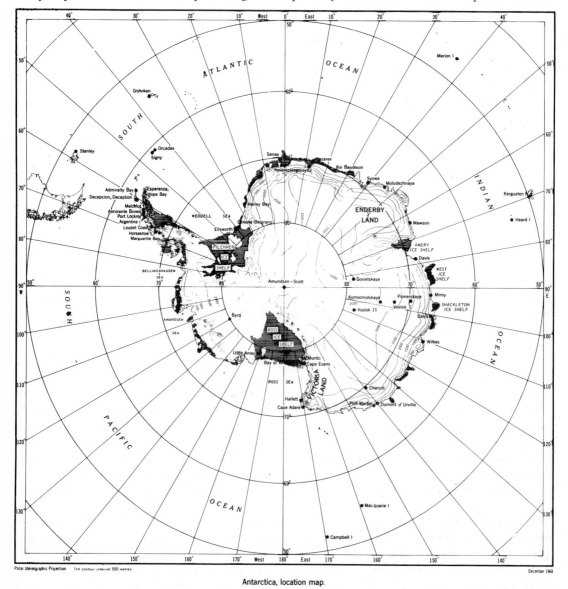

Antarctica, location map.

Fig.2a.1. Locations of meteorological stations (from Phillpot, 1968).

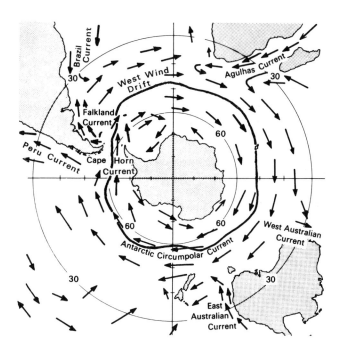

Fig.2a.2. The main features of the surface circulation of the oceans south of Lat 30°S (from Pittock *et al.*, 1978). The position of the Antarctic Convergence ( − ) is also shown.

in area to have any significant impact on the broad-scale atmospheric behaviour, but their local surface climates are very different from that of their surrounding environment and are of great biological significance (*see* Chapter 3a).

Before looking at characteristic values of various climatic parameters, we need to consider whether site differences between stations can bias the data. Mean annual statistics on surface wind, air temperature and total cloud cover for 34 stations on the Antarctic continent (Mather, 1969) show that there are some important variations between groups of stations. For instance between groups of coastal stations there are differences in surface wind speed and air temperature. Stations on ice-shelves are colder than nearby ones which are on rock; as an example, the mean annual temperature at the ice-shelf station Lazarev ( − 15.7°C) is 4.7°C lower than at Novolazarevskaya only 80 km distant, but in an area with extensive surface rock.

The surface wind behaviour is very complex, but from an examination of mean wind speed against mean temperature at the continental stations Mather (1969) concluded that 'the mildest climate is encountered in the Wilkes, Davis, Oasis and Syowa group'. However, if total cloud cover is also considered, Davis is marginally better than the other stations. In general, the mildest climates experienced on the Antarctic continent are those of the Greater Antarctica coast in areas of extensive exposed rock, on off-shore islands or peninsulas, but Dumont d'Urville is an important exception to this.

We turn now to a closer examination of total cloud cover, surface air temperature, surface wind and precipitation. Cloud estimates in the Antarctic are particularly difficult and subjective, but a general picture can be drawn for average total cloudiness in January (summer) and July (winter). Mean cloudiness over the oceans is high; it is a little lower over the continental coastline and there is a marked gradient in cloud amount in the area of steep ice slopes. There is no doubt that Lesser Antarctica is a

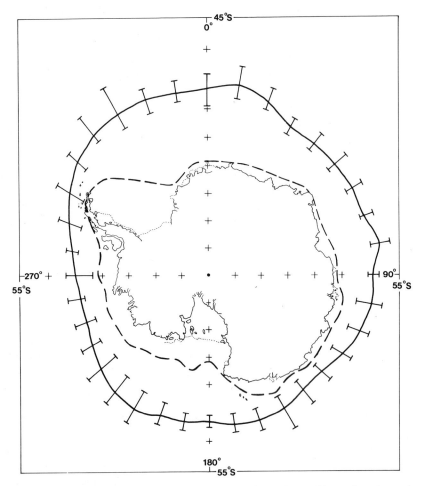

Fig. 2a.3. Mean maximum ice extent (— in September), calculated independently at each 10° of longitude, and range bars indicating the smallest and largest maxima. Mean minimum ice extent (— —) in February — March (from Jacka, 1981).

more cloudy region than Greater Antarctica and that minimum cloud cover centres over the high plateau, particularly in winter (Fig. 2a.4).

The surface air temperature differences between summer and winter (Fig. 2a.5) averaged around the latitude circles of 50, 55, 60 and 65°S, are 3.6, 4.3, 6.5 and 11.8°C respectively (Schwerdtfeger, 1970), the large difference at 65°S reflecting the effects of the sea-ice cover. Over the high plateau in the continental interior the temperature difference is more than 30°C.

Isotherm charts such as Fig. 2a.5 have limitations. Firstly, since the surface air temperature is height-dependent, the temperature pattern is very much a reflection of the ice topography. Secondly, these temperature fields represent 'long-term' values, but in any given year significant departures from the long-term averages may be found at any station, particularly on the continent. Thirdly, for any local consideration of temperature patterns the nature of the recording station sites must be considered.

Although not apparent from Fig. 2a.5 the mean temperature on the continent decreases slightly with latitude, for a given altitude. An analysis of mean annual temperatures from eleven coastal ice-shelf stations shows this to be 0.66°C/degree latitude. However, because temperature depends so strongly on

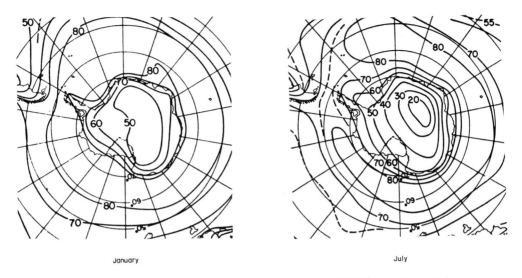

Fig. 2a.4. Total cloudiness (per cent) over the Antarctic in January and July (from van Loon, 1972).

height, and the highest elevation (exceeding 4 km) is not at the South Pole but near 82°S, 75°E, the lowest values will normally occur near this 'pole of low temperature'.

The mean diurnal temperature variation is generally small on the subantarctic islands, being about 2 – 3°C in summer and 1°C or less in winter, whilst on the continent there is a pronounced variation in spring and summer (2 – 6°C on the coast, 7 – 10°C in the interior), but no marked variation (about 1°C only) in winter.

Temperature extremes on the subantarctic islands vary rather widely. At stations in lower latitudes in the South Atlantic and Indian Ocean sectors temperatures in excess of 20°C have been reached, whilst at Heard Island a maximum of 14.4°C, 2°C higher than at Macquarie Island (in nearly the same latitude), has been recorded. Minimum temperatures appear rather more consistent, being around – 10°C at the lower latitude island stations. In the maritime Antarctic Orcadas Station, near latitude 61°S, is interesting because it shows clearly the effect of sea-ice cover — a winter minimum of – 40.1°C having been recorded in August. On the Antarctic continental coast extreme maxima in summer may exceed + 9°C in areas with considerable exposed rock, and winter minima – 30 to – 40°C. In the higher latitudes of the Ross and Weddell Sea areas, much lower temperatures are found (Ellsworth – 56.7°C in October), and in the continental interior a temperature of – 89.6°C was recorded at Vostok in July 1983.

Surface temperature inversions are a common feature at Antarctic continental coastal stations, but they vary in intensity and depth for no obvious reason. At inland stations however, and outside the two summer (December – January) months, they increase in strength to reach a maximum over the high plateau through the winter (Fig. 2a.6).

Mean surface wind patterns in the Antarctic are determined by four principal features of the mean surface pressure field. Variable pressure gradients at about 60°S off the coast of Greater Antarctica and 70°S off the Lesser Antarctic coast together with the strong oceanic gradient from 35°S to 60°S largely determine the average surface wind speeds shown in Fig. 2a.7 for January and July for the seas around Antarctica. The pressure gradient across the steep coastline is too complex to allow a general determination of wind speed, and for both this area and the continental interior we must rely on direct observations from the scattered meteorological stations.

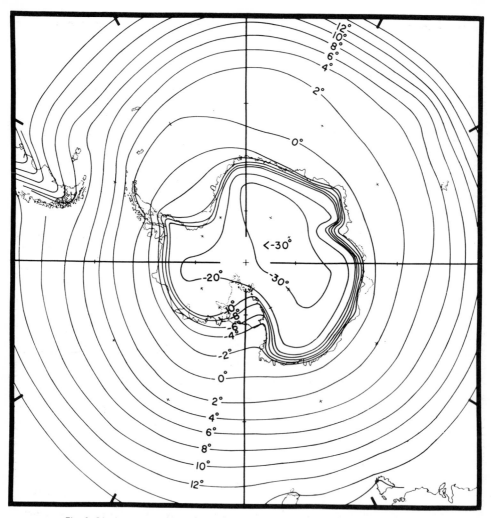

Fig. 2a.5A. A mean monthly surface isotherm (°C) in January (after Taljaard *et al.*, 1969).

Observations on some of the subantarctic islands demonstrate some aspects of the wind regime associated with the strong pressure gradient in latitudes 40°S − 60°S. The daily wind speed variation is very small; even through the year there is no marked variation, the minimum occurring in summer and the maximum tending to occur in the spring period, August to October. On South Georgia sudden local katabatic winds are frequent, whilst at Marion Island extremely violent storms sometimes occur, at Iles Kerguelen there is a much higher percentage (3%) of storms exceeding 25 m s$^{-1}$ than at the other island stations.

Although of great significance, the continental surface wind regime has not yet been completely explained. A general picture of wind direction (Mather and Miller, 1967) shows air flowing down-slope and crossing the line-of-fall at an angle of about 45° to the left of the fall-line. The airflow in the continental interior has been described (Schwerdtfeger, 1970) as 'equilibrium' flow — an approximate balance of pressure gradient, Coriolis and frictional forces — which becomes established under the strong and persistent temperature inversion. 'Equilibrium' flow can extend over a long time

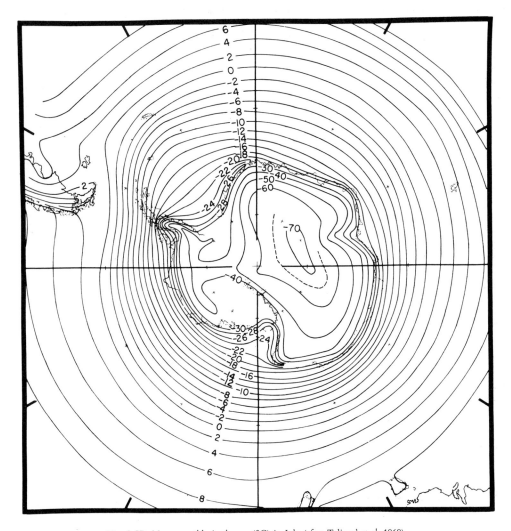

Fig. 2a5B. Mean monthly isotherms (°C) in July (after Taljaerd *et al*, 1969).

interval — of the order of weeks — with some speed variation resulting from the effect of superimposed pressure gradients. When the 'equilibrium' air-flow moves from the continental interior down the steep ice-slope to the coastline, it becomes a 'katabatic' or 'gravity' type wind. In the coastal sector at the foot of the ice slope, variations in wind strength may be due either to changes in the katabatic wind strength or to local pressure gradients, or both.

Wind speed varies considerably around the continental coastline. Understandably the wind regime at 'oases' varies from that observed at the foot of the ice-slope, but on one exceptional section of the coast, near Dumont d'Urville, surface winds attain very high speeds and persist for very long periods. At Cape Denison (1912 − 13) and Port Martin (1950 − 1) the mean annual wind speed for the total four years of observation was 18.5 m s$^{-1}$ ....' by far the highest wind speed near sea-level anywhere on earth...' (Loewe, 1974).

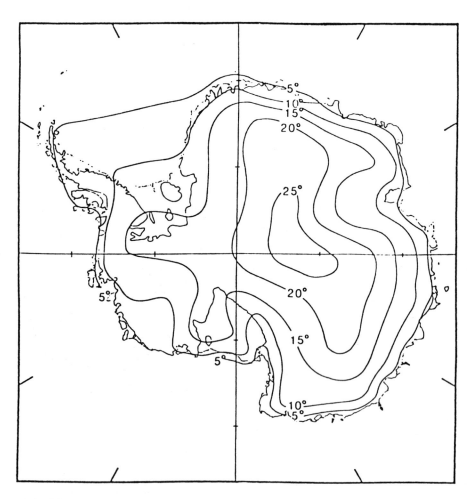

Fig. 2a.6. Isopleths of the average strength (°C) of the surface temperature inversion over the Antarctic continent in winter, June – August (from Phillpot and Zillman, 1970 with slight modification by Schwerdtfeger, 1970).

A feature of these surface winds is the remarkable suddenness with which speeds may vary, increasing within a few minutes from very light or near-calm conditions to as high as 50 m s$^{-1}$ or falling from gale force to near-calm just as quickly. Variations in surface pressure, temperature and humidity occur with the wind changes. This peculiar behaviour, with which high and eddying walls of drift-snow may be observed, has been named 'Loewe's Phenomenon'. It has been extensively discussed and illustrated in relation to Dumont d'Urville station, where it occurs about 10 times a year (Valtat, Gilbert and Magniez, 1960). It seems that the katabatic wind phenomenon may be a much more significant circulation feature on the Antarctic coastline than has been accepted or recognized so far. It also appears that this effect is observed at other coastal stations near the foot of the ice-slope of East Antarctica, but is not so well-defined and therefore not well documented.

Observations from ships show that wind speed drops considerably only a few kilometres off the coast. However the passage of depressions, particularly near stations on the coast, leads to very strong winds

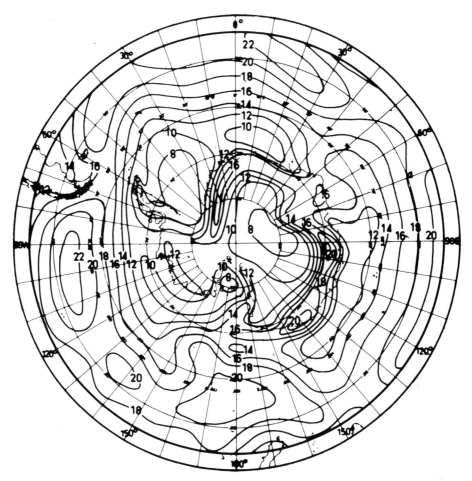

Fig. 2a.7. Average scalar wind speed (knots) in January and July (prepared by Zillman, shown by Schwerdtfeger, 1970).

with speeds of 50 m s$^{-1}$ or more being experienced. By contrast in the calmer continental interior, the South Pole station (Amundsen-Scott) has reported a maximum gust of 24 m s$^{-1}$.

Measurements of precipitation are not very reliable over the entire area south of 50°S. On the sub-Antarctic islands strong winds, frequent snow falls and the atypical locations of some of the observing stations affect the usefulness of the data. Over the oceans only ships' observations can be used and these are, in the main, available only for the summer months, whilst on the continent — where precipitation is almost entirely in solid form — direct measurement is considered totally unreliable.

On the sub-Antarctic islands the highest mean annual rainfall exceeds 200 cm at Marion Island, and for the remainder is mainly between about 100 and 150 cm (Loewe, 1957). Rainfall is fairly evenly distributed through the year, spring generally being the driest season. The number of days with precipitation is high, values ranging from 154 days per year at South Georgia to more than 300 days per year in the Australian — New Zealand sector. In general rainfall intensity shows relatively small variation throughout the year with mixed precipitation in liquid and solid form — or a rapid alternation of these — frequently occurring. Snowfalls have been recorded in summer at most stations. Over the sea, variations between different areas have been found to be no greater than variations from month to month in the same area.

On the Antarctic continent the present mass balance of the Antarctic Ice Sheet is important in assessing 'precipitation' (Bull, 1971). The mass balance is the difference between 'accumulation' by which material is added to the glacier, ice sheet or ice shelf, and 'ablation' by which mass is lost. The value calculated refers only to the upper surface since the rates of accumulation or ablation at the bottom of the ice sheet or ice shelves are not known.

Apart from direct measurements with snow gauges (which are considered unreliable) methods used for estimating local balance are: direct measurements using stakes and other markers; interpretation of snow stratigraphy revealed in walls of pits and in snow and ice cores; studies of stable and radioactive isotopes, and studies of microparticle concentrations. Over almost all of Greater Antarctica above 2 km surface elevation, the annual balance is less than 10 g cm$^{-2}$, and within this region there is an area of 2 million km$^2$ in which the balance is less than 5 g cm$^{-2}$. Much of the continent must therefore be classified as amongst the most arid areas in the world (Fig. 2a.8). There are significant spatial and temporal variations in the surface mass balance, and it is unlikely, given the techniques available, that these can be easily resolved.

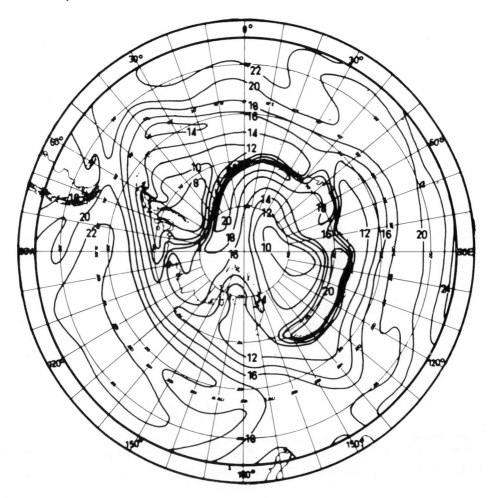

Fig. 2a.7. *cont.*

## 2a.2. THE ANTARCTIC CONTINENTAL CLIMATE — PAST AND FUTURE

The Antarctic continent is the principal heat sink of the global climate system. An understanding of its climate in the past, and a soundly-based assessment of its likely future trend, would clearly be most advantageous. Collecting meteorological and climatic data is not always easy (Fig. 2a.9) but the information is likely to be of considerable use in both the short and long terms.

Techniques generally used for the study of past climates, such as the analysis of meteorological records, the study of histories, old records, diaries etc., together with studies of geological changes, plant and animal fossils, sea level variations and so on, can be extended on the Antarctic continent to include the study of ice cores. The types of information these can give are: measurements of the ratios of stable isotopes of oxygen or hydrogen; gas content (samples of atmospheric gases being trapped in the ice during the process of compaction from snow to ice); particulate matter and chemical content (soluble and insoluble aerosols of various types), all of which can be used to derive information about past environments.

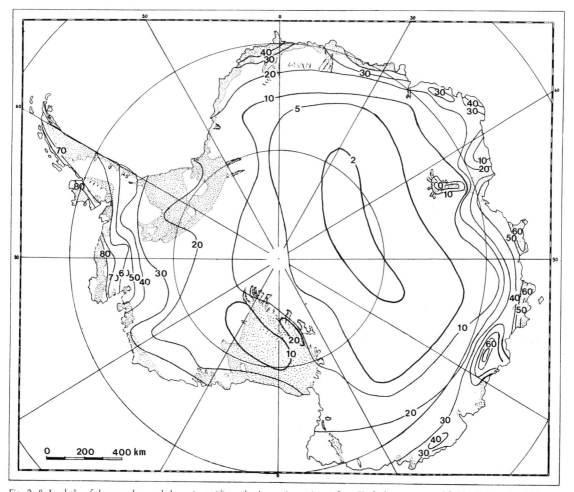

Fig. 2a.8. Isopleths of the annual snow balance (g cm$^{-1}$) on the Antarctic continent (from Kotlyakov, 1974, modified by Budd and Smith).

Fig. 2a.9. Rime ice on instruments interferes with the collection of accurate data. (Photo: British Antarctic Survey)

Measurement of the ratios of stable isotopes of oxygen (or hydrogen) in the ice cores underlies an important technique having wide application, because the isotope ratio depends on the temperature at which a particular process takes place — such as the evaporation of water, the formation of snow, ice, or of marine sediments made up of micro-fossil material.

Difficulties are encountered in the practical application of the analysis technique. For an ice cap such as the Antarctic, the isotope ratio reflects the climatic temperature at the time and place of snow deposition, but surface temperature decreases with both latitude and (more particularly) with height, hence any ice elevation changes resulting from ice sheet movement may cause isotope ratio variations similar to climatic change effects. This emphasizes the need for ice flow models (Jenssen, 1978), whilst Robin (1981) noted, in addition, the problems of determining a mean condensation temperature for the formation of snow, and the 'noise' attendant on relationships determined between isotopic composition and temperature.

When the technique is applied to marine sediment analysis, it is not simply a surface temperature that is being derived. Although the original organism lived in the near-surface waters, the solvent effect of the water, and the exposure time on the sea floor need to be allowed for. However, despite these problems, the technique is an important one, and is the basis of the CLIMAP (Climate: Long-range Interpretation, Mapping and Prediction) Project, a multi-institutional project for the study of climatic change.

Whichever technique is used, it is essential to keep in mind that variations with time in the meteorological elements are evident regardless of the period over which the analysis is made. Furthermore, variations are observed when space-scales are changed, hence the determination of any trend or change in climate is a combined space-time problem. In addition to establishing the particular nature of past climatic regimes, the underlying reason for them also needs to be found.

Time scales of interest may extend from those of a yearly scale, through scales of decades, centuries, of millenia, or epochs.

Daily observations quite frequently show positive surface temperature anomalies in one area, and negative temperature anomalies in another, with the overall result that over the entire area temperature conditions would be described as 'average'. Sometimes temperature anomalies may persist in particular areas for some time and show a trend, as occurred in the winter temperature data at two Antarctic stations (Sanae and McMurdo) almost 180° in longitude apart over the period 1955 − 69 (Williams and van Loon, 1978). The trends were of almost the same magnitude but opposite in sign. In another study (Phillpot, 1968), a comparison was made of two particular years when surface air temperatures over quite a large sector of the Antarctic continent showed significant variation from normal, but no very clear association could be established between mean surface temperatures and the mean tropospheric circulation features. The temperature differences could be explained, however, by considering the origin of, and the path followed by, the air, i.e. whether from off the continent or the ocean. These examples illustrate the need for caution in suggesting the possibility of 'climatic change' over a particular area, and indeed, even of drawing inferences of probable mean atmospheric circulation patterns from a given mean surface temperature field.

On a scale of decades, an examination of the meteorological and sea-ice data from Orcadas (60.7°S, 44.7°W), the only station in the high southern latitudes with a reasonably long period of observation (since 1903), has shown no convincing evidence of any climatic change since the beginning of this century. On the continent, data from Amundsen's Framheim base in 1911 − 12, the various Little America stations and the IGY, do not support the hypothesis of any significant climatic trend in that area (Schwerdtfeger, 1970).

On a time scale of about a century, ice core analysis of relatively recent temperature behaviour has been interpreted to suggest a slight degree of warming, supporting views that the warming trend observed in the Northern Hemisphere between about 1880 and 1940 was of global extent. On the question of trends in snow accumulation on the ice cap over the last 200 years the evidence appears inconclusive.

Moving to the time scale of millenia, temperature trends revealed in the ice cores obtained from five deep boreholes in Antarctica show that considerably lower temperatures were experienced around 8000 y BP (Before Present). An extreme occurred around 18,000 y BP when climatic temperatures were about 5 − 7°C lower, with atmospheric dust content several-fold greater and atmospheric carbon dioxide up to 50% less than at present (Young, 1981a). Additional evidence derived from the analysis of marine sediments (the CLIMAP project) indicates that the Northern Hemisphere temperature minimum event also occurred around 18,000 y BP, although the Southern Hemisphere temperature led the Northern Hemisphere ice volume record by about 3000 y. Furthermore, an equatorial Pacific Ocean core shows evidence of cycles at periods of about 40,000 and 22,000 y over the past 800,000 y BP (Shackleton, 1978).

Just as ice core studies have shown that relatively short periodicity can be detected in data extending over the last 100 − 200 year period, so too can longer term variations be detected in data appropriate to millenia. For example, on the oxygen isotope profile for one of the Antarctic deep ice-cores (Dome C), a rather high and uniform mean value corresponding to present surface conditions can be fitted down to a depth equivalent to about 11,000 y BP. The detailed core variations, however, show that different periods from 11,000 to 8000 y BP (relatively warm), 8000 to 4000 y BP (relatively cold) and 4000 y to present (moderately warm with a slight negative trend) can be identified. Two points emerge here, firstly the warmer and cooler conditions for shorter periods are contained within a longer 'uniform' regime, and secondly, this record suggests significant differences between the two hemispheres over the last 11,000 y (Lorius et al., 1981), as does the marine sediment analysis.

On the time scale of epochs, a generalized scheme of climatic events for the high southern latitudes based on studies ranging from glacial marine sediments to ice-core analyses, is shown in Table 2a.1. A suggested 'scenario' is roughly this: for some 200 million years prior to the onset of the present glaciation (which probably occurred around the mid-Oligocene epoch some 26 − 30 million years BP) the world's oceans were characterized by low thermal gradients giving warm water to high latitudes; continental fragmentation then opened the Atlantic Ocean, the southern polar oceans became dominant sources of cold water and were distributed more uniformly around Antarctica as a result of the deepening of the sea-way between Australia and Antarctica; the Arctic seas were opened up as a cold water source (about 14 million years BP); the Panama Isthmus was closed (about 3 million years BP) and the modern scheme of oceanic circulation was essentially established (Frakes, 1981).

In summary therefore: significant variations in past climate can be detected for short periods from within the longer-term climatic records deduced from ice-core analyses; no attempts have yet been made to explain even relatively short time scale variations in the Antarctic; various interpretations are possible from the complex ice-core data; major climatic events in the two hemispheres did not coincide completely, and there is some evidence of cyclicity in the major changes.

The three major cycles thought to affect earth climates (on the time scale of concern here) are those arising from the eccentricity of the earth's orbit around the sun (period about 93,000 y); the precession of the equinoxes (period about 21,000 y), and the obliquity of the ecliptic of the earth (period about 41,000 y). The third cycle is very significant for high latitude zones including Antarctica, and through affecting global albedo by growth of ice there, may be the driving force for recent global changes (Frakes, 1981).

To completely understand and explain past climate would demand a knowledge of all the factors, periodic and aperiodic, which control climate, but clearly an understanding of events on the time scale of the major ones, such as that experienced around 18,000 y BP would be extremely valuable, and would significantly assist the problem of assessing the Antarctic climate likely to be experienced at some future time.

Some progress is being achieved on this, and at a symposium in 1979 (Allison, 1981) a model was proposed for the growth and retreat of Northern Hemisphere ice sheets, suggesting that glacial maxima occurred around 110,000, 60,000 and 20,000 y BP. It was also shown that sea level changes, induced by the Northern Hemisphere ice sheets, could be the primary driving force for changes in the Antarctic ice sheet (Budd and Smith, 1981; Budd, 1981b).

Recognition of the need to improve our understanding of climatic behaviour has been stimulated, not only by the rising importance of food, water resources, energy, health and environmental planning on a planet with a continually increasing population and a need to raise significantly the living standard for a very large proportion of the present population, but also because during the last century or so, human activities have begun to significantly affect conditions important for the maintenance of a balanced climate state. The most important of these activities is the increased use of fossil fuels, which have caused a steady rise in the concentration of carbon dioxide in the global atmosphere, and led to the recognition of the '$CO_2$ − Climate problem' as one of the most pressing importance in atmospheric science.

Using comprehensive numerical models, simulation of the atmospheric response to a doubling of the atmospheric $CO_2$ concentration (likely within about the next half-century or so) has revealed rising surface air temperatures, which could be as high as 10°C in high latitudes, but are likely to be rather smaller in the Southern Hemisphere (Tucker, 1981). This alarming prospect may however be partly due to inadequacies in the models.

Whilst an atmospheric warming of the order of even 10°C would have little effect on the deep ice sheet of Greater Antarctica, the behaviour of the Lesser Antarctica sheet is harder to predict. Budd (1980) drew attention to preliminary results by Young (1981b) which showed that even after several hundred years, a 5°C warming at the atmosphere − ice sheet interface would have only a slight effect on the ice flow, but the problem here is that we do not know whether the ice sheet is retreating or not. At present it appears that it may be stable (Budd, 1980), and if so the problem is not so serious, but, because of the long lead-time involved, this is an important point to be resolved.

Any major change in the volume of the Antarctic ice sheet would produce catastrophic problems for man throughout the rest of the world. On this crowded planet we must make the best possible use of all our resources. An understanding of the Antarctic climate may help to ensure that disaster does not catch us unprepared.

TABLE 2a.1. The Generalized Time Scale of Climatic Events for High Southern Latitudes (From Frakes, 1978)

| Year $\times$ 10$^6$ BP | Climatic event |
|---|---|
| 37 | Sharp drop in surface ocean temperature ($4-5°C$) |
| 26 | Glaciers reach sea-level in Ross Sea |
| 15 — 5 | Pronounced drop in surface ocean temperatures. |
|  | Rapid build-up of ice in Greater Antarctica |
|  | Valley-carving state in Trans-Antarctic Mountains (Taylor V) |
| 5.5 — 3 | Antarctic Polar Front shifts 300 km northward. |
|  | Ross IceShelf expands to edge of continental shelf. |
| 4.7 — 4.3 | Substantial build-up of ice in polar regions. Possible glacial surge. |
| 4.25 — 3.95 | Marked warming of surface ocean temperatures ($8-10°C$). Possible melting of the ice sheet in Lesser Antarctica and marine invasion of Wright Valley. |
| 3.95 — 3.35 | Lengthy and marked cooling. |
| 3.35 — 0.7 | Fluctuations in ocean surface temperatures (period: 170,000 years). |
| 3.5 — 2.7 | Spillover of ice from Greater Antarctica (Taylor IV) |
| 2.1 — 1.6 | (Taylor III) |
| 1.6/1.2 — >49,000 | (Taylor II) |
| 34,800 — 9490 | (Taylor I) |
| 0.7 — present | Wider fluctuations in ocean surface temperatures (period: 110,000 years). |

# REFERENCES

Allison, I. (Ed.) (1981) *Sea Level, Ice and Climatic Change. Proc. Canberra Symp. 1979*. IAHS Publ. No. 131.

Budd, W.F. (1980) The importance of the polar regions for the atmospheric carbon dioxide concentrations. *Carbon Dioxide and Climate: Australian Research. Proc. Canberra Symp. 1980)*. Ed. G.I. Pearman, pp. 115 — 28. Australian Academy of Science, Canberra.

Budd, W.F. (1981a) The role of Antarctica in Southern Hemisphere weather and climate. *Antarctica: Weather and Climate*. Compiled by N.W. Young, p.20. Roy. Met. Soc. Australian Branch, University of Melbourne.

Budd, W.F. (1981b) The importance of ice sheets in long term changes of climate and sea-level. *Sea Level, Ice and Climatic Change. Proc. Canberra Symp. 1979*. Ed. I. Allison, pp. 441 — 71. IAHS Publ. No. 131.

Budd, W.F. and Smith, I.N. (1981) The growth and retreat of ice sheets in response to orbital radiation changes. *Sea Level, Ice and Climatic Change. Proc. Canberra Symp. 1979*. Ed. I. Allison, pp. 369 — 409. IAHS Publ. No. 131.

Bull, C. (1971) Snow accumulation in Antarctica. In *Research in the Antarctic*. Eds. L.O. Quam and H.D. Porter, pp. 367 — 421. American Association for the Advancement of Science. Washington D.C.

Frakes, L.A. (1978) Cenozoic climates: Antarctica and the Southern Ocean. In *Climatic Change and Variability: a Southern Perspective*. Eds. A.B. Pittock, L.A. Frakes, D. Jenssen, J.A. Peterson and J.W. Zillman, pp. 53 — 69. Cambridge University Press, Cambridge.

Frakes, L.A. (1981) The history of Antarctic climates over 'geologic' time-scales. *Antarctica: Weather and Climate*. Compiled by N.W. Young, p. 10. Roy.Met.Soc. Australian Branch, University of Melbourne.

Jacka, T.H. (1981) Antarctic temperature and sea ice extent studies. *Antarctica: Weather and Climate*. Compiled by N.W. Young, p.10. Roy. Met. Soc. Australian Branch, University of Melbourne.

Jenssen, D. (1978) Climatic and topographic changes from glaciological data. In *Climatic Change and Variability: a Southern Perspective*. Eds. A.B. Pittock, L.A. Frakes, D. Jenssen, J.A. Peterson and J.W. Zillman, pp. 77 — 81. Cambridge University Press, Cambridge.

Kotlyakov, V.M., Barkov, N.I., Loseva, I.A. and Petrov, V.N. (1974) New map of the accumulation on the Antarctic ice sheet. Data of Glaciological Studies, Chronicle Discussion, pp. 248 — 55. Acad. of Sciences of the USSR and Institute of Geography, Publ. No. 24.

Loewe, F. (1957) Precipitation and evaporation in the Antarctic. In *Meteorology of the Antarctic*. Ed. M.P. Van Rooy, pp. 71 — 90. Weather Bureau, Pretoria.

Loewe, F. (1974) Considerations concerning the winds of Adélie Land. *Z. Gletscherk. Glazialgeol.* 10, 189 – 97.

Lorius, C., Merlivat, L., Duval, P., Jouzel, J. and Pourchet, J. (1981) Evidence of climatic change in Antarctica over the last 30,000 years from the Dome C ice core. In *Sea Level, Ice and Climatic Change. Proc. Canberra Symp. 1979.* Ed. I. Allison, pp. 217 – 25. IAHS Publ. No. 131.

Mather, K.B. (1969) The pattern of surface wind flow in Antarctica. *Pure Appl. Geophys.* 75, 332 – 54.

Mather, K.B. and Miller, G.S. (1967) The problem of the katabatic winds on the coast of Terre Adélie. *Polar Rec.* 13, 425 – 32.

Phillpot, H.R. (1968) A study of the synoptic climatology of the Antarctic. *IAMRC Tech. Rept. No.12.* Commonwealth Bureau of Meteorology, Melbourne.

Phillpot, H.R. and Zillman, J.W. (1970) The surface temperature inversion over the Antarctic continent. *J. Geophys. Res.* 75, 4161 – 9.

Pittock, A.B., Frakes, L.A., Jenssen, D., Peterson, J.A. and Zillman, J.W. (Eds) (1978) *Climatic Change and Variability: a Southern Perspective.* Cambridge University Press, Cambridge.

Robin, G. de Q. (1981) Climate into ice: the isotopic record in polar ice sheets. *Sea Level, Ice and Climatic Change. Proc. Canberra Symp. 1979.* Ed. I. Allison, pp. 207 – 16. IAHS Publ. No. 131.

Schwerdtfeger, W. (1970) The Climate of the Antarctic. In *World Survey of Climatology, Vol.14.* Editor-in-Chief H.E. Landsberg, pp. 253 – 361. Elsevier Publishing Company, Amsterdam.

Shackleton, N.J. (1978) Some results of the CLIMAP project. *Climatic Change and Variability: a Southern Perspective.* Eds. A.B. Pittock, L.A. Frakes, D. Jenssen, J.A. Peterson and J.W. Zillman, pp. 69-76. Cambridge University Press, Cambridge.

Taljaard, J.J., van Loon, H., Crutcher, H.L. and Jenne, R.L. (1969) *Climate of the Upper Air: Southern Hemisphere 1 Temperatures, Dew Points and Heights at Selected Pressure Levels.* NAVAIR 50-1C-55 Washington D.C.

Tucker, G.B. (1981) *The $CO_2$ – Climate Connection — A Global Problem from an Australian Perspective.* Australian Academy of Science, Canberra.

Valtat, B., Gilbert, J. and Magniez, R. (1960) Le 'phenomene de Loewe' en Terre Adélie. *La Meteorologie* 57, 77 – 112.

van Loon, H., Taljaard, J.J., Sasamori, T., London, J., Hoyt, D.V., Labitzke, Karin, Newton, C.W. (1972) Meteorology of the Southern Hemisphere. In *Meteorological Monographs 13, No.35.* Ed. C.W. Newton, pp. 263. American Meteorological Society, Boston.

Williams, J. and van Loon, H. (1978) Trends of winter temperature in the Southern Hemisphere. *Climatic Change and Variability: a Southern Perspective.* Eds. A.B. Pittock, L.A. Frakes, D. Jenssen, J.A. Peterson and J.W. Zillman, pp. 191 – 4. Cambridge University Press, Cambridge.

Young, N.W. (1981a) Climate from ice core studies. *Antarctica: Weather and Climate.* Compiled by N.W. Young, p.9. Roy. Met. Soc. Australian Branch, University of Melbourne.

Young, N.W. (1981b) Responses of ice sheets to environmental changes. *Sea Level, Ice and Climatic Change. Proc. Canberra Symp. 1979.* Ed. I. Allison, pp. 331 – 60. IAHS Publ. No. 131.

Zwally, H.J., Parkinson, C.L., Carsey, F.D., Gloerson, P., Campbell, W.J. and Ramseier, R.O. (1979) Seasonal variation of total antarctic sea ice area, 1973 – 75. *Antarct. J. U.S.* 14, 102 – 103.

CHAPTER 2b

# Physical Geography — Geological Evolution

Institute of Polar Studies and Department of Geology and Mineralogy, Ohio State University, Columbus, Ohio 43210, U.S.A.

## CONTENTS

## 2b.1. INTRODUCTION

To an earth scientist, Antarctica differs from other continents only in the present strong influence of ice on geological processes. The ice sheet, furthermore, plays an important role in any attempt to understand the evolution of the continent because less than two per cent of the surface area is formed of exposed rock and is thus accessible for geological study. That two per cent is distributed principally around the periphery of the continent, in the Transantarctic Mountains and the Ellsworth Mountains (Fig. 2b.1). Over most of the continent ice forms a cover that in places is more than 4000 m thick. If the ice were to melt, East Antarctica would be largely above sea level and the sub-glacial Gamburtsev Mountains would become a range over 3000 m high; West Antarctica would form an archipelago with deep channels and basins between the various islands (Fig. 2b.2). After isostatic rebound, that is recovery from the depression of the land caused by the weight of ice, there would be some increase in land above sea level in East Antarctica but little change in West Antarctica. Geological knowledge of the

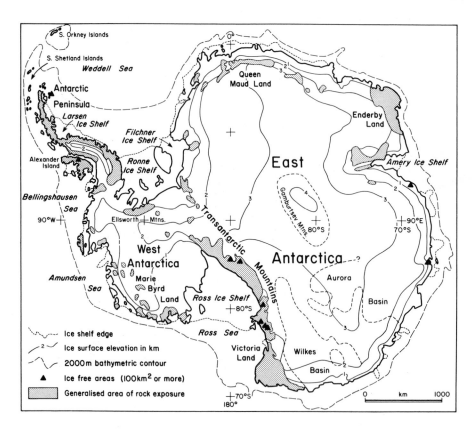

Fig.2b.1. Generalized distribution of ice and rock in Antarctica. The location of the sub-glacial Gamburtsev Mountains and the sub-glacial Wilkes and Aurora basins are also indicated.

ice-covered regions is limited to the sorts of information that can be derived by remote sensing and other geophysical techniques. Thus, sub-glacial morphology has been mapped over much of the continent; based on contrasting magnetic signatures of the rocks, magnetic provinces have been defined in a few regions; and some information on structure and stratigraphy has been obtained for certain areas by seismic methods.

## 2.b.2. ANTARCTICA AND PLATE TECTONICS

In terms of global plate tectonics (Wilson, 1972; Wyllie, 1976), in which the surface of the earth is divided into plates defined by seismically active zones or belts, Antarctica is part of a larger Antarctic Plate bounded for the most part by the mid-ocean ridges in the Pacific, Indian and southern-most South Atlantic Oceans. The Antarctic Plate itself consists of the continent plus the ocean floor out to those mid-ocean ridges (Fig. 2b.3). In the South American — Scotia Sea — Antarctic Peninsula region the boundary of the Antarctic Plate is less well defined: the Scotia Sea region includes two or more small microplates; the South American Plate is being destroyed at the South Sandwich Islands trench; and the Antarctic Plate is being driven at a very slow rate beneath the Andes south of 44°S.

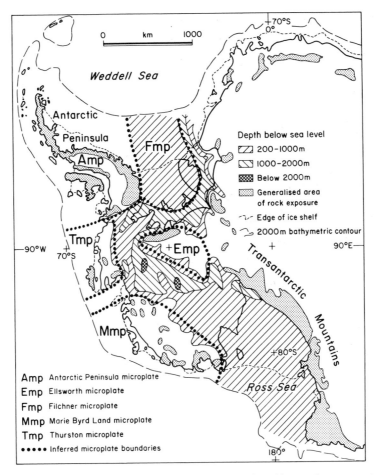

Fig. 2b.2. Sub-glacial topography of West Antarctica. The outlines of possible microplates are superimposed.

Although there is debate over whether the mechanisms of plate tectonics operated during the Precambrian and Early Palaeozoic (Table 2b.1), from the standpoint of Antarctica and the biogeography of the Southern Hemisphere, the Late Palaeozoic to the present constitutes by far the most important period. For this time span, the geologic history can be interpreted in terms of plate tectonics, and in particular in terms of active plate margins — those plate margins at which oceanic crust may be driven, or subducted, beneath continental (or oceanic) crust. At such margins various geological terrains can be recognized (Fig. 2b.4): a magmatic arc along which there is volcanic activity and associated emplacement of magma at depth to form plutons; a fore-arc region between the arc and the oceanic trench, where sediment accumulation may be thick and the sediments rich in volcanic detritus, both eroded off the arc and as air-fall debris from explosive eruptions; and a back-arc terrain in which sedimentary sequences may have much volcanic debris in the earlier stages of accumulation but often pass upward into sequences without volcanic debris and to terrestrial rather than marine deposits. Deformation (folding and faulting) accompanies the processes at active plate margins and affects both the fore-arc and back-arc terrains. The evolution of an active plate margin frequently includes orogeny (mountain building), a term that is used, more loosely, to encompass deposition of thick sedimentary

sequences, deformation of the rocks, emplacement of plutons and uplift that leads to the development of erosion surfaces on the roots of the orogen.

Antarctica consists of two contrasting geological provinces. East Antarctica is principally a Precambrian craton (a stable continental area consisting largely of igneous and metamorphic rocks) with, along its Transantarctic Mountains flank, a Lower Palaeozoic orogenic belt that is overlain by flat-lying Upper Palaeozoic to Lower Mesozoic sedimentary rocks of the Beacon Supergroup and the intrusive and extrusive rocks of the Ferrar Supergroup. In contrast, West Antarctica consists of a series of orogenic belts of Early Palaeozoic to Late Mesozoic — Early Cenozoic age.

## 2.b.3. ANTARCTICA IN GONDWANA

Although F. B. Taylor (1910) first advanced the idea of continental drift, it was Alfred Wegener (1912) who drew together evidence from a variety of sciences and postulated the former existence of a supercontinent (Pangaea) consisting of Laurasia and Gondwana in the northern and southern hemispheres, respectively. After Wegener, the hypothesis of continental drift found its greatest exponent in Alex du Toit who placed the idea on a solid geological footing. Du Toit (1937) based his argument on geological and palaeontological data from the southern continents and peninsular India: the occurrence of distinctive floras and faunas together with similarities in the stratigraphic record; matching of tectonic fabrics (style and orientation of belts of folding and faulting) across what are now disjunct boundaries; and provenance studies — that is identification of the rock types in the source terrains of sedimentary sequences. The study of the palaeomagnetism of rocks during the 1950s,

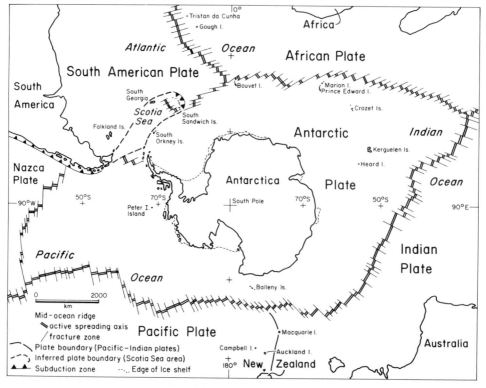

Fig. 2b.3. The Antarctic Plate.

brought a new tool to bear on the question, and provided strong support for the hypothesis. In the following decade marine geophysical and geological research led to the sea-floor spreading hypothesis which gives an explanation for the evolution of the ocean floors; at the same time it provided a mechanism for continental drift which, if not conclusive, very strongly supported the concept of Gondwana. The discovery of Triassic land vertebrate fossils in Antarctica in 1969 clinched the argument for many biologists and palaeontologists.

Palaeomagnetism offers convincing evidence for the existence of Gondwana from the late Precambrian through to early Jurassic time (McElhinny, 1973). It does not, however, give precise geometric arrangements for the supercontinent. A more precise assembly may be obtained from the study of the linear sea-floor magnetic anomalies and 'back-tracking', that is closing up the oceans by superimposing anomalies of the same age, to the initiation of the first or oldest recognizable magnetic lineation (Fig. 2b.5).

The fits between Africa and South America (West Gondwana) and between Antarctica, Australia and India (East Gondwana) have been satisfactorily established. However, serious problems remain with the region of the 'southwestern South Atlantic Ocean', that is between South America and the Antarctic Peninsula, and with the 'Pacific' margin of Gondwana from the Antarctic Peninsula to New Zealand.

TABLE 2b.1. Geological Time Scale

| Era | Period | Epoch | | Age in Ma |
|---|---|---|---|---|
| CENOZOIC | Quaternary | | Holocene | |
| | | | Pleistocene | |
| | | | | 2 |
| | | | Pliocene | |
| | | Late | | 5 |
| | | | Miocene | |
| | | | | 25 |
| | Tertiary | | Oligocene | |
| | | | | 38 |
| | | Early | Eocene | |
| | | | | 55 |
| | | | Palaeocene | |
| | | | | 65 |
| MESOZOIC | Cretaceous | | | 144 |
| | Jurassic | | | |
| | | | | 213 |
| | Triassic | | | |
| | | | | 248 |
| | Permian | | | |
| | | | | 286 |
| | Pennsylvanian | | | |
| | | | | 320 |
| PALAEOZOIC | Mississippian | | | |
| | | | | 360 |
| | Devonian | | | |
| | | | | 408 |
| | Silurian | | | |
| | | | | 438 |
| | Ordovician | | | |
| | | | | 505 |
| | Cambrian | | | |
| | | | | 590 |
| PRECAMBRIAN | | | | 4500 |

Assigned ages simplified from Harland et al., 1982

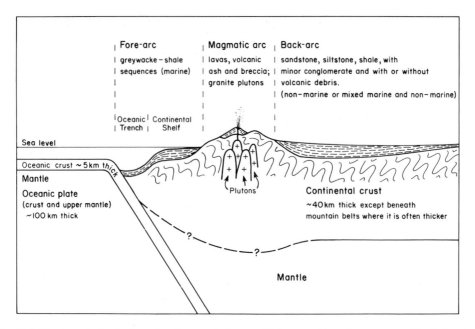

Fig. 2b.4. Diagrammatic sketch of an active plate margin; in general it is comparable to the west coast of South America.

The solution appears to lie in that part of Antarctica consisting of a number of blocks or microplates that have been rearranged into their present geography, subsequent to break-up of the super-continent (Figs. 2b.2 and 2b.6).

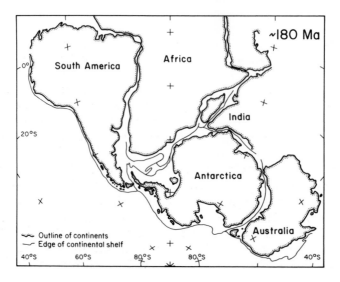

Fig. 2b.5. Reconstruction of Gondwana (after Norton and Sclater, 1979). Note that reconstruction permits orientation and latitude to be established, but not longitude. Note also the overlap of the Antarctic Peninsula and the South American continent; this is geologically unacceptable and constitutes part of the argument for West Antarctica being formed of several microplates.

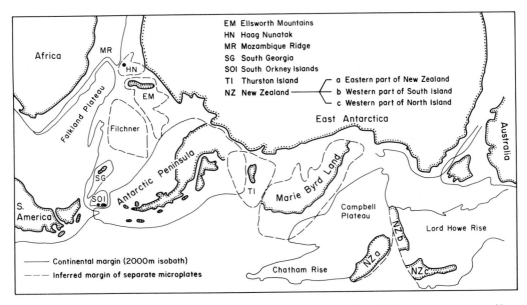

Fig. 2b.6. Reconstruction of the 'Pacific' margin of Gondwana (modified from Dalziel and Elliot, 1982), illustrating the space problem. It is one of many that might be proposed and is used here because of the need for a frame of reference for discussion of the geological evolution of the continent. Alternatives to Fig. 2b.6 which do not violate the known geology, will not substantially alter the interpretation of the geological history. Note that significant areas of the present Ross Sea – Ross Ice Shelf area have been omitted; eastern New Zealand has been separated from the two western parts along the Alpine Fault.

## 2b.3.1. Early evolution of the continent

The Precambrian history of the continent is represented mainly by metamorphic and igneous rocks in East Antarctica that range in age from as old as 3.8 Ga to as young as 480 Ma. Most of East Antarctica is an old craton similar to that of Western Australia or southern Africa. The sparseness of the outcrops and the limited work done so far have not permitted the development of a coherent geological history. Nevertheless it would seem unlikely that East Antarctica differs from other cratons. Although evidence of Precambrian life has been discovered in other old cratons and in rocks more than 3.0 Ga old, as yet no record exists of early life in Antarctica; this is probably because no rocks of the types that are fossil-bearing elsewhere have been found so far.

The Late Precambrian to Early Palaeozoic history (Fig. 2b.7A) is contained largely in the Transantarctic Mountains where extensive outcrops occur of greywacke-shale sequences, volcanic rocks, carbonate and clastic sediments, and plutonic rocks (Elliot, 1975*; Laird, 1981). These rocks record an active plate margin, with the Late Precambrian events culminating in the Beardmore Orogeny and the Cambrian events in the Ross Orogeny. Only sparse microfossil assemblages have been recovered from some of the Upper Precambrian rocks. The Cambrian sedimentary rocks, in contrast, have yielded abundant invertebrate remains. The carbonate rocks include extensive reef-like deposits made up of

*Access to much of the Antarctic geological literature can be obtained through this publication and the proceeding volumes of the 1977 and 1982 Antarctic Geology Symposia (Craddock, 1982; Oliver *at al.*, in press).

archaeocyathids that are closely related to Australian faunas. Trilobite faunas are widespread and other fossil groups found include bivalves, gastropods, brachiopods, echinoderms and conodonts. The most interesting fossils are the monoplacophorans (Fig. 2b.8), forms thought by some to be ancestral to the cephalopods. Conditions during Cambrian time were suitable for deposition of great thicknesses of limestone, and a warm tropical climate can be inferred. This agrees with palaeomagnetic data that points to the central Transantarctic Mountains lying in mid to low latitudes.

The mid-Palaeozoic history is poorly documented. Devonian plutons occur in north Victoria Land and sparsely in Marie Byrd Land (Fig. 2b.7A). The bulk of the sedimentary rocks of this age are marine sandstones and in the principal outcrop area in south Victoria Land contain a sparse Devonian flora which includes lepidodendron, sigillaria, and plant microfossils. Devonian fish have also been recovered from the uppermost beds; jaws, skulls, teeth, spines and dermal plates of acanthodian, crossopterygian and rhiphidistian fish have been identified and include *Bothriolepis* which has affinities to Greenland and other northern hemisphere examples. Only in the Ohio Range is a more diverse invertebrate fauna present. It is principally a brachiopod fauna (Fig. 2b.8) that has affinities with the Malvinokaffric Fauna of the South Atlantic; bivalves and other groups are also present. A very sparse brachiopod fauna with similar affinities occurs in the Ellsworth Mountains. Fossil wood and plant microfossils have been recovered from the Devonian strata of the Pensacola Mountains, and Devonian shales with plant megafossils are known from a single outcrop in Marie Byrd Land (Grindley and Mildenhall, 1981). On

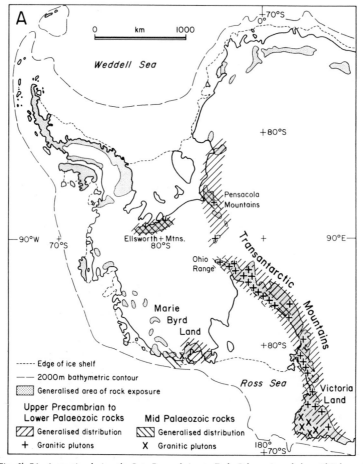

Fig. 2b.7A. Antarctica during the Late Precambrian to Early Palaeozoic and the mid-Palaeozoic.

the basis of these rather meagre data, it would appear that a magmatic arc, possibly part of a plate margin, crossed north Victoria Land and Marie Byrd Land. On the continental flank of the arc a broad shallow sea existed but the sedimentary record is not obviously related to an adjacent plate margin.

## 2.b.3.2. Antarctica during Gondwana time: the Late Palaeozoic and Early Mesozoic

Rocks formed during this time span occur quite widely in West Antarctica, the Antarctic Peninsula and the Transantarctic Mountains (Fig. 2b.7B; Dalziel and Elliot, 1982). Along the Peninsula there are extensive outcrops of thick greywacke-shale sequences which locally include shallow-water marine sediments containing Triassic invertebrates. Metamorphic rocks that represent similar sediments plus cherts, basaltic volcanic rocks and limestones are known from the South Orkney and South Shetland Islands. In scattered places in the Antarctic Peninsula and West Antarctica, granitic plutons of this age crop out. By far the most extensive deposits, however, are the flat-lying sedimentary rocks of the Victoria Group in the Transantarctic Mountains. The Victoria Group, ranging in age from early Permian to Triassic — Jurassic, forms the Gondwana sequence of Antarctica and has great similarities

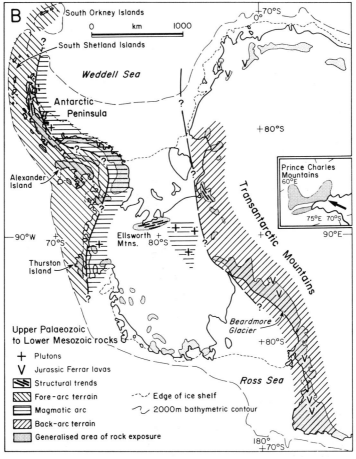

Fig. 2b.7B. Antarctica during the Late Palaeozoic to Early Mesozoic. Limited outcrops of Permian coal measures occur on the margin of the Prince Charles Mountains; location marked by arrow in inset.

Fig. 2b.8. a. *Proplina* (Class Monoplacophora); X 1.6; Ellsworth Mountains. Photo courtesy of John Pojeta (to be described by G. Webers, J. Pojeta and E. Yochelsen).

b. *Dicroidium odontopteroides*; X 1.5; Mt. Bumstead, Transantarctic Mountains.

c. *Araucarioxylon*; X 0.4; Fremouw Peak, Transantarctic Mountains. Collected by the late J.M. Schopf.

d. *Maorites seymourianus*; X 0.5; Seymour Island, Antarctic Peninsula. Collected by C. Macellari.

e. *Carcharodon auriculatus* (shark tooth); X 0.8; Seymour Island, Antarctic Peninsula. From Welton and Zinsmeister (1980) by permission of the Natural History Museum of Los Angeles County.

f. *Pleurothyrella antarctica* (Phylum Brachiopoda); X 1.2; Ohio Range, Transantarctic Mountains. From Doumani *et al.* (1965) by permission of the American Geophysical Union.

g. *Struthioptera camachoi*; X 0.7; Seymour Island, Antarctic Peninsula. From Zinsmeister (1977) by permission of the *Journal of Paleontology*.

h. *Oreochima*; X 1; Storm Peak, Transantarctic Mountains. From Schaeffer (1972) by courtesy of the American Museum of Natural History.

with Gondwana sequences elsewhere and particularly that in southern Africa. Above the glacial strata (tillites) at the base, the sequence is made up almost entirely of fluvial sedimentary rocks with interbedded volcanic ash and breccia in the upper part.

The geologic record for the Late Palaeozoic to Early Mesozoic demonstrates the existence of an active plate margin. The greywacke-shale sequences and other rocks of the Antarctic Peninsula represent the fore-arc region between the magmatic arc and the oceanic trench, a region of high sediment input, submarine slope instability to cause turbidity currents and hence the deposition of greywacke, and deep water environments. This fore-arc terrain was backed by a magmatic arc now represented by scattered plutonic rocks in the Antarctic Peninsula and West Antarctica and by the abundant contemporaneous volcanic ash in the Victoria Group and the Gondwana sequence of South Africa. The post-glacial Permian and Triassic Victoria Group rocks clearly form part of an extensive non-marine sedimentary basin (part of the back-arc terrain) that stretched from the flank of the magmatic arc onto the old Precambrian craton.

The Triassic invertebrates that allow age determinations for the Antarctic Peninsula rocks include the bivalves *Neoschizodus*, *Bakevelloides*, *Balantioselena* and *Waagenoperna* which have close similarities with forms from other parts of the circum-Pacific, and gastropods allied to forms in South America (Thomson, 1982). The paucity of fossils in the marine rocks is counterbalanced by the Gondwana floras and faunas from the Victoria Group and its equivalents. The Permian macroflora is dominated by the Gymnospermous *Gangamopteris* and *Glossopteris* floras with the former predominant in the beds immediately above the glacials and the latter predominant in the upper coal-bearing rocks (Schopf and Askin, 1980). Minor components of the flora include lepidophytes, cordaiteans, and sphenopsids such as *Phyllotheca*. The leaves are occasionally accompanied by seeds and rare fructifications. Fossil wood is widely preserved and normally displays well developed growth rings. Perhaps of greatest importance is the fossil peat found near the head of the Beardmore Glacier; this material is analogous to the Carboniferous coal balls of the northern hemisphere (Schopf, 1970).

The Triassic floras are not nearly so common but, as with the Permian floras, are dominated by a single genus, in this case the corystosperm *Dicroidium* (Fig. 2b.8; Townrow, 1967). Conifer, cycad and arthrophyte remains accompany it. *Dicroidium* has also been found at Livingston Island in the South Shetland Islands (Lacey and Lucas, 1981). Fossil wood, much of which shows good growth rings (Fig. 2b.8), is abundant locally and includes logs up to 22 m long. A Triassic permineralized deposit, similar to the Permian peat just mentioned, has also been found near the Beardmore Glacier (J. M. Schopf, 1978). Fragmentary plant material of Jurassic age, mainly ferns with subordinate conifers and cycads, is present in the sedimentary interbeds in the basalt sequence. Plant microfossils are widespread and provide the best tool for biostratigraphic zonation.

The lower Triassic rocks have yielded an excellent *Lystrosaurus* zone fauna (Hammer and Cosgriff, 1981). The therapsid *Lystrosaurus* dominates the fauna and at least three species, *L. murrayi*, *L. curvatus*, and *L. mccaigi*, of the nine that comprise the genus have been identified in the Antarctic material. In addition, *Thrinaxodon liorhinus*, a carnivorous therapsid, *Myosaurus gracilis*, an herbiverous therapsid, and *Procolophon trigoniceps*, a member of the anapsida, are present. On the generic level the eosuchian *Prolacerta* has been tentatively identified. Cynodontid reptile and thecodent remains compose the rest of the reptilian fauna. Labyrinthodont amphibians are an important part of the total fauna and include two new species, the brachyopod *Austrobrachyops jenseni*, and the lydekkerinid *Cryobatrachus kitchingi*. A single tooth, possibly of the lungfish genus *Ceratodus*, is the only fish material recovered so far. The fauna shows strong correspondence with that of the *Lystrosaurus* Zone of South Africa, both in the species present and their relative proportions.

Fossil fish occur in the sedimentary interbeds in the Jurassic lavas (Schaeffer, 1972) and the majority belong to the order Pholidophoriformes and to a new actinopterygian genus *Oreochima* (Fig. 2b.8). The closest affinities are with Australian faunas.

Invertebrates occur very sparsely in the Gondwana rocks except in the sedimentary interbeds associated with the Jurassic lavas. The Permian rocks have yielded conchostracans in the Ohio Range and a beetle wing in the Ellsworth Mountains. No invertebrates have been found in Triassic strata; however, the Jurassic beds have yielded conchostracans, notostrachans, isopods, syncarids, and ostracods (all fresh water crustaceans), and four orders of insects (Ball *et al.*, 1979). These crustaceans have strong affinities with other southern hemisphere faunas.

Broad land connections are implied by the vertebrate faunas: no barrier existed to free interchange between Antarctica and South Africa, and it can reasonably be assumed that the same situation existed between Antarctica and Australia, India and South America.

Antarctic Gondwana was dominated geologically by the evolution of the active plate margin along the 'Pacific' border, and climatically by a change from glacial to non-glacial temperate conditions. The glacial conditions, so well documented in the tillites that occur across Gondwana, are attributed to a polar position, but the arrangements of land and sea differed greatly from those of today. The glacial environment appears to have terminated abruptly and was replaced by climatic conditions suitable for abundant plant growth and its accumulation eventually to form coal. Together with well-developed growth rings in fossil wood, it suggests a strongly seasonal temperate climate. The homogeneity of the Permian Gondwana floras — most specimens in all assemblages can be referred to a small number of species — point to a climatically uniform floral province. Conditions changed little during the Triassic, the vegetation again being locally abundant and forming coal swamps.

## 2b.4. GONDWANA BREAK-UP

The Lower Jurassic lavas that cap the Beacon rocks are part of the Ferrar Supergroup, a related set of lavas, sills (horizontal sheets of igneous rock emplaced, normally, within a sedimentary sequence) and massive plutonic rocks of which the Dufek Intrusion is the most important.

The Ferrar, which may well have overlapped the late stages of sedimentation and volcanism associated with the Gondwana active plate margin, marks the first geological event, dated at about 180 Ma (Early Jurassic), related to the dispersal of the supercontinent (Kyle *et al.*, 1981).

The lavas, sills and plutons, almost all of basaltic composition, are confined to the Transantarctic Mountains and their extension along western Queen Maud Land. Their emplacement was accompanied by rifting across the whole continent 'Pacificward' of the present Transantarctic Mountains, and within 20 Ma the first new sea-floor, dated at about 160 Ma, had been generated between East and West Gondwana in what is now the Weddell Sea (LaBrecque and Barker, 1981). The former wide land junctions that had existed for hundreds of millions of years were now breaking down or would do so in a short geological time span. The unity of the supercontinent was broken but there remained many pathways, albeit some narrow, others extensive, for floral dispersal and faunal migration right up to the mid-Cenozoic.

### 2b.4.1. Antarctica during Gondwana dispersal: mid-Mesozoic to mid-Cenozoic time

The emplacement of the Ferrar Supergroup marks the initial stage, in Antarctica, of the fragmentation of Gondwana. Excluding the late Cenozoic, rocks post-dating this event are found only along the Antarctic Peninsula, the coast of West Antarctica (Fig. 2b.9A), and on the continental shelves. Sedimentary rocks on the western flank of the Antarctic Peninsula are Late Jurassic to Early

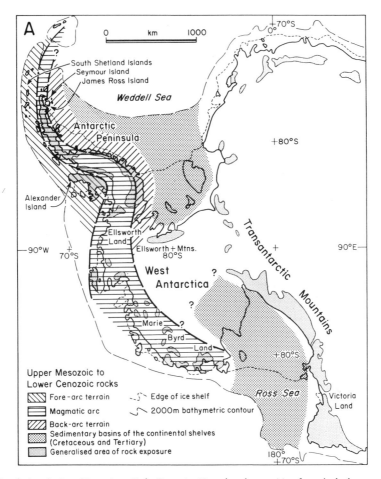

Fig. 2b.9A. Antarctica during the Late Mesozoic to Early Cenozoic. Note that the transition from the back-arc terrain of the Antarctic Peninsula into the continental shelf deposits is gradual rather than abrupt as might be inferred from the illustration.

Cretaceous in age. In the South Shetland Islands the limited outcrops consist largely of greywacke-shale sequences containing much volcanic debris. In contrast, the Alexander Island sequence, the Fossil Bluff Formation, consists of interbedded volcanic and sedimentary rocks in the lower part overlain by a thick sequence of shallow marine clastic sediments with a significant volcanic content and containing an abundant invertebrate fauna; the uppermost part is non-marine and plant-bearing. There are, in addition, a few small outcrops of marine beds scattered along the west coast of the Peninsula. To the east and southeast lies a chain of Upper Jurassic and Cretaceous granitic plutons that form the backbone of the Antarctic Peninsula; thick sequences of lavas, tuffs and breccias are associated with this belt of plutons. The string of plutons can be traced throughout the Antarctic Peninsula, into Ellsworth Land and thence into Marie Byrd Land. Mesozoic sedimentary rocks on the Weddell flank of the Peninsula crop out principally from eastern Ellsworth Land north to about latitude 72°S and in the James Ross Island area. In the southern Peninsula these rocks are fossiliferous marine shales and sandstones of Middle and Late Jurassic age, whereas in the James Ross Island area they are Cretaceous marine conglomerates, sandstones and shales with locally an abundant invertebrate fauna. Marine and non-marine clastic sediments ranging in age from Palaeocene to Early Oligocene overlie the Cretaceous strata of the James

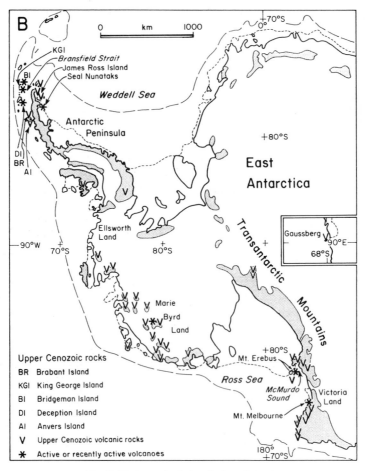

Fig. 2b.9B. Antarctica during the Late Cenozoic.

Ross Island region. Other Lower Cenozoic rocks are largely volcanic and confined to the South Shetland Islands and other offshore islands along the northwestern part of the Antarctic Peninsula.

The marine sediments of the Antarctic Peninsula region contain abundant and varied invertebrate faunas. Molluscs, principally bivalves and ammonites, dominate the faunas; gastropods, nautiloids, belemnites, echinoderms, crinoids, asteroids, brachiopods, bryozoans, corals, annelids, crustaceans and foraminifera have also been recovered (Thomson, 1977) and some are important biostratigraphically. On Alexander Island the Fossil Bluff Formation contains ammonites, mainly perisphinctids, belemnites (16 species of Belemnopseidae), and bivalves such as *Inoceramus*, *Buchia*, and members of the Trigonid and Grammatodontid families. The faunas of late Jurassic to earliest Cretaceous age show affinities to the Indo-Pacific faunas, whereas the younger faunas show a high degree of provincialism. On the Weddell flank of the southern Peninsula the Jurassic faunas are dominated by bivalves but ammonites, belemnites and crinoids are also important; strong affinities exist with New Zealand faunas. Ammonites are the principal component of the rather sparse faunas in the South Shetlands. In the James Ross region the faunas are again dominated by ammonites, mainly kossmaticeratids (Fig. 2b.8), but a wide variety of other molluscs including *Inoceramids*, which are important for dating, are also present. Molluscs are the main phylum represented in the Eocene faunas of Seymour Island and consist of 34 gastropod, 23 bivalve and two nautiloid taxa; the most common genera are the bivalve, *Cucullaea* and the gastropods *Antarctodarwinella* and *Struthioptera* (Fig. 2b.8). The faunas in the overlying Lower Oligocene beds are markedly different; only seven species out of a total of over 100 invertebrate taxa reported are common to both parts of the section and the diversity is much reduced (Zinsmeister and Camacho, 1982). The dominant species are now the gastropods *Struthiolarella* and *Perissodonta*, and the bivalve *Cucullaea* but a species different from that in the Eocene rocks. The only vertebrate fossils recovered so far from the Cretaceous beds are plesiosaurs, mososaurs, fish vertebrae and shark teeth. The Lower Cenozoic (Palaeogene) beds on the other hand have yielded numerous shark teeth (Fig. 2b.8), abundant penguin bones, and subordinate reptile (turtle), fish and whale (*Zeuglodon*?) remains, and most important of all, fragments of three jawbones of the extinct marsupial genus *Polydolops* (Woodburne and Zinsmeister, 1982), previously known only from Patagonia. The fossil penguins, represented by four genera including a giant form up to 135 cm tall, have strong affinities with the New Zealand fossil faunas.

Plant fossils, particularly leaf and wood material, are widespread in the Mesozoic rocks. Leaf remains are principally of pteridophytes, cycads and conifers such as those recovered from Upper Jurassic beds at Hope Bay and Lower Cretaceous rocks on Alexander Island. The latter also contain fossilized stumps of coniferous trees preserved in growth position (Jefferson, 1982). Angiosperm leaves have been collected from Lower Cretaceous beds on Alexander Island (T. H. Jefferson, personal communication) and from Lower Tertiary beds on the same island and on Seymour Island. Plant microfossils are present in the James Ross Island basin rocks and in the Cenozoic rocks the most common forms are podocarpaceous conifer pollen and *Nothofagidites* (forms similar to Nothofagus) (Askin and Elliot, 1982). Dinoflagellate cysts have also been recovered.

The Late Mesozoic to Early Cenozoic history of the Pacific margin of Antarctica records an active plate margin. The sedimentary sequences of the western flank of the Antarctic Peninsula constitute the fore-arc terrain of a magmatic arc that is documented in the abundant plutonic and volcanic rocks of the Peninsula and their correlatives scattered through to Marie Byrd Land. The sedimentary sequences on the Weddell flank of the Peninsula form the back-arc terrain which includes in the James Ross Island area a sequence that passes upward into deltaic rocks. This Andean active plate margin indicates consumption of oceanic crust beneath the 'Pacific' rim of Antarctica, but it is only the Cenozoic history that can be related to the sea-floor anomaly pattern of the southeastern Pacific.

The major part of the geological history of the continent during the Late Mesozoic and Early Cenozoic is not recorded in any exposed rocks, but is surely present on the continental shelves and in West Antarctica.

Basins that contain sediment several kilometres thick have been identified by geophysical surveys in the Ross Sea, Weddell Sea and the Amery Ice Shelf region, and have been inferred for other parts of the continental shelf. The ages of these basins are uncertain although likely to be late Mesozoic to Cenozoic; the oldest recovered sediment, from off north Victoria Land, is Early Cretaceous (Domack *et al.*, 1980). The sedimentary basins of the Ross and southern Weddell Seas project southward beneath the ice shelves and possibly contain great thicknesses of post break-up strata. This leads into the problem of the evolution of West Antarctica.

The geology of the Ellsworth Mountains, a deformed Palaeozoic sequence clearly related in its upper part to the Gondwana glacial beds and *Glossopteris*-bearing strata of the Transantarctic Mountains, and having a tectonic fabric aligned at right angles to that of the similar-aged rocks in the Pensacola Mountains and to that of the younger Andean plate margin, led to the suggestion that they are part of a rotated and displaced continental block (Dalziel and Elliot, 1982). This implies the fragmentation of the Gondwana active plate margin, and possible thinning of the continental crust by rifting processes. The crustal thickness and sub-glacial depths make parts of West Antarctica analogous to features such as the Campbell Plateau to the east of New Zealand and to the Falkland Plateau east of southern South America; both are sunken blocks of continental crust resulting from continental break-up. The inferred sub-glacial sedimentary basins in West Antarctica, therefore, probably date from sometime in the Jurassic. This low area through West Antarctica would have been a shallow seaway across the continent (Transantarctic Seaway) prior to the opening of the Drake Passage in mid-Cenozoic time, and possibly a means by which surface circumpolar waters circulated long before the full circumpolar deep water circulation was formed in late Cenozoic time.

The late Mesozoic to early Cenozoic was a time of diminishing land connections in the southern hemisphere (Fig. 2b.10). After the initial rifting and separation between Antarctica and South America, recorded in the Weddell Sea magnetic anomalies, land connections were possibly maintained via the microplates of the Antarctic Peninsula and West Antarctica. Southern Africa had separated from Antarctica by 145 Ma and from South America at about 128 Ma. India separated from Antarctica at about 105 Ma; however, anomalies remain in the vertebrate distribution if the sea floor data are presumed to be correct (Colbert, 1981). New Zealand had separated from Australia by about 80 Ma and from Antarctica by 75 Ma. An elongate shallow seaway existed during Mesozoic time between Antarctica and Australia, and the oldest sea-floor magnetic anomaly is dated at 80 Ma (Cande and Mutter, 1981). However this was not the final break because it was not till the South Tasman Rise cleared the coast of Antarctica at $30-25$ Ma that deep water flow could have been initiated. The relations between the Antarctic Peninsula and South America are not constrained yet by sea-floor data prior to 29 Ma, the time of initial opening of the Drake Passage. Before that time barriers to faunal exchange of invertebrates may have existed; certainly a major oceanographic boundary was present in the early Cenozoic because of the lack of correspondence between the faunas of that age from Seymour island and Patagonia. The opening of the Drake Passage in the late Cenozoic completed the physical isolation of the continent.

The climate during the Late Mesozoic and Early Cenozoic was equable and probably warm temperate, as indicated by the common occurrence of broad angiosperm leaves and logs from large trees, and certainly seasonal, as shown by the well-defined growth rings in these logs. Oxygen isotopic studies on marine organisms, however, suggest a gradual cooling of the surrounding ocean during the Late Cretaceous and Early Tertiary.

## 2b.4.2 Late Cenozoic history

The advent of glacial conditions and the development of alkaline volcanic provinces (Fig. 2b.9B) conveniently separate the geological events of the last 25 to 30 Ma from those of the Late Mesozoic and Early Cenozoic. Sedimentation on the continental shelves continued throughout the Late Cenozoic though exposures of Upper Cenozoic rocks are very sparse. Pliocene to Pleistocene marine deposits are known at Cockburn Island and in the South Shetlands, and contain foraminifera, bivalves, brachiopods, crustacea and bryozoa, many of which are closely allied to living forms.

Most information on the late Cenozoic is derived from cores from the continental shelves and the deep ocean floor around the continent. Glacially-related sediments dominate the record on all the continental shelves of Antarctica. The geologic record in the deep ocean basins out to the Antarctic Convergence is characterized by alternating biosiliceous and ice-rafted sediment with, locally, thick wedges of sediment derived from the continent.

The late Cenozoic is also characterized by volcanism (Elliot, 1975). The volcanic provinces of the Transantarctic Mountains, Marie Byrd Land and Ellsworth Land are dominated by large stratovolcanoes; a few are still active such as Mt. Erebus, or recently active (Mt. Melbourne). The composition of these volcanoes is indicative of continental rifting. A single recently active volcano is located at Gaussberg at 90°E. Late Cenozoic volcanism is widespread on the northern Antarctica

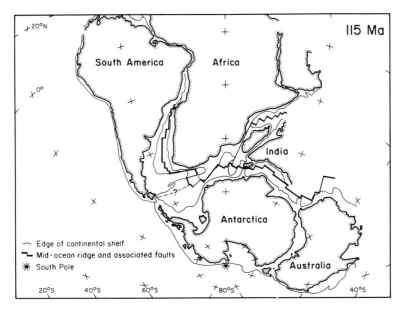

Fig. 2b.10. Stages in the dispersal of the Gondwana continents (after Norton and Sclater, 1979). No account is taken here of the possibility of rearrangement of West Antarctica. Note that only a very limited separation of Antarctica and Australia had occurred by 53 Ma, and also that South America and the Antarctic Peninsula were most likely still abutting each other at 39 Ma. Key for all figures in reconstruction for 115 Ma. The position of the south Pole is that inferred for the time given in each of the diagrams.

Peninsula. In the South Shetland Islands it includes the active or recently active centres of Deception, Penguin and Bridgman Islands, and the slightly older lavas of the south coast of King George Island.

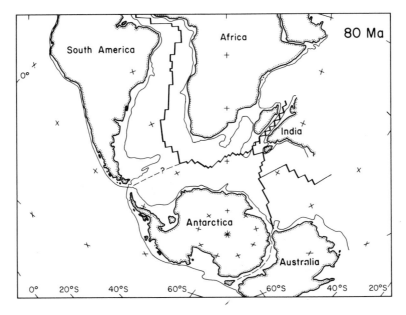

Fig. 2b.10. *cont.*

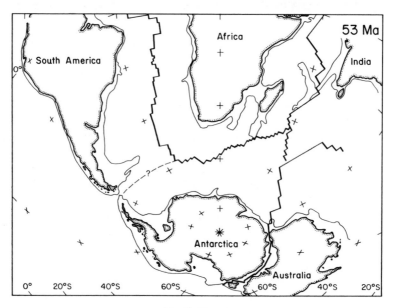

Fig. 2b.10. *cont.*

These rocks are related to the opening of the Bransfield Strait trough within the last 2 Ma. Probable late Cenozoic lavas and pyroclastic rocks occur on Anvers and Brabant Islands. The James Ross Island Volcanic Group on the east coast is 6.0 Ma and younger, and includes still active centres on the Seal Nunataks. The South Sandwich Islands constitute an active volcanic arc above a subduction zone and are no older than about 8 Ma.

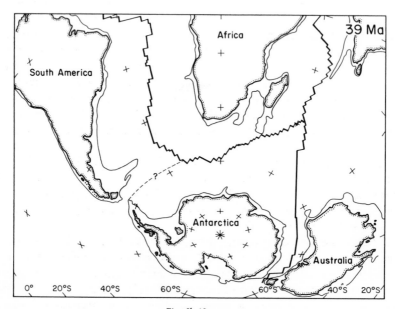

Fig. 2b.10. *cont.*

## 2b.5. PALAEOCLIMATE AND FORMATION OF THE CONTINENTAL ICE

The palaeoclimate of the late Cenozoic shows, with the exception of a warming interval in the middle Miocene (23 – 16 Ma), a progressive cooling that is a continuation of a trend inferred from oxygen isotopic studies for the Early Cenozoic and part of the Late Cretaceous (Shackleton and Kennett, 1975). The abrupt faunal change from the lower (Eocene) to the upper (Oligocene) part of the Seymour Island section occurred at about 38 Ma and may reflect changing circulation or a palaeoclimatic event; no other evidence exists from land exposures in Antarctica though some of the undated glacial deposits at high elevations in the Transantarctic Mountains could date from the early stages of glaciation and could be pre-Miocene in age. The faunal change is placed in a broader Southern Hemisphere context by Zinsmeister (1982) who discusses the biogeographic evolution of the molluscan faunas in terms of continental drift. In the marine environment the cooling appears more gradual although there are abrupt jumps in the isotopic record (Kennett, 1978).

The sudden isotopic change at about 38 Ma has traditionally been interpreted as a major cooling of Antarctic surface waters with the first formation of sea ice, and is coincident with a southern hemispheric erosional event recorded in deep sea cores. The clearing of Antarctica by the South Tasman Rise at about 30 – 25 Ma as Australia moved northward initiated deep water circulation between those two continents. The Drake Passage opened at about 28 Ma with the establishment of full deep water circulation around Antarctica after about 23 Ma, thus completing the physical isolation of the continent. Before this, shallow water circulation may well have been possible in the Drake Passage region, and it is likely that such circulation existed across West Antarctica as well. The second major oxygen isotopic change occurred at about 15 Ma and has been regarded as the time of most rapid growth of the East Antarctica ice sheet, the time when it built out to the edges of the continent. Evidence from Antarctica itself for the development of the ice sheets is sparse and ideas on this topic are controversial. Because the West Antarctic ice sheet is grounded below sea level and therefore requires colder conditions for development than a land-based ice sheet, it is generally accepted that the ice sheet in East Antarctica was initiated first, and probably as mountain glaciation that expanded out from the Transantarctic Mountains and the Gamburtsev Mountains to cover the whole of East Antarctica. Ice-rifted debris in cores from the Ross Sea is the first indication of substantial ice reaching sea level and this occurred at about 25 Ma. Large ice shelves such as exist today may not have formed till the end of the Miocene when there was an abrupt 5° northward migration of the limit of ice-rafted debris in the Southern Ocean. The traditional view is that the ice started accumulating at about 30 Ma, and by 25 Ma was calving substantial numbers of bergs from West Antarctica into the Ross Sea region. The East Antarctic ice sheet attained its present dimensions by about 15AMa. However, it was not until about 7 Ma that the climate at sea level became cold enough for ice shelves to form; this set the stage for the local ice caps grounded above sea level in the archipelago of West Antarctica to expand into the marine basins within the archipelago, coalesce with the East Antarctic ice sheet, and become grounded at great depth below sea level.

This traditional view has been recently challenged by Matthews and Poore (1980). They argue that the isotopic jump at 38 Ma reflects a major increase in global ice volume and that the existence of significant continental ice before that time should not be discounted. They interpret the 15 Ma isotopic event as a change in bottom water temperature alone. This and other aspects of Southern Hemisphere glaciation are reviewed by Mercer (1983).

The fluctuations of the ice sheet since the inferred full development about 15 Ma ago are poorly known. From the Dry Valleys of South Victoria Land there is evidence that ice overrode the Transantarctic Mountains twice; the first time was probably before the Late Miocene and certainly before 4 Ma, the radiometric age of basalt cones in the glacially-carved Dry Valleys. Other evidence of possible fluctuations is largely derived from proxy data gathered in Patagonia and from deep sea cores, and suggests ice expansion and cool periods in the interval 7 to 4.6 Ma, about 3.6 Ma and repeatedly after about 2.5 Ma. It seems likely that the major ice changes in Antarctica have been associated with changes in the grounding line of the ice: if sea level falls (for instance, due to larger ice volume in the northern hemisphere) the ice fronts will advance. The greatest advances and retreats would have occurred where the continental shelf is widest, that is the Ross and Weddell embayments which today contain the large ice shelves. Those ice shelves and that part of the West Antarctic ice sheet grounded below sea level may well have disappeared in the interval 3.4 to 2.5 Ma and also during the last Interglacial (Sangamon or Eem) *ca* 125,000 years ago. The record in the Dry Valleys shows four separate periods of ice expansion in the last 1.2 Ma, the most recent maximum being *ca* 17,000 to 21,200 BP. Ice was still grounded in McMurdo Sound 8300 BP and did not reach its present grounding line near the head of the Ross Ice Shelf till about 6700 BP.

## 2b.6. ECONOMIC GEOLOGY: POTENTIAL RESOURCES

This survey of the geological history of the Antarctic continent would be incomplete without mention of the non-living resources (Splettstoesser, in press). There are no known economically exploitable mineral or hydrocarbon resources in Antarctica, and it is doubtful whether any resource now known would constitute an economic deposit even if located in a less harsh environment closer to centres of industry.

One major banded iron formation, of the type that occurs around Lake Superior, has been found in the Prince Charles Mountains (Fig. 2b.11) and aeromagnetic studies show that it extends sub-glacially for as much as 180 km. The grade of ore is such that it would be marginal even under the best conditions. There are, in addition, a number of places in East Antarctica where iron ore occurs in lenses or pods, but these are of limited extent. Magnetite layers are known in the Dufek Intrusion, a layered basic intrusion which can be compared with the Bushveld complex in South Africa. Magnetite forms seams a metre or more thick and the grade of ore is about 50% Fe. In all probability chromite occurs at depth in the lower, and unexposed, part of the intrusion. Whether platinum group metals also occur at depth is another matter; they cannot be presumed to do so. For the mining geologist, the Dufek is unquestionably the prime target in Antarctica: whether it has either chromite seams sufficiently thick or platinum-bearing horizons of sufficient grade to be economically exploited is in the realm of speculation. Only by drilling will this ever be known.

Vein and disseminated copper sulphide mineralization is present in the Andean plutons of the southern Antarctic Peninsula. Associated minerals include sulphides of iron, molybdenum, lead, zinc and silver. Similar mineralization has been reported from many localities in the northern Peninsula. The mineralization is like that associated with porphyry copper deposits of the Andes; however, as yet no well-documented porphyry-type deposit has been reported from Antarctica. Furthermore, all known occurrences are uneconomic even without the penalties imposed by Antarctica.

Numerous other metallic and non-metallic minerals occur in trace amounts; none can be regarded as anything but a mineral occurrence, in other words an interesting find. Doubtless, if Antarctica were in a temperate climate, much prospecting would have been done by now.

Coal beds occur throughout the Victoria Group and its equivalents in the Transantarctic Mountains and elsewhere. Most occur in Permian coal measures; a minor amount is Triassic in age. Known coal beds are up to 10 m thick but more commonly are 0.5 to 1.0 m in thickness; seldom can they be traced for more than 2 km in outcrop. The coals that have been analysed are mainly high ash, high sulphur, high rank bituminous coals. The depositional environment for the coal, a non-marine floodplain, is such that most coal beds will be lenticular. Even allowing for the lack of a regional assessment of the true

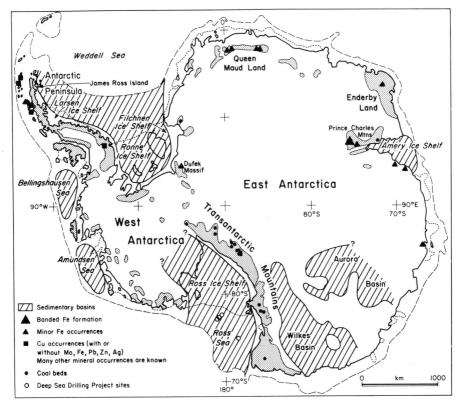

Fig. 2b.11. Mineral occurrences and late Mesozoic — Cenozoic sedimentary basins in Antarctica.

extent of the coals and the lack of adequate sample analyses, it is difficult to see how these coals, which in no way compare with those of North America in extent or thickness, could ever be economically exploited.

The possibility or probability of natural gas and liquid hydrocarbons in the sedimentary rocks around Antarctica has been discussed and commented on frequently in recent years. All estimates of the reserves around Antarctica are quite without the scientific foundation that would make them credible. The occurrence of gas in cores taken from the Ross Sea by the Deep Sea Drilling Project has been used as an argument for deposits being present. Most of the gas consisted of methane, which can be formed in any sedimentary rock containing organic debris. The higher homologues of methane are better indicators, but the amount and proportion of compounds of higher molecular weight were very small and thus of uncertain significance. With one exception, all analogies drawn so far with oil-bearing basins in formerly adjacent continents lack adequate supporting data. The one exception is the late Mesozoic — early Cenozoic sedimentary basin of the James Ross Island area; this basin is analogous in some respects, but not all, to the Magallanes basin of Tierra del Fuego. It differs in that the reservoir rock, the Springhill Sandstone, is not known to occur in the James Ross region; it also differs in its tectonic history which may reflect, or be reflected in, the heat input into the rocks and thus the maturation of the organic material.

This short summary is not very encouraging; however, some points need to be kept in mind. Ninety-eight per cent of the continent is ice-covered and to all intents and purposes not accessible. There is no reason to believe that Antarctica, as a continent, differs from the other southern continents; it is reasonable to presume that metallogenetic provinces such as those of the Precambrian cratons of South Africa, India and western Australia, and the fold belts along the Pacific rim, also exist in Antarctica. It would be most surprising if there were no metallic resources in a continent the size of Antarctica. Furthermore, sedimentary basins undoubtedly exist around the continental shelves and in West Antarctica; it would be equally surprising if hydrocarbons are totally absent.

At present, we do not know what the metallic, non-metallic and hydrocarbon resources might be. Whether any resource could ever be economically exploited considering the cost penalty for working in so remote an area remains uncertain. It is not improbable that factors other than economics will dictate whether or not exploitation occurs.

# REFERENCES

Askin, R.A. and Elliot, D.H. (1982) Geologic implications or recycled Permian and Triassic polynomorphs in Tertiary rocks of Seymour Island, Antarctic Peninsula. *Geology* 10, 547 – 51.

Ball, K.W., Borns, H.W., Hall, B.A., Brooks, H.K., Carpenter, F.M. and Delevoryas, T. (1979) Biota, age, and significance of lake deposits, Carapace Nunatak, Victoria Land, Antarctica. In *IV International Gondwana Symposium: Papers (volume 1)*. Eds B. Lasker and C.S. Rajo Rao, pp. 166 – 75. Hindustan, Delhi.

Cande, S.C. and Mutter, D.H. (1982) A revised identification of the oldest sea floor spreading anomalies between Australia and Antarctica. *Earth Planet. Sci. Lett.* 58, 151 – 60.

Colbert, E.H. (1981) The distribution of tetrapods and the break-up of Gondwana. In *Gondwana V*. Eds M.M. Cresswell and P. Vella, pp. 277-82. A.A. Balkema, Rotterdam.

Craddock, C. (1982) *Antarctic Geoscience*. Madison, University of Wisconsin Press.

Dalziel, I.W.D. and Elliot, D.H. (1982) West Antarctica: Problem child of Gondwanaland. *Tectonics* 1, 3 – 19.

Domack, E.W., Fairchild, W.W. and Anderson, J.B. (1980) Lower Cretaceous sediment from the East Antarctic continental shelf. *Nature, Lond.* 287, 625 – 6.

Doumani, G.A., Boardman, R.S., Rowell, A.J., Boucot, A.J., Johnson, J.G., McAlester, A. Lee, Saul, John, Fisher, D.W. and Miles, R.S. (1965) Lower Devonian Fauna of the Horlick Formation, Ohio Range, Antarctica. In *Geology and Paleontology of the Antarctic, Antarctic Research Series (Volume 6)*. Ed. Jarvis B. Hadley, pp. 241 – 81. American Geophysical Union, Washington, D.C.

Du Toit, A.L. (1937) *Our Wandering Continents*. Oliver and Boyd, Edinburgh.

Elliot, D.H. (1975) Tectonics of Antarctica: A review. *Am. J. Sci.* 275A, 45 – 106.

Gindley, G.W. and Mildenhall, D.C. (1981) Geological background to a Devonian plant fossil discovery, Ruppert Coast, Marie Byrd Land, West Antarctica. In *Gondwana V*. Eds M.M. Cresswell and P. Vella, pp. 23 – 30. A.A. Balkema, Rotterdam.

Hammer, W.R. and Cosgriff, J.W. (1981) *Myosaurus gracilis*, an anomodont reptile from the Lower Triassic of Antarctica and South Africa. *J. Paleont.* 55, 410 – 24.

Harland, W.B., Cox, A.V., Llewellyn, P.G., Pickton, C.A.G., Smith, A.G. and Walters, R. (1982) A geologic time scale. Cambridge Earth Sciences Series, 131 pp.

Jefferson, T.H. (1982) Fossil forests from the Lower Cretaceous of Alexander Island, Antarctica. *Palaeontology* 25, 680 – 700.

Kennett, J.P. (1978) Cenozoic evolution of Antarctic glaciation, the circum-Antarctic Ocean, and their impact on global paleoceanography. *J. Geophys. Res.* 82, 3843 – 60.

Kyle, P.R., Elliot, D.H. and Sutter, J.F. (1981) Jurassic Ferrar Super-group tholeiites from the Transantarctic Mountains, Antarctica, and their relationship to the initial fragmentation of Gondwana. In *Gondwana V*. Eds M.M. Cresswell and P. Vella, pp. 283 – 7. A.A. Balkema, Rotterdam.

LaBrecque, J.L. and Barker, P.F. (1981) The age of the Weddell Basin. *Nature, Lond.* 290, 489 – 92.

Lacey, W.S. and Lucas, R.C. (1981) The Triassic flora of Livingston Island, South Shetland Islands. *Bull. Br. Antarct. Surv.* 53, 157 – 73.

Laird, M.G. (1981) Lower Palaeozoic rocks of Antarctica. In *Lower Palaeozoic of the Middle East, Eastern and Southern Africa, and Antarctica*. Ed. C.H. Holland, pp. 257 – 314. John Wiley & Sons, Inc., New York.

Matthews, R.K. and Poore, R.Z. (1980) Tertiary – $^{18}$0 record and glacio-eustatic sea-level fluctuations. *Geology* 8, 501 – 4.

McElhinny, M.W. (1973) *Palaeomagnetism and Plate Tectonics*. Cambridge University Press, Cambridge.

Mercer, J.H. (1983) Cenozoic glaciation in the Southern Hemisphere. *Ann. Rev. Earth Planet. Sci.* 11, 99 – 132.

Norton, I.O. and Sclater, J.G. (1979) A model for the evolution of the Indian Ocean and the breakup of Gondwanaland. *J. Geophys. Res.* 84, 6803 – 30.

Schaeffer, B. (1972) A Jurassic fish from Antarctica. *Amer. Mus. Novit.* 2495, 1 – 17.

Shackleton, N.J. and Kennett, J.P. (1975) Paleotemperature history of the Cenozoic and the initiation of Antarctic glaciation: oxygen and carbon isotope analyses in DSDP sites 277, 279, and 281. In *Initial Reports of the Deep Sea Drilling Project* 29, 743 – 56.

Schopf, J.M.(1970) Petrified peat from a Peruvian coal bed in Antarctica. *Science* 169, 276 — 70.

Schopf, J.M. (1978) An unusual osmundaceous specimen from Antarctica. *Can. J. Bot.* 56, 3083 — 95.

Schopf, J.M. and Askin, R.A. (1980) Permian and Triassic floral biostratigraphic zones of southern land masses. In *Biostratigraphy of Fossil Plants*. Eds D.L. Dilcher and T.N. Taylor, pp. 119-52. Dowden, Hutchinson & Ross, Stroudsburg, Pa.

Splettstoesser, J.F. (in press) Mineral resource potential of Antarctica. University of Texas Press, Austin.

Taylor, F.B. (1910) Bearing of the Tertiary Mountain belt on the origin of the earth's plan. *Bull. Geol. Soc. Am.* 21, 179 — 226.

Thomson, M.R.A. (1977) An annotated bibliography of the palaeontology of Lesser Antarctica and the Scotia Ridge. *N.Z. J. Geol. Geophys.* 20, 865 — 904.

Thomson, M.R.A. (1982) Mesozoic paleogeography of West Antarctica. In *Antarctica Geoscience*. Ed. C. Craddock, pp. 331 — 7. University of Wisconsin Press, Madison.

Townrow, J.A. (1967) Fossil plants from Allan and Carapace Nunataks, and from the Upper Mill and Shackleton Glaciers, Antarctica. *N.Z. J. Geol. Geophys.* 10, 456 — 73.

Wegener, A. (1912) Die entstehung der Kontinente. Petermanns Mittelungen 1912, 185 — 95. [See also Wegener, A. (1929). The origin of continents and oceans. Translated by John Biram (1966) from the 6th (1929) German edition. Dover, New York.]

Welton, B.J. and Zinsmeister, W.J. (1980) Eocene Neoselachians from the La Meseta Formation, Seymour Island, Antarctic Peninsula. *Contributions in Science* 329, 1 — 10. Natural History Museum of Los Angeles County.

Wilson, J.T. (1972) *Continents adrift; readings from Scientific American*. San Francisco. W.H. Freeman and Company.

Woodburne, M.O. and Zinsmeister, W.J. (1982) The first fossil land mammal from Antarctica. *Science* 218, 284 — 6.

Wyllie, P.J. (1976) *The Way the Earth Works*. 296 pp. John Wiley & Sons, Inc., New York.

Zinsmeister, W.J. (1977) Note on a new occurrence of the Southern Hemisphere Aporrhaid gastropod *Struthioptera* Finlay & Marwick on Seymour Island, Antarctica. *J. Paleont.* 51, 399 — 404.

Zinsmeister, W.J. (1982) Late Cretaceaous — Early Tertiary molluscan biogeography of the southern circum-pacific. *J. Paleont.* 56, 84 — 102.

Zinsmeister, W.J. and Camacho, H.H. (1982) Late Eocene (to possibly Earliest Oligocene) molluscan fauna of the La Meseta Formation of Seymour island, Antarctica Peninsula. In *Antarctica Geoscience*. Ed. C. Craddock, pp. 299 — 304. University of Wisconsin Press, Madison.

CHAPTER 2c

# Physical Geography — Soils

## G. G. C. CLARIDGE and I. B. CAMPBELL

Soil Bureau, Department of Scientific and Industrial Research, Private Bag, Lower Hutt, New Zealand

## CONTENTS

Ice-free areas occupy some 2% of the Antarctic continent and occur in the Antarctic Peninsula, along the Trans-Antarctic mountains, and in other scattered coastal situations, where ice sheets have retreated or where glaciers have diminished in volume and disappeared from valleys. In these, mostly moraine-covered bare ground, weathering and soil formation on the exposed rocks and glacial deposits has commenced.

Although many deposits lack an identifiable biological component the products of weathering in Antarctica have long been regarded as soils in which the influence of living organisms is very small (Claridge, 1965). Their development and properties are influenced by chemical and physical processes identical with those in soils of temperate regions. They may thus be properly called soils.

## 2c.1. SOIL-FORMING FACTORS

The climate of Antarctica is for the most part cold and arid. In so far as soil formation is concerned, the most important features are the low temperatures, which result in a very slow rate of chemical processes, a very low level of biological activity, and more importantly a very low level of moisture availability. Since moisture is only available for chemical processes when it exists in a liquid form, the length of time during which soil temperatures are above freezing point is particularly relevant. In regions where soil temperature rarely rises above freezing, soil moisture is almost completely absent and chemical weathering processes are minimal, while in regions where there is more pronounced thawing, the soils may be quite wet, and leaching and chemical weathering are more evident.

In many situations, especially in the rigorous environments of the Trans-Antarctic Mountains, biological activity is confined to a very few micro-organisms, and in the most extreme situations the soil may be practically sterile. The rates of weathering and soil formation in Antarctica are extremely slow and the soils are consequently extremely fragile with respect to environmental changes.

Most soils are formed on glacial moraines. Throughout much of the Trans-Antarctic Mountains, moraines are formed largely from comminuted sandstone and larger fragments of the harder and more resistant dolerites (the two most common rock types found); consequently the proportion of fine particle-size material, largely quartz sand, can be relatively high in the soil parent material. On the other hand, moraines formed largely from granite, metamorphic rocks or indurated sediments tend to be much coarser in texture. On exposed bedrock, the soils are frequently limited to residual occurrences of accumulations of fine-textured material occurring either as a thin veneer or within cracks and crevices of the rock.

Although weathering rates are extremely slow in Antarctica, the time scale available for soil formation is very long. It has been shown that some surfaces exposed by glacial retreat have been exposed to weathering processes for as much as 2−4 million years (Armstrong, 1978; Fleck et al., 1972). In other places it can be shown by extrapolation that the soils may be very much older than this, having formed on surfaces that have been exposed to weathering for periods much longer than 4 million years under climatic conditions very similar to those of the present day (Campbell and Claridge, 1975, 1978).

## 2c.2. SOIL ZONES

Soil moisture status is largely a function of climate and is dependent on soil temperature. In those regions where temperatures are lowest, snowfall is minimal because of low humidity, and little thawing takes place of snow or of ice within soil profiles (Fig. 2c.1). In the warmer coastal situations where humidity is greater, there is generally more snowfall, and the higher temperatures result in the presence of more ice or water within the soil. In these situations the soils may be moist or even wet for short periods when thawing occurs. From place to place, local differences are found caused by diurnal or seasonal changes as well as local site influences. Moisture availability is reflected in the depth of permafrost or frozen ground, which varies from close to the surface to more than 1 m. The soils of Antarctica fall into three major soil zones, based on climate and moisture availability: the dry valleys and bare ground expanses of the Trans-Antarctic Mountains, the oases of coastal Greater Antarctica, and the maritime Antarctic Peninsula.

### 2c.2.1. Trans-Antarctic mountains

In terms of temperature and aridity, the soils of the Trans-Antarctic Mountains are formed in the most extreme environment known on earth. Although surface temperatures may rise well above freezing for short periods, the permafrost always remains close to the surface. Precipitation, in the form

Fig.2c.1. Soil on a moraine covered bench in the Aagaard Range, above the Wright Valley. Note the absence of melting of the light snow cover.

of snow, is very slight and moisture available for soil development is derived either from small amounts of melting snow, or from existing ice within the soil. Moisture may move within the soil in the vapour phase. These soils have been extensively investigated (Campbell and Claridge, 1981) and the processes of soil formation are relatively well known. In general, a well-developed soil, i.e. one that has been exposed to weathering for a considerable period of time (more than 4 million years), of this region shows a number of quite distinctive features.

The soil surface is usually covered by a stone pavement, or lag gravel, consisting of the more resistant rocks, often in older soils stained and polished. The size of the boulders protruding through the stone pavement is a reflection of soil age, as in the old soils all but the most resistant rocks may be eroded to ground level. Below this stone pavement, there is often a thin (1 − 4 cm) horizon of fine gravel and sand which accumulated through fretting of the surface rocks.

The soil proper generally consists, especially if formed from moraine or till, of oxidized and stained finer-textured material (up to 4% clay) with accumulations of salts, either as a separate horizon or as discrete small scattered accumulations. The upper horizons are the most strongly oxidized, and staining diminishes downward, but may be detected to depths of a metre or more in the oldest soils.

Throughout the soil profile, although more particularly near the surface, 'ghosts', or the crumbled constituents of more easily decomposed rocks may be found. These are frequently fragments of sandstone boulders, which disintegrate easily to form a mass of loose sand. Coarse-grained granite also crumbles readily into its constituent minerals, due to the weathering of the less stable constituents, the disintegrating effect of salt films in intergranular spaces, and even the freezing of thin moisture films. Resistant rocks such as fine-grained dolerite may be fresh and angular in the soil profile, while similar rocks on the soil surface are almost completely destroyed.

Most soils are underlain by permafrost. In young soils this permafrost is at a shallow depth, whereas in older soils it is deeper. Where the frozen ground layer is near the soil surface, some melting may occur when soil temperature rises during summer. When it occurs at greater depths, i.e. below 1 m, it is usually permanently frozen.

There is little chemical weathering in these soils. The ferromagnesian minerals of basic igneous rocks, especially those on rock surfaces, begin to decompose first, together with the glassy intergranular matrix of basalts and dolerites. This releases iron which precipitates on the surfaces of more stable mineral strains, producing a characteristic red-brown colour. When sufficient moisture is available, smectites will be formed; otherwise the weathering products remain amorphous, consisting of very small glassy fragments. Sandstone rocks contain relatively abundant clay-size mica, which becomes the dominant clay mineral of soils that have received contributions from such rocks. During weathering the micas tend to hydrate, often as far as vermiculites. Where the pH is high, in particular in saline situations, smectites may also be formed from micas.

Soil moisture status is largely dependent on soil temperature and moisture is derived either from small amounts of snow melt or from existing ice within the soil. Where the soils are dry (i.e. where no ice is present in the profile) temperature changes do not significantly influence the soil. However, where ice crystals are present, the soil may become moistened when the temperature rises. In these situations, the soil has a frosted appearance before thawing occurs, with ice crystals occurring between the soil grains.

In situations where the soil can receive greater amounts of moisture than usual, such as north-facing slopes at low altitude in coastal regions where moisture can be obtained from the melting of summer snowfalls, the soils may be quite wet during the warmest part of the year. Such soils often show little development of profile features since they are often young, and are also disturbed by freeze – thaw processes, resulting in considerable mixing and overturning of the soil material. Because of their high moisture content, they also tend to be a more favourable habitat for plant life, and mosses, lichens, algae, etc. may be found at the surface. There is little, if any observable accumulation of humus however, and the plant life appears to have little influence on the soil.

Characteristic features of these soils are accumulations of salts, occurring as encrustations or surface coatings under stones, or concentrated as horizons within the profile. The nature of the salt deposits and their chemistry depend on the moisture status of the soil and the source of supply of the salt (Claridge and Campbell, 1975; Keys and Williams, 1981).

## 2c.2.2. Coastal Greater Antarctica

The soils of coastal Greater Antarctica, which includes the Bunger Oases (100°E), the Vestfold Hills (78°E), and Enderby Land (45°E) (MacNamara, 1969), are formed in a much warmer environment than those of the bare ground areas of the Trans-Antarctic Mountains. During the summer air temperatures may rise as high as 10°C, while soil temperatures at 1 m may be as high as 4°C. As a consequence of these higher temperatures the soils may have more available moisture, deeper penetration of weathering, and to some extent, a greater degree of weathering.

The soils are generally dry, and salt accumulations are abundant especially on the surface. In some of the soils, relatively large amounts of clay are present, and there is some evidence of clay movement. Because of the increased temperatures, mosses and lichens are locally more abundant than in the soils of the Trans-Antarctic Mountains, and they exhibit a detectable influence on the soils where they are present. Beneath the vegetation, there may be 2 − 3 cm of relatively organic-rich sandy loam over up to 10 cm of sand with traces of humic material.

The soils appear to be considerably younger than the oldest soils of the Trans-Antarctic Mountains, as they have formed largely on surfaces exposed by retreat of grounded ice in the last 10,000 years. Most of them are grey in colour, but the soils of Enderby Land may be somewhat older as they have developed reddish colours. Because of the higher temperatures and increased moisture availability, the soils are subjected to considerable frost stirring, and do not appear to have developed distinct soil horizons.

## 2c.2.3. Maritime Antarctica

The northern part of the Antarctic Peninsula and associated islands, north to 60°S, are exposed to a cold moist climate that does not have the extremes that prevail elsewhere in Antarctica. There are small daily fluctuations only, and frequent drizzle during the summer months. Temperatures may be above zero at any time during the year, rising to as high as 9 or 10°C during January or February. Precipitation may be as high as 400 mm. Soil formation thus proceeds under cold oceanic conditions, with high moisture and regular freeze − thaw activity. The soils found can be grouped in two main classes − mineral or mainly organic.

In mineral soils the parent material is always subject to seasonal cryoturbation. Soil horizons do not develop and the soils are very youthful in appearance. Such soils have been described from Signy Island (Holdgate et al., 1967), Elephant Island (O'Brien et al., 1979) and the South Shetland Islands (Everett, 1976).

Chemical weathering is much more active than elsewhere in Antarctica, with some mineral soils containing more than 10% of clay-size material. The soils have significant exchange capacity and levels of exchangeable cations. Some may contain considerable amounts of organic carbon and nitrogen, and begin therefore to resemble more closely the soils of more temperate regions in their chemical properties. Because of greater moisture availability, the soils are leached of soluble salts and in most cases calcium and magnesium are the dominant cations of the exchange complex, whereas in most Antarctic soils where soluble salts are present, the dominant cation is sodium. Nitrogen input is usually from guano produced by the penguins and other seabirds. Biological nitrogen fixation is very low.

Where the ground surface is stable, mosses, lichens and algae are abundant, forming in some cases an almost continuous plant cover. Organic matter content in the soils beneath the vegetation can be very high, although there is no development of a true A horizon. Decomposition of the organic matter is slow, leading to the accumulation of peat up to 3 m deep (Fig. 2c.2) much of which is permanently frozen (Smith, 1979). Soil-mixing microfauna are not present and consequently there is little incorporation of organic matter into the material beneath, the few soil-dwelling animals that are present being concentrated in the upper organic layers.

The soils of the Maritime Antarctic region therefore are very different from the soils of the cold deserts as characterized by those of the Trans-Antarctic Mountains and resemble more closely the soils of the polar deserts of the north polar regions.

Fig.2c.2. *Polytrichum* peat bank on Signy Island, South Orkney Islands. (British Antarctic Survey).

## 2c.3. SIGNIFICANCE OF SOIL STUDIES

### 2c.3.1. Soils and the history of Antarctica

Surfaces in the McMurdo Sound region have been dated radiometrically and it has been possible to show that some of these on which soils are forming have been exposed to weathering, much like that prevailing at the present day, for 2 – 4 million years. From soil data a sequence of soil weathering stages of increasing development (Fig. 2c.3) was related to surfaces of known age (Campbell and Claridge, 1975, 1978). The youngest of these surfaces was considered to be less than 50,000 years old while the oldest was probably 10 – 15 million years old. By making allowances for variations in soil development it is possible to correlate weathering stages in different parts of Antarctica and produce a detailed chronology even in the absence of radiometrically dated surfaces.

Soil studies have therefore contributed to a knowledge of the detailed chronology of glacial events in Antarctica. It can be deduced from a study of the features of the older soils of the Trans-Antarctic Mountains, that they have been subjected to a climate very similar to that prevailing at the present time since Miocene times, without the interposition of any periods of significantly higher temperatures.

Fig.2c.3. Landscape in the ice-free area of McMurdo Sound, showing moraine-filled valley floor, with moraines of different ages.

## 2c.3.2. Nature and significance of salts

A unique feature of Antarctic soils is the diversity of the nature and distribution of salts (Fig. 2c.4). A study of the composition of the water-soluble salts in soils from widely-separated areas along the length of the Trans-Antarctic Mountains, in coastal regions and in low-lying dry valleys has shown that the salts can arise from several different sources. For example, we have already shown (Claridge and Campbell, 1968, 1977) that nitrate salts occur widely in soils in inland high altitude areas close to the polar plateau. Towards the coast, nitrates diminish while sodium chloride increases. This distribution is explained in terms of atmospheric circulation, with nitrate being derived from the tropical oceans via the upper atmosphere and polar snows, and sodium chloride mainly from coastal precipitation from local oceans. Other salts may be derived from the weathering of rocks, while in coastal situations the salts in precipitation may largely be derived from local marine sources. The salts are redistributed and precipitated in the soil by processes related to the soil environment; if enough moisture exists, the salts may be leached out of the soils into saline lakes or hollows.

Detailed studies of the salts in soil sequences, including those in buried soils, may reveal evidence of past variations in climatic conditions through major changes in the accumulation rates or distribution patterns of salts, while further investigations of the nitrate salts may be useful in detecting evidence of changes in global atmospheric circulation patterns. It is also obvious from the stability of the salts in the soils of Antarctica that widespread global pollution can accumulate and persist in these soils.

Fig.2c.4. Soil on moraines on the floor of Bull Pass, Wright Valley, showing a salt horizon.

### 2c.3.3. Relationship with soils of the hot deserts

The desert-like nature of Antarctica has long been recognized and the soils described as cold desert soils. There are close relationships in many soil properties between soils from cold and hot desert areas. There are parallels with respect to salts (including nitrate occurrences), the disintegration of surface rocks, the type of lichens and algae found on and in rocks, the formation of surface soil crusts, and the presence of pale coloured horizons and red colours in old soils.

### 2c.3.4. Influence of man on soils

As is the case in other desert regions, environmental conditions in Antarctica markedly limit both weathering and biological activities, and as a consequence the soils are much simpler than those of more temperate lands. Their very simplicity renders them more fragile and therefore more susceptible to permanent damage.

One of the most important influences on the soil ecosystem has been the direct impact of organic and inorganic material resulting from man's activities. Over the years, vast quantities of organic matter in the form of foodstuffs, building and packaging materials, have been transported to Antarctica, and much of this has been abandoned. Also of importance are the products of combustion, most of which finally accumulate and remain in the soil. The introduced organic materials all have an exceedingly long life and remain in the soil because of the slow rate of decomposition and absence of effective leaching.

As well as changes in ecosystems through disturbance of plant and animal life and additions of organic material to the soils, man has the capacity to alter local conditions and hence the soil in various other ways. The scraping of snow for water, or the smoothing of the ground surface, for example, causes less snow to be trapped and results in increased local warming and thawing through greater exposure of soil and rock surfaces. Dust from mechanical operations also accumulates on snow surfaces and accelerates thawing; the resulting increase in available moisture causes soil activation and patterned ground movements. The effects of direct mechanical disturbance of the soil by digging or scraping are permanent, and no recovery of the soil to its natural undisturbed state is possible because of the exceedingly slow rate of soil development. Care therefore needs to be taken to ensure that soils are not altered, and that they are protected from local activities including physical disturbance, accumulation of atmospheric pollutants, and other pollutants including radioactive, soluble and non-decomposable organic or inorganic materials and the accidental introduction of new soil organisms. In some areas it is already difficult to find uncontaminated or undisturbed sites for conducting important baseline studies.

## REFERENCES

Armstrong, R.L. (1978) Late Cenozoic McMurdo Volcanic group and dry valley glacial history, Victoria Land, Antarctica. *N.Z. J. Geol. Geophys.* 21, 685 – 98.
Campbell, I.B. and Claridge, G.G.C. (1975) Morphology and age relationships of Antarctic soils. *Bull. R. Soc. N.Z.* 13, 83 – 8.
Campbell, I.B. and Claridge, G.G.C. (1978) Soils and late Cenozoic history of the upper Wright Valley area, Antarctica. *N.Z. J. Geol. Geophys.* 21, 635 – 43.
Campbell, I.B. and Claridge, G.G.C. (1981) Soil research in the Ross Sea region of Antarctica. *J. Roy. Soc. N.Z.* 11, 401 – 410.
Claridge, G.G.C. (1965) The clay minerology and chemistry of some soils from the Ross Dependency, Antarctica. *N.Z. J. Geol. Geophys.* 8, 165 – 220.
Claridge, G.G.C. and Campbell, I.B. (1968) Origin of nitrate deposits. *Nature* 217, 428 – 30.
Claridge, G.G.C. and Campbell, I.B. (1977) The salts in Antarctic soils, their distribution and relationship to soil processes. *Soil Sci.* 123, 377 – 84.
Everett, K.R. (1976) A survey of the soils in the region of the South Shetland Islands and adjacent parts of the Antarctic Peninsula. *Rep. Inst. Polar Stud.* 58, 1 – 44.
Fleck, R.J., Jones, L.M. and Behlins, R.E. (1972) K – Ar dates of the McMurdo volcanics and their relation to the glacial history of Wright Valley. *Ant. J. U.S.* 7, 244 – 6.
Holdgate, M.W., Allen, S.E. and Chambers, M.J.G. (1967) A preliminary investigation of the soils of Signy Island. *Bull. Br. Antarct. Surv.* 12, 53 – 71.
Keys, J.R. and Williams, K. (1981) Origin of crystalline cold desert salts in the McMurdo region, Antarctica. *Geochim. Cosmochim. Acta* 45, 2299 – 2309.
MacNamara, E.E. (1969) Soils and geomorphic surfaces in Antarctica. *Biul. Peryglac.* 20, 299 – 320.
O'Brien, R.M.G., Romans, J.C.C. and Robertson, L. (1979) Three soil profiles from Elephant Island, South Shetland Islands. *Bull. Br. Antarct. Surv.* 47, 1 – 12.
Smith, R.I. (1979) Peat forming vegetation in the Antarctic. In *Proceedings of the International Symposium on Classification of Peat and Heathlands, Hyytiala, Finland* September 17 – 21, 1979. International Peat Society, Helsinki, pp.58 – 67.

# CHAPTER 3

# Terrestrial Habitats — Introduction

## D. W. H. WALTON and W. N. BONNER

British Antarctic Survey, Natural Environment Research Council, High Cross, Madingley Road,
Cambridge CB3 0ET, U.K.

Anyone, with more than a passing interest in the Antarctic, quickly realises that all its visitors over more than two centuries have complained about the lack of land. The Continent may be the highest in the world, and contain some of the world's ice, but its tiny proportion of bare ground puts it at the bottom of the table for biological diversity in terrestrial habitats.

The two flowering plants, found only in the Maritime Antarctic area, are surprising more for their occurrence than for the lack of phanerogamic diversity. The remoteness of the land from sources of propagules and the short time most of it has been free of ice clearly militate against a highly diversified group of phanerogams. Diversity is restricted to the cryptogams and, as Royce Longton shows, there is sufficient to allow a considerable development of specific associations. These cryptogams are subjected to the most severe environmental stresses of any of the Antarctic flora and fauna and may show physiological or biochemical characteristics unnecessary for survival in more temperate climates.

The invertebrate fauna is generally less environmentally stressed and much less diverse than the Antarctic flora. Lauritz Sømme, in reviewing our generally fragmentary knowledge of microarthropod life cycles and ecology has highlighted the ability of many species to survive harsh conditions either by super cooling or by anhydrobiosis. Much remains to be discovered about the nature of these mechanisms, the triggers that control them and their significance in species survival.

In a land covered by ice, water is generally very scarce. Yet in some localised areas lakes and pools are important features. They range from the very saline, never unfrozen lakes of Victoria Land through to the oligotrophic pools on Signy Island, free of ice for nearly three months. Some are shallow and freeze to the bottom, others are deep and provide a protected and buffered environment for life. Because their flora and fauna are limited and many remain as sealed systems for much of the year, Antarctic lakes make ideal research environments. Julian Priddle, in reviewing examples from the range of lake types, has drawn particular attention to the present microbiological studies and their relationship to biogeochemical cycling within both saline and freshwater systems.

Microbiology, although not reviewed in this volume, has contributed considerably to Antarctic science. From the early studies of Pirie and Eklöf three major areas have developed. The most substantial has been an interest in the microbiology of Man and in this field there have been many expedition studies. An associated but much less developed area has been the microbiology of Antarctic vertebrates. The second major area, that concerned with soil microbiology and decomposition, was given a major impetus in the 1960s by American studies in the Dry Valleys under NASA contracts. Victoria Land was chosen as an analogue area to Mars and it was suggested that detailed studies on microbial life there could provide adequate testing of Martian sampling systems. The emphasis in this area has now moved over to microbiology and decomposition processes with much of the recent work having been done in the more diverse and complex soils of the Maritime Antarctic and the subantarctic islands.

The third major area is the freshwater field, already touched on. Complementary to all these areas are studies of the air spora, but except for a detailed study on pollen dispersed to the subantarctic islands, work in this field has been limited and fragmentary so far. Recently research has begun into the role of micro-organisms in rock weathering and the establishment of vegetation.

These apparently simple terrestrial and freshwater systems offer exciting possibilities to ecologists. Perhaps in these it will prove possible to attempt realistic models of the functioning of complete ecosystems, without the inevitable black boxes of unresolved transfers found in all temperate models. Certainly current studies on competition and interactions between organisms show considerable promise for theoretical developments. Perhaps it will also be possible in these nutrient poor environments to quantify nutrient transfers, not only within terrestrial systems but from the sea to the land and back, via freshwater, to the sea.

# CHAPTER 3a

# Terrestrial Habitats — Vegetation

## R. E. LONGTON

Department of Botany, The University, Reading  RG6 2AS, U.K.

## CONTENTS

## 3a.1. ORIGIN AND DISTRIBUTION OF THE FLORA

Alone amongst all the continents the Antarctic has an almost wholly cryptogamic flora. Yet the fossil record indicates that parts of the continent were more richly vegetated during the Late Palaeozoic, and the Late Cretaceous and Tertiary periods (*see* Chapter 2b). The Tertiary flora disappeared as the climate

73

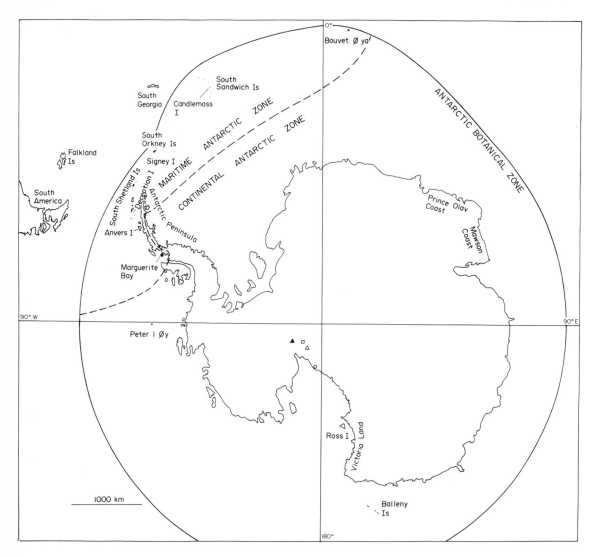

Fig. 3a.1. Antarctica. The heavy line encloses the Antarctic botanical zone (Skottsberg, 1960). The dashed line indicates the approximate boundary of the maritime and continental regions (Holdgate, 1964; Smith, in press). Southernmost known localities are indicated for angiosperms − (●), bryophytes − (○), blue-green algae − (▲), lichens − (△) and for bacteria and yeasts − (□) from data in Cameron *et al.* (1971), Wise and Gressitt (1965) and Smith (1982).

deteriorated with the Pleistocene glaciation producing continuous ice cover over the continent, its off-lying islands, and some subantarctic islands, except perhaps for isolated nunataks. Subsequent recolonization of areas left bare by retreating ice on a now very isolated continent, must have been impeded by the ocean barrier, with a minimum width of c. 900 km (Fig. 3a.1.).

The present native terrestrial flora of the Antarctic zone (Fig. 3a.1) includes at least 200 species of lichens, 85 mosses, 28 macrofungi, 25 hepatics and two angiosperms, together with many species of algae. Both flowering plants occur in southern temperate regions and neither is thought to have survived in Antarctica throughout the Pleistocene. Phytogeographical relationships of most of the cryptogamic flora are confused, with inadequate taxonomic work and meagre distribution data. For the lichens one hypothesis (Dodge, 1965,; 1973) suggests the present taxa are derived principally from species which survived the Late Cenozoic glaciations on nunataks, and subsequently spread to all parts of the continent with simultaneous evolution of numerous local endemics. Another hypothesis however (Lindsay, 1977) recognized substantial cosmopolitan and bipolar elements which may have recolonized areas after glaciation, as well as a significant assemblage of endemics. Most of the bryophytes are widespread at lower latitudes in the southern hemisphere, some are cosmopolitan, and a bipolar element is represented by several of the ecologically important species. Like the flowering plants, most of the bryophytes are thought to be post-glacial immigrants, survival in the Antarctic throughout the Pleistocene being most likely among the few endemics (Robinson, 1972).

Both the moss and lichen floras contain species that are widespread throughout the Antarctic zone and others that are apparently restricted to either the maritime or the continental region (Fig. 3a.1). The greatest diversity of plant species, and the most extensive stands of vegetation, occur on the west coast of the Antarctic Peninsula, its off-lying islands, and other islands in the South Atlantic. This maritime Antarctic region, which has its origins in parts of Western Gondwana, has a strong floristic relationship with South Georgia and southern South America, and it is from these areas that its present flora may be principally derived.

## 3a.2. GENERAL FEATURES AND CLASSIFICATION OF THE VEGETATION

Over 97% of the continental land surface remains covered by perennial snow and ice. Bare areas available for plant colonization occur principally in coastal regions, on islands, on inland nunataks, and locally in inland dry valleys. Antarctic vegetation is of significant ecological interest despite its limited extent, and it has attracted scientists from several nations during the past 25 years. Their studies have confirmed that maritime Antarctic vegetation contrasts strongly with that in continental Antarctica (Fig. 3a.1), primarily in response to differences in climate and water availability. Mean monthly air temperatures in the former region reach 0°C to 2°C in summer and seldom fall below − 15°C in winter; precipitation is frequent, ranging from 25 − 100 cm water equivalent per year, some falling as rain in summer. At most continental Antarctic sites, mean monthly air temperatures remain below 0°C in summer and generally fall below − 25°C in winter: sub-freezing temperatures, combined with infrequent summer precipitation falling exclusively as snow, severely limit the availability of liquid water in most habitats. Cloud cover is generally less frequent than in the maritime region. In favourable maritime Antarctic localities there are substantial closed stands of mosses, lichens and algae: liverworts and macrofungi are widespread, though seldom abundant, and the two native species of flowering plants are dominant locally to 68°S. The more sparsely vegetated continental Antarctic supports occasional stands dominated by mosses, lichens and algae, and more widespread communities formed almost exclusively by micro-organisms. The former are best developed at coastal sites, being largely replaced by microbial communities inland and at high altitudes. However, mosses and lichens have been recorded on nunataks to beyond lat 84°S and 86°S respectively (Fig. 3a.1).

Several general habitat features, in addition to geographical isolation and low temperature, are responsible for the limited vegetation (Table 3a.1). The lack of water in continental Antarctica acts both directly and indirectly, for example through its influence on soil salinity. The soils are immature, with little incorporation of organic matter, and are deficient in certain mineral nutrients, notably available nitrogen, in areas beyond marine influence. Site aspect at high latitudes has a major effect on radiation receipt, whilst lack of winter snow cover can expose plants to both lethal winter temperatures and inadequate free water in summer.

TABLE 3a.1. Ecological Factors Determining the Distribution of Life in Antarctic Dry Valleys and the Luxuriance of Vegetation Elsewhere in the Antarctic Zone. (Modified from Cameron 1972)

| Favourable | Unfavourable |
| --- | --- |
| N − S orientation | E − W orientation |
| Northern exposure | Southern exposure |
| Gentle, north-facing slopes | Flat or south-facing slopes |
| High solar radiation | Low solar radiation |
| Micro-climate above freezing during the day | Micro-climate below freezing |
| Low mean wind speed | High mean wind speed |
| Northerly winds | Southerly winds |
| Shelter from wind | Exposure to wind |
| High humidities | Low humidities |
| Slow or impeded drainage | Rapid drainage |
| Lengthy duration of available water through presence of snow and ice fields, glaciers, lakes and melt streams | Short duration of available water |
| Translucent pebbles | Opaque pebbles |
| Non-saline soils with balanced ionic composition | Saline soils with unbalanced ionic composition |
| Marine influence through wind-blown spray and excrement from birds and seals | No marine influence |
| Well developed winter snow cover | Winter snow cover sparse or absent |
| Stable soils | Soils subject to solifluction and frost-induced movement |

TABLE 3a.2. Outline of a Classification of Vegetation in the Antarctic Zone

Antarctic herb tundra formation
    Grass and cushion chamaephyte subformation
Antarctic cryptogam tundra formation
    Crustaceous lichen subformation
    Fruticose and foliose lichen subformation
    Short moss turf and cushion subformation
    Tall moss turf subformation
    Bryophyte carpet and mat subformation
    Moss hummock subformation
    Alga subformation
    Snow alga subformation

The classification has been developed by Holdgate (1964), Longton (1967) and Gimingham and Smith (1970). Associations and sociations are listed by Longton (1979a) with additions by Seppelt and Ashton (1978) and Kanda (1981).

A classification of Antarctic macrophytic vegetation has been developed (Table 3a.2). The communities are grouped in two formations defined by the dominance of flowering plants or cryptogams, while the eight subformations of the cryptogamic group are differentiated by the growth form of community dominants. Associations and sociations are recognized in each subformation based on floristic composition. There is gradation between the groups at all levels, but the scheme provides an effective framework for describing vegetation in the Antarctic zone.

Fig. 3a.2. Grass and cushion chamaephyte subformation. Mats of the grass *Deschampsia antarctica* on soil on a rocky, north-facing slope on Signy Island. The disc is 5 cm in diameter.

## 3a.3. VEGETATION IN THE MARITIME ANTARCTIC

### 3a.3.1. Antarctic herb tundra formation

The two native flowering plants are widely distributed, though seldom abundant, throughout the maritime region. *Deschampsia antarctica*, a grass, forms low mats (Fig. 3a.2) and *Colobanthus quitensis* (Caryophyllaceae) occurs in compact cushions up to 25 cm wide. Both are most frequent at low altitudes as occasional components of essentially cryptogamic communities on level ground, or more typically on the warmer sites of north or west facing slopes. *D. antarctica* is the more widespread, and usually the more abundant species. Plants may coalesce to form continuous swards several metres wide on rock ledges and stony ground. *C. quitensis* and mosses such as *Drepanocladus uncinatus* are also abundant in some of these occasional stands of phanerogamic vegetation. Soils showing general similarities to temperate brown earths, although lacking earthworms, have developed in places beneath the grass (Allen and Heal, 1970). Small populations of the introduced temperate grasses *Poa annua* and *P. pratensis* have survived for several years in the South Shetland Islands but are now extinct. Several species

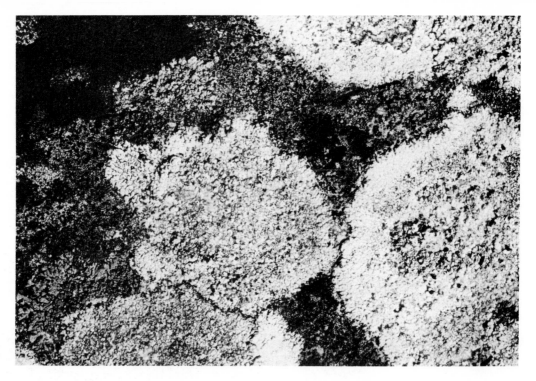

Fig. 3a.3. Crustaceous lichen subformation, with *Caloplaca* and *Verrucaria* spp. (Photo — D. W. H. Walton)

persisted for one or more years in experimental introductions of temperate, sub-Antarctic and Arctic flowering plants to maritime Antarctic sites, but, with the exception of *Poa flabellata*, permanent establishment appeared unlikely.

### 3a.3.2. Antarctic cryptogam tundra formation

*Crustaceous lichen subformation*

All eight subformations of the Antarctic cryptogam tundra formation are found in the maritime Antarctic (Table 3a.2). The crustaceous lichen subformation is dominated by crustose, dwarf foliose and other small lichens which grow adpressed to the substratum (Fig. 3a.3). Many of the crustaceous lichens appear to prefer substrata subjected to salt spray or nitrogen enrichment by birds, and thus the present subformation is particularly characteristic of coastal cliffs. Other communities occur extensively on dry, rocky substrata inland and on the shingle of raised beaches.

Vertical zonation of crustaceous lichens is often evident on coastal rocks. *Verrucaria* spp predominate from immediately above to slightly below the high-water mark. The principal species are brown or black in colour, and thus contrast strongly with the species of *Caloplaca* and *Xanthoria* that form striking splashes of yellow and orange higher on the cliffs. Each community occupies a characteristic position in

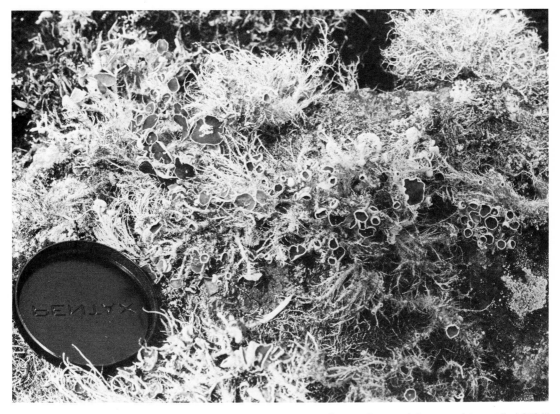

Fig. 3a.4. Fruticose and foliose lichen subformation. The fruticose lichen *Usnea fasciata* and crustose lichens on rock in an upland fellfield community on Signy Island. The disc is 5 cm in diameter.

relation to sea spray, birds' nests and other factors, with the moss *Orthotrichum crassifolium* and the fruticose lichen *Ramalina terebrata* important associates in some stands. Other lichens, including nitrophobous species of *Buellia*, *Lecanora*, *Lecidea*, *Pertusaria*, *Rhizocarpon* and *Placopsis*, become abundant on exposed inland cliffs up to at least 600m.

*Fruticose and foliose lichen subformation*

Macrolichen communities are also widely and extensively developed from sea level to at least 500 m alt., principally on inland boulders, cliffs and scree, and in fellfields on finer mineral soils in dry or exposed situations. *Usnea antarctica* is abundant in many communities, and other fruticose species, including *Himantormia lugubris*, *Usnea fasciata* and *U. sulphurea*, and foliose species of *Umbilicaria* are also important (Fig. 3a.4). Crustaceous species form an understorey beneath many stands of macrolichens, particularly in the driest habitats where examples of the present subformation intergrade with vegetation dominated by crustaceous forms. Conversely, *Usnea* spp also occur in association with small cushion-forming mosses, which become co-dominant with macrolichens under relatively moist conditions in communities which merge into examples of the next subformation (Fig. 3a.5).

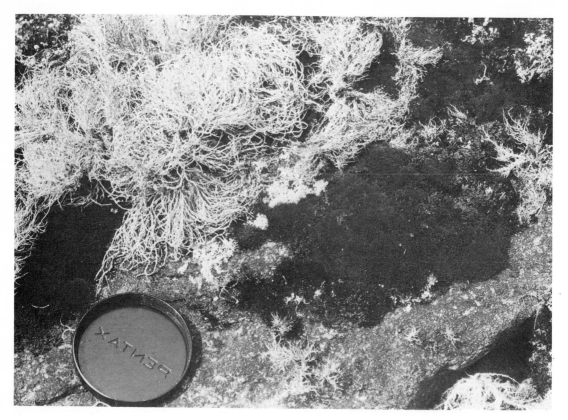

**Fig. 3a.5.** Fruticose and foliose lichen subformation. *Usnea antarctica* with *Andreaea* sp. and other bryophytes and lichens on an exposed boulder on Signy Island. Lettering on the disc is 5 mm tall. Reproduced from Longton (1967) by courtesy of the Royal Society.

## Short moss turf and cushion subformation

Short turf- and small cushion-forming acrocarpous mosses predominate in the driest habitats which support abundant bryophytes. Short turfs are composed of parallel, erect shoots, typically densely packed and under 2 cm tall. Small cushions develop a compact hemispherical form which may coalesce to form continuous areas of undulating moss cover (Fig. 3a.6). The dominant species are distributed in relation to physical and chemical features of the underlying soil or rock.

*Andreaea* spp. are abundant on many acidic substrata, occurring on rock faces, scree slopes and more finely divided mineral soil. The development of winter snow cover appears to be an important factor determining the relative abundance of mosses and lichens in the open, cryptogamic fellfields which occur in such habitats, the former preferring relatively sheltered situations where snow accumulates. Other mosses, and locally hepatics, are co-dominant with *Andreaea* spp. in some communities, and lichens may also be prominent associates growing both on mineral ground and epiphytically on the mosses. Species of *Grimmia* and *Tortula* replace *Andreaea* spp. on base rich substrata, such as the local marble outcrops on Signy Island, and occur in small stands around deposits of limpet shells left by gulls on acid soils. Species of *Bryum*, *Ceratodon* and *Pohlia* are particularly characteristic of recent, porous volcanic substrata on the

Fig. 3a.6. Short moss turf and cushion subformation. *Tortula* spp. with *Bryum algens* and *Grimmia antarctici* on soil by a marble outcrop on Signy Island. The disc is 5 cm in diameter. British Antarctic Survey photograph reproduced from Gimingham and Smith (1970) by courtesy of Academic Press.

South Sandwich and South Shetland Islands. These mosses are often extensively encrusted by lichens, as on scoria-covered plains on Candlemas Island where abundant short turfs of *Pohlia nutans* are so heavily colonized by *Lepraria* sp. that the vegetation assumes the pale greenish-white colour of the lichen. On Deception Island, species of *Bryum* and *Ceratodon* form irregular, elongate cushions up to 5 cm wide, which are impregnated with volcanic ash and hollow in the centre, possibly as a result of frost action. Physiognomically different vegetation develops locally on ground disturbed by solifluction, frost-heave and animals. *Pottia austro-georgica* colonizes these areas, probably by spores, in short, open turfs with cover only 10 — 15%. Short turf- and cushion-forming mosses also comprise the principal component of communities found in rock crevices, although other growth forms, including fruticose and foliose lichens and bryophyte mats, are also well represented.

*Tall moss turf subformation*

The tall moss turf subformation includes some of the most striking and distinctive stands of Antarctic vegetation. The dominant species form turfs of parallel erect shoots, which may be loosely aggregated as in *Polytrichum alpinum*, or more densely packed, sometimes with abundant interlacing rhizoids below, as in *Chorisodontium aciphyllum* and *Polytrichum alpestre*. The latter have given rise to peat deposits often up

Fig. 3a.7. Tall moss turf subformation. The peat bank in the foreground is 140 cm deep and is composed of *Chorisodontium aciphyllum* with lichens on the more exposed parts of the surface. Near the N.W. coast of Signy Island. British Antarctic Survey photograph reproduced from Gimingham and Smith (1970) by courtesy of Academic Press.

to 2m deep. Tall turf-forming mosses are most abundant on gently to moderately sloping ground and overlie boulders, scree or more finely divided, but generally well drained substrata. The deeper peats retain a reservoir of water, permanently frozen below 20 – 30 cm from the surface. Moreover, the tall moss turfs are generally extensive only below 150 m alt., thereby avoiding sites most strongly exposed to desiccating winds.

Stands of *Polytrichum alpestre* and *Chorisodontium aciphyllum* range from discrete turfs to continuous banks of either or both species, which extend continuously for up to 50 m, or even more, particularly on well insolated north and west facing slopes (Fig. 3a.7). *C. aciphyllum* prefers wetter substrata than *P. alpestre* and thus tends to predominate towards the base of the more extensive banks and on gently sloping ground (Fig. 3a.8) where it forms oval, dome-shaped islands of peat. Associated bryophytes include other turf-forming mosses, and hepatics growing as individual stems among the mosses or as small mats on the surface. Epiphytic lichens are abundant on the larger banks, and basidiomycete fruiting bodies may be found.

*Polytrichum alpinum*, although widespread, is particularly characteristic of recent volcanic substrata. Its colonies range from compact mounds c. 20 cm in diameter and impregnated with volcanic ash, to large, loose mounds up to 1 m wide and 30 cm deep which merge to form more extensive continuous stands. The colonies expand by means of radiating subterranean rhizomes, and thus are frequently circular in outline. *Pohlia nutans* is usually present, and sometimes co-dominant with *Polytrichum alpinum*, and tall turfs of *Campylopus* spp are found locally on ground influenced by volcanic heat and moisture.

Fig. 3a.8. Tall moss turf and bryophyte carpet and mat subformations. The turf of *Chorisodontium aciphyllum* with occasional *Polytrichum alpestre* (top right) appears to be invading the carpet of *Calliergon* cf. *sarmentosum* and *Drepanocladus uncinatus* (bottom-left). On wet, level ground on Signy Island. The disc is 5 cm in diameter. Reproduced from Longton (1967) by courtesy of the Royal Society.

### Bryophyte carpet and mat subformation

This subformation occurs extensively in wet habitats on level or gently sloping ground at low altitudes, often around lakes or on slopes subject to melt water seepage (Fig. 3a.9). The characteristic components are species of pleurocarpous mosses in the general *Brachythecium*, *Calliergidium*, *Calliergon* and *Drepanocladus* (Fig. 3a.8), which commonly form in the maritime Antarctic carpets of closely packed, parallel erect or ascending shoots arising from a prostrate layer of older stems. In drier habitats these mosses occur as more typical compact mats of almost prostrate shoots, some individual species such as *Drepanocladus uncinatus* showing growth form variation in relation to habitat moisture. These hydric mosses in places overlie soligenous peat up to 15 cm deep. Other bryophytes, notably *Cephaloziella varians*, occur as associates: the alga *Prasiola crispa* is locally frequent, but lichens are generally rare. A characteristic feature of many carpets are expanding rings of white moribund shoots (Fig. 3a.10) associated with fungal infection. This subformation also includes a distinctive series of hepatic-

Fig. 3a.9. Bryophyte carpet and mat and moss hummock subformations. A variety of carpet forming mosses on a moist seepage slope give way to hummocks of *Brachythecium austro-salebrosum* by a small melt stream in the centre. Factory Cove, Signy Island. The boulder at top-centre is c. 75 cm wide.

dominated communities confined to relatively warm, moist ground around fumaroles on the South Sandwich Islands. These range from shallow compact mats of *Cephalozia badia* to luxuriant stands of *Cryptochila grandiflora* and other species, which may occur as mat, carpet or even tall turf growth forms. Thallose mats of *Marchantia* spp become dominant locally, both near fumaroles on the South Sandwich Islands, and on moist north facing slopes in other localities.

### Moss hummock subformation

The remaining major bryophyte community type, the moss hummock subformation, is confined to the margins of melt water streams, wet rock ledges subject to dripping water, and other flushed habitats in contact with flowing water (Figs. 3a.9 and 3a.11). The growth forms have been described as a tall compact cushion (*Bryum algens*), a tall loose cushion (*Brachythecium austro-salebrosum* and *Tortula excelsa*), or a deep undulating carpet (*Drepanocladus uncinatus*) (Smith, 1972). Lichens are again rare. The large cushions may coalesce to form a closed hummocky stand, but the communities are seldom extensive. This vegetation type commonly intergrades with stands of the bryophyte carpet and mat subformation, from which it is not always clearly defined.

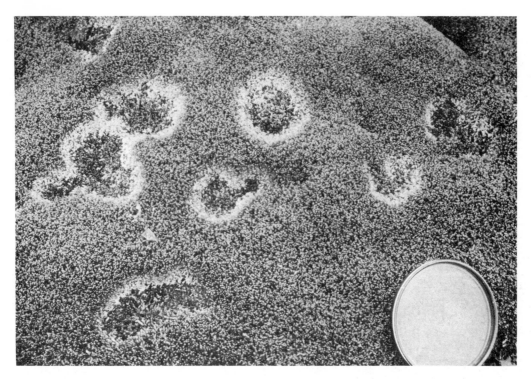

Fig. 3a.10. Rings of white moribund shoots associated with an unidentified ascomycete in a carpet of *Drepanocladus uncinatus* on Signy Island. The disc is 5 cm in diameter. The distribution and development of rings in Antarctic mosses is described by Longton (1973a).

## Alga subformation

The green foliose alga *Prasiola crispa* is one of the most widespread and conspicuous plant species in coastal regions. Its thalli are abundant over large areas subjected to trampling and manuring by seals and penguins. In areas surrounding elephant seal wallows and penguin rookeries, algal cover may reach 75%. Few other macrophytes can tolerate the conditions of mechanical disturbance, low pH and high concentrations of N, Na and P which characterize such sites. Bryophytes are rare except around the periphery of the *Prasiola* zone, but a variety of crustaceous lichens colonizes the larger boulders. The latter include *Mastodia tesselata*, a lichenized form of *Prasiola*. The most intensively disturbed ground within the penguin colonies is devoid of visible vegetation.

Other terrestrial algae are abundant locally, notably on the South Shetland Islands. Here a foliose form of *Nostoc commune* forms rosettes up to 5 cm in diameter, spaced almost equidistantly from each other some 10 − 20 cm apart, on clay soil saturated by almost stagnant, non-nitrogenous meltwater. More continuous stands of this alga occur elsewhere on moist base rich soils, especially on Signy Island.

## Snow algae subformation

During the main period of thaw in mid to late summer, firn snow coloured red, green or occasionally yellow by aggregations of unicellular algae, including species of *Chlamydomonas*, *Raphidonema* and *Ochromonas*, is widespread in northern coastal areas. The coloration may extend continuously over

Fig. 3a.11. Moss hummock subformation. Hummocks of *Brachythecium austro-salebrosum* by the melt stream shown in Fig. 3a.9. The disc is 5 cm in diameter. Reproduced from Longton (1967) by courtesy of the Royal Society.

several hundred square metres. Species composition, abundance and productivity of these cryoplankton communities were discussed by Fogg (1967), who noted that their rapid appearance during summer is likely to be caused by accumulation of cells during snow melt and ablation rather than by high productivity.

### 3a.3.3. Extent of the major vegetation types

The extent of plant cover, and the relative importance of the vegetation types described above, varies widely within the maritime Antarctic. Few quantitative data are available, except for Signy island, a small relatively well vegetated member of the South Orkney Island. Some 30% of its total area of *c.* 20 km² is glaciated. Of the remainder, 12% supports stands of the bryophyte carpet and mat, and less extensively the moss hummock subformations and 6% communities in the tall moss turf subformation, particularly banks of *Polytrichum alpestre* and *Chorisodontium aciphyllum*. The fruticose and foliose lichen subformation, and *Andreaea*-dominated stands of the short moss turf and cushion subformation, extend over half of the exposed land surface, but these vegetation types are often open with cover less than 50%. Approximately 26% of the exposed surface is either bare mineral substrate or carries stands of the crustaceous lichen subformation, and a further 5% is so strongly influenced by birds and seals that it supports at most scattered stands of *Prasiola crispa*. The estimated cover of phanerogam-dominated vegetation is less than 0.001% (Tilbrook, 1970).

Cover and diversity of cryptogamic vegetation tend to decrease from the South Orkney Islands (Lat. 60 – 61°S) and South Shetland Islands (Lat. 61 – 63°S) southwards along the west coast of the Antarctic Peninsula, while the east coast is, in general, sparsely vegetated, principally by lichen communities, and appears to be continental Antarctic in character. In the west, plant cover remains extensive south to the Argentine Islands – Anvers Island area (Lat. 65°S) where the range of communities is comparable with that on Signy Island, but such important species as *Chorisodontium aciphyllum* and *Himantormia lugubris* are uncommon and apparently near the southern limit of their ranges. Further south, most of the cryptogamic subformations (Table 3a.2) have been recorded around Marguerite Bay (Lat. 68 – 69°S), but they are only locally distributed among extensive areas of arid substrata devoid of macrophytes. Even here, however, luxuriant and extensive bryophyte vegetation may be encountered in unusually moist, sheltered, well-insolated sites, but the number of species is very much reduced. The two native flowering plants also occur locally around Marguerite Bay, where they reach the southern limit of their known range in a region that is clearly near the southern extremity of the maritime Antarctic. Deception Island, Candlemas Island and several other recent volcanic islands are also less extensively vegetated than the South Orkney Islands, from which they differ in other respects such as the rarity or absence of *Andreaea* spp., *Polytrichum alpestre* and *Chorisodontium aciphyllym*, and the greater abundance of *Polytrichum alpinum*.[*]

## 3a.4. VEGETATION ON CONTINENTAL ANTARCTICA

### 3a.4.1. Lichen subformations

The general impression presented by most ice-free terrain in continental Antarctica is of bare rock, gravel and scree: only locally does cryptogamic vegetation form a prominent feature of the landscape. The few detailed vegetation accounts relate principally to sites in Victoria Land (Longton, 1973b; Rudolph, 1963), the Ingrid Christensen and Mawson Coasts (Pickard, 1982; Seppelt and Ashton, 1978) and the Prince Olav Coast (Kanda, 1981). Information on other areas can be inferred from more generalized botanical accounts and floristic studies. The occasional stands of macrophytic vegetation are referable to the crustaceous lichen, fruticose and foliose lichen, short moss turf and cushion, and alga subformations.

Lichen communities are more widespread and variable than those dominated by mosses, particularly on inland nunataks. Thus Filson (1966) recorded 26 lichen species and only two mosses from Mac-Robertson Land. Crustaceous lichens in the genera *Buellia*, *Caloplaca*, *Rhizocarpon* and *Xanthoria* are prominent in much of the sparse vegetation, but fruticose and foliose species of *Alectoria*, *Usnea* and *Umbilicaria* are locally abundant. In some of the driest habitats supporting epilithic lichens the plants occur only as crustose forms on the sides or lower surfaces of pebbles. The lichens commonly extend

[*]Information on the vegetation of specific localities may be found in the following papers, which have been drawn upon freely in preparing the present account: Allison and Smith (1973), Follman (1965), Gimingham and Smith (1970), Lindsay (1971), Longton (1967), Longton and Holdgate (1979), Smith (1972), Smith and Corner (1973) and Smith and Gimingham (1976).

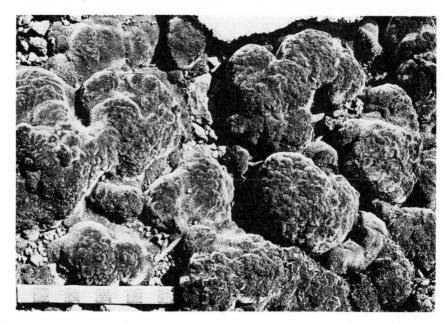

Fig. 3a.12. Short moss turf and cushion subformation. Cushions of *Bryum algens* beside a melting snow drift at Mawson. The scale is in cm. Reproduced from Seppelt and Ashton (1978) by courtesy of Blackwell Scientific Publications.

below the uppermost layer of stones, and algae may become frequent locally under the conditions of increased humidity in the sub-surface region. It is in such habitats that translucence of pebbles (Table 3a.1) becomes important.

## 3a.4.2. Short moss turf and cushion subformation

Lichen communities occur both on rock and on more finely divided sands and gravels, but short moss turfs and cushions have been recorded principally on the latter. No tall turf-, carpet- or hummock-forming mosses have been recorded and only one liverwort, a species of *Cephaloziella*, known from a number of sites. Floristic composition varies between the widely separated localities so far studied, but species of *Bryum*, *Ceratodon*, *Grimmia* and *Sarconeurum* are the principal dominant bryophytes.

Plant abundance in the moss communities is variable: thus on Ross Island *Bryum antarcticum* locally gives 85% cover on sandy soil in moist gullies, whereas cover values are less than 5% in more typical moss communities of dry stony ground. In some cases the mosses occur as short turfs having shoot apices flush with substrate, projecting stones affording additional protection from wind. Elsewhere they form compact, often elongate and contorted cushions up to 10 cm wide and 5 cm high. It is clear, however, that these dimensions are attained by persistence rather than rapid growth, as the green, photosynthetic layer commonly extends only to 1−2 mm. Some moss colonies at Mawson may originate as small, mobile cushions with green shoots all over the surface. They eventually become attached to the substrate, and the lower shoot apices die. The larger cushions commonly have a mineral core, suggesting that the sedentary phase of colony development involves cryoturbic disturbance (Figs. 3a.12 and 3a.13). The surface of bryophyte colonies in some communities becomes extensively colonized by lichens and in the wetter habitats by algae, especially blue-green forms, which may inhibit moss growth.

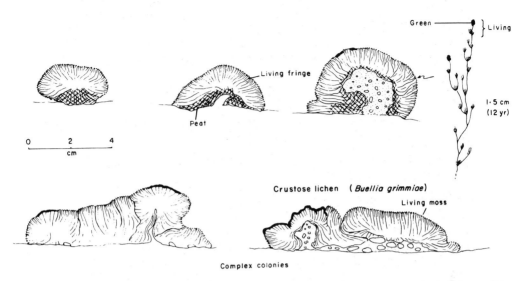

Fig. 3a.13. Suggested sequence of cushion and turf development in *Bryum algens* at Mawson. Peat areas are crosshatched. Crustose lichens are shown as black strips on the surface of the cushions. Reproduced from Seppelt and Ashton (1978) by courtesy of Blackwell Scientific Publications.

### 3a.4.3. Alga subformation

Algal vegetation is, at least locally, more variable and extensive at continental than at maritime Antarctic sites with communities widely distributed in coastal regions. *Nostoc* is particularly abundant in several localities. Algae are also prominent in some seasonally wet habitats in and beside shallow melt streams and around pools, the genera present including *Oscillatoria*, *Phormidium* and *Ulothrix* (Koob, 1967). It is possible that their abundance is related to the absence of carpet-forming mosses which occupy comparable habitats in the Maritime Antarctic.

### 3a.4.4. Microphytic vegetation

The proportion of the ice-free land surface supporting vegetation visible to the naked eye declines with increasing altitude and distance inland, until the ground surface appears lifeless. Apparently barren terrain also predominates under conditions of unusually low precipitation in ice-free dry valleys in coastal regions of Southern Victoria Land. However, auto- and hetrotrophic bacteria, algae, micro-fungi and lichens exist in many areas lacking macrophytes, and in places they are capable of supporting arthropods and other soil animals. Incorporation of their communities into the classification outlined in Table 3a.2 would require recognition of a third formation, to include populations of snow algae, and other assemblages of exclusively microscopic plants. Microphytes occur both in gravel soils and beneath the surface of rocks.

The richest soil-based communities occur in coastal regions (Claridge *et al.*, 1971), but bacteria and yeasts extend to at least 87°21'S (Fig. 3a.1). The poleward limit of soil biota is thought to be determined by the occurrence of soil temperatures above 0°C failing to penetrate to sufficient depths to liberate ice-bound water (Janetschek, 1970). Beyond this limit, in much of Antarctica's central ice-plateau region, there is no known life. The soil microbial systems of South Victoria Land have been studied intensively with the objective of developing techniques for detecting extra-terrestrial life under comparable, although more extreme, conditions of cold and aridity. The results have shown that under the severe regime operating in the dry valleys, many soils are sterile, while others support small populations of heterotrophic bacteria that are not unusually tolerant of either cold or salinity. It has been suggested that many such populations represent air-borne contaminants from other parts of the world, a balance being maintained between influx and mortality, although there is evidence of an indigenous soil microflora even in this inhospitable region (Vishniac and Hempfling, 1979). The introduction and dispersal of microbial contaminants, an inevitable consequence of human activities in the Antarctic, can seriously jeopardize the study of these, the most simple of all terrestrial ecosystems.

Recent studies (Broady, 1981; Friedmann, 1982) have suggested that the most widespread biotic communities in continental Antarctica comprise endolithic micro-organisms occurring within rocks in two principal habitats: chasmoendoliths in minute fissures, and cryptoendoliths which penetrate between the crystals of relatively porous rocks such as granite, granodiorite, sandstone and marble. A number of primary producers occur as endoliths, including unicellular and filamentous green, blue-green and yellow-green algae and lichens. They are commonly associated with colourless bacteria. The Antarctic endolithic communities contrast with those reported from hot deserts, in which procaryotes predominate (Friedmann, 1980). Water supply is primarily by snow melt absorbed in the surface layers of rock, where relative humidity in the air between the rock particles has been shown to remain considerably higher than that in the external air for several days following summer snowfall.

Chasmoendolithic algae appear to be widespread in coastal regions of continental Antarctica and have been recorded also in the maritime Antarctic and in the dry valleys. They grow both in fissures running perpendicular to the rock surface and under thin flakes parallel with the surface, occurring beneath 20% of the rock surface at some sites. Species zonation has been reported, with blue-green algae in the innermost vegetated region where light intensity and oxygen concentration are likely to be low, and green algae, fungi and lichens nearer the surface.

Cryptoendolithic lichens are the dominant form of life in the dry valleys of South Victoria Land. The rocks commonly show an abiotic surface crust $1-2$ mm thick inside which occur successive zones appearing black (*c.* 1 mm wide), white (2−4mm) and green (2 mm) as seen in rock sections. The cementing material between the rock particles is apparently dissolved by substances released from the organisms, resulting in the surface crust periodically peeling off by a process of biogenic exfoliative weathering. The three inner zones each support filamentous fungi and unicellular green algae including species of *Trebouxia*, the algae being abundant in the inner, green zone and scarce in the white zone. Although the mycobiont (fungal partner) lacks the organized thallus found in epilithic lichens, there is ample evidence of a symbiotic lichen association between the two organisms. The fungal hyphae are closely associated with the algal cells, being adpressed to their surface or penetrating them by means of haustoria; centric bodies occur in the mycobiont cytoplasm, as in other lichens, and lichen substances such as norstilic acid are produced. Moreover, where cryptoendolithic lichens become exposed on the rock surface by exfoliative weathering they may, under favourable conditions, produce small organized thalli bearing reproductive structures such as apothecia and pycnidia. The genera *Acorospora*, *Buellia* and *Lecidea* have been recognized on this basis, all strongly represented in the macroscopic lichen flora of Antarctica. It seems likely that the cryptoendolithic lichens represent highly adaptive forms which evolved from more typical lichens in response to deteriorating Antarctic environments since the Tertiary (Friedmann, 1982).

## 3a.5. VEGETATION PATTERN AND DYNAMICS

### 3a.5.1. Patterns in plant communities

Antarctic vegetation provides striking examples of zonation, on both local and regional levels. The sequence: barren central plateau — micro-organism communities of inland nunataks and the dry valleys — sparsely developed moss and lichen communities of the coastal and slope regions of continental Antarctica — more luxuriant and varied cryptogamic vegetation in the maritime Antarctic, can be regarded as an irregularly concentric zonation pattern occurring on a regional scale in response to increases in water availability and temperature outwards from the South Pole.

Smaller scale examples of concentric zonation are associated with gradients of temperature and moisture surrounding fumarole vents on the South Sandwich Islands. In the example illustrated in Fig. 3a.14 the vents are lined with low mats of *Cephaloziella* sp. and other bryophytes, including *Pohlia nutans* in a form having slender, prostrate stems and small, distantly spaced leaves. The vents are surrounded by carpets of *Cryptochila grandiflora*, with a range of bryophyte associates including tall turfs of *Campylopus* spp. and *Pohlia nutans*. The third zone is formed principally by tall, loose turfs of *Pohlia nutans* and *Polytrichum alpinum*, and this is surrounded by short, compact turfs of *Pohlia nutans* encrusted with *Lepraria* sp. and other lichens giving way gradually to almost bare volcanic ash and scoria. This example illustrates the wide ecological amplitude shown by a number of taxa in species-poor Antarctic vegetation, as *Pohlia nutans* is prominent in all four zones occurring in three different growth forms. It is also significant that several of the fumarole bryophytes, including *Cryptochila grandiflora* and *Campylopus* spp., are widespread in these highly localized sites of volcanic heat and moisture on the South Sandwich Islands but are otherwise unknown from the Antarctic zone. Other examples of zonation occur on

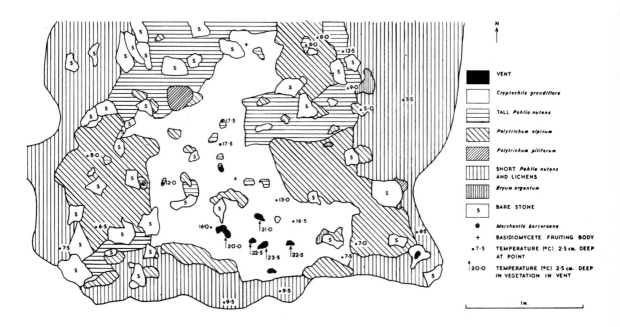

Fig. 3a.14. Map of the vegetation surrounding one group of vents at a fumarole on Candlemas Island. The figures indicate temperatures (°C) recorded as spot readings 2.5 cm deep in the soil and vegetation. Reproduced by courtesy of the British Antarctic Survey from Longton and Holdgate (1979).

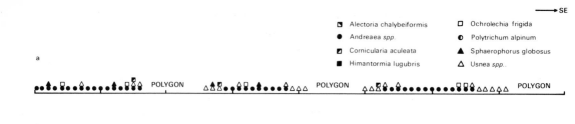

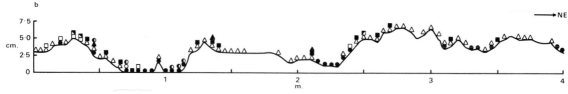

Fig. 3a.15. Line transects through stands of *Andreaea* spp. and lichens on Signy Island. The distribution of species is related to micro-topography on (a) level ground disrupted by three soil and stone polygons and (b) undulating ground subjected to irregular frost heaving. Reproduced by courtesy of the British Antarctic Survey from Smith (1972).

coastal cliffs, on raised beaches, around melt streams and, on a small scale, among endoliths (page 90) and in response to minor topographic features. Seppelt and Ashton (1978) described how rock steps only 1 cm high in the largely barren terrain near Mawson Station shelter on the lee side a narrow zone of *Caloplaca elegans* giving way to *Buellia frigida* farther from the step, the total width of the vegetated zone being only 2 – 3 cm.

Vegetation is commonly influenced by several environmental factors which vary independently of each other, either at random or along intersecting gradients. Such situations give rise to irregular mosaic patterns in the distribution both of communities and of species within a vegetation unit. Smith (1972) described examples from Signy Island where he was able to relate the distribution of sociations dominated by lichens and small cushion-forming mosses on exposed stony ground to variation in exposure, moisture, winter snow cover, soil accumulation, substrate pH and other factors. Further examples of vegetation pattern are attributable to physical or biotic disturbance. Zonation near penguin colonies and seal wallows has already been alluded to, and cryoturbation can also give rise to striking vegetation patterns as shown in examples from Signy Island (Figs. 3a.15 and 3a.16).

The transect in Fig. 3a.15a crosses three frost heave polygons in which instability has prevented plant establishment. Bands of *Usnea* spp. occur on the peripheral stone fringes, with *Andreaea* spp. prominent among the mosses and fruticose lichens on the finer, relatively stable inorganic soil between the polygons. Figure 3a.15b shows a less regular pattern in which lichens are abundant on frost-induced stone hummocks, giving way to bryophytes on finer, moister material in intervening depressions. A

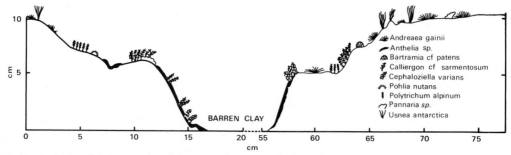

Fig. 3a.16. Distribution of species across the marginal ramps of a narrow soil and gravel stripe on Signy Island. Reproduced by courtesy of the British Antarctic Survey from Smith (1972).

profile across soil and stone stripes on more steeply sloping ground is illustrated in Fig. 3a.16. The central region of wet clay is moving downslope by solifluction faster than the surrounding stripes of gravel and stone. The former is largely barren centrally, but bryophytes, including a hepatic in the genus *Anthelia*, become abundant under moist conditions along the less mobile margins. Other bryophytes and lichens colonize drier, relatively stable stones between the soil stripes. Solifluction appears, in some localities, to inhibit the development of vegetation over extensive areas, but in many instances the factors controlling the distribution of Antarctic vegetation remain to be elucidated, and it is not always easy to distinguish between the effects of spatial variation under the present environment and patterns associated with plant succession.

## 3a.5.2. Plant succession

It is commonly assumed that ecological succession in deglaciated regions of the Antarctic is initiated by algae and soil micro-organisms, with mosses and lichens appearing later (Llano, 1965). While there is as yet no direct evidence for this sequence, it seems clear that micro-organisms are the primary colonists in areas where climatic conditions do not permit the growth of mosses and lichens. Recently deglaciated rocks in the maritime Antarctic often show a trimline at distances up to 2.5 m from the retreating ice. Rock surfaces between the ice-edge and the trimline lack colonizing lichens, whereas beyond this boundary there may be a progressive increase in species diversity, in colony diameter of the first established crustose lichens, and in thallus size in species of *Usnea* and *Umbilicaria*. However, mosses, notably *Drepanocladus uncinatus*, become established on accumulations of moist mineral soil even within the lichen trimline. A similar pattern has been observed on moraines, where the first macrophytic colonizers appear to be lichens on rock surfaces and mosses, or even grass in pockets of soil. In neither case has it yet been determined when micro-organisms become established.

These observations emphasize that a range of life forms may act as pioneer macrophytic colonizers depending on the nature of the substrate and other factors, there being little evidence of rigid temporal succession in this respect. Smith (1972) summarized his concept of community development and interrelationships on Signy Island as follows: '......when a particular habitat becomes available for plant establishment it is colonized only by the species typical of a sociation favouring the conditions provided by that habitat. In most cases one or a few species are more successful than the rest and will form the basis of the community, usually as the dominant or co-dominant components. Various sociations which lie adjacent to one another show little evidence of serious competition or the displacement of one by the other, but merely form a heterogeneous zone where the stands merge'. This position was supported by an examination of plant remains below all the widespread bryophyte communities in non-fumarolic areas of Candlemas Island, the results suggesting that each vegetation type had developed independently of the others (Longton and Holdgate, 1979). Cyclic succession does occur in the maritime Antarctic, however, as shown in a community on Candlemas Island dominated by *Polytrichum alpinum*. A common sequence here appears to be growth and subsequent erosion of a *P. alpinum* hummock, colonization of the resulting patch of bare ground by *Ceratodon* sp., and eventual re-establishment of *P. alpinum* in the *Ceratodon* cushions.

Other authors have contended that directional succession is more prevalent in Antarctic vegetation than is suggested by Smith, occurring due to fluctuations in climate, and inter-specific differences in rates of establishment, as well as through vegetation-induced changes in edaphic conditions. The process from initial colonization of maritime Antarctic moraines by crustose lichens to a climax of macrolichens and moss cushions may take at least 200 years (Lindsay, 1978). An example of seral succession involving peat-forming mosses is considered below (page 94). Thus it is not yet obvious how far present Antarctic vegetation is developing either along the lines of the classic xerosere, or as a relatively wide range of climax units distributed in response to local environmental variation. It is important to resolve this issue in order both to understand the factors underlying present vegetation patterns and to formulate management practices aimed at maintaining the diversity of biotic communities. An unusual situation may occur in the soil systems of continental Antarctica for here, an inverse relationship can be anticipated between soil development and the diversity and abundance of the associated biota, due to progressive increase in salinity with time under conditions of extreme aridity. Ugolini (1977) suggests that salt accumulation by mosses, followed by death and erosion of the plants, may be important in periodically reversing this process.

### 3a.5.3. Development of moss peat banks

One of the few well-established cases of seral succession in Antarctic vegetation involves the extensive banks of *Polytrichum alpestre* and *Chorisodontium aciphylum*, a vegetation type with no direct counterpart outside Antarctic regions (Fenton and Smith, 1981). The stands may be initiated either through colonization by *P. alpestre* of cushion-forming mosses on slopes, or by *C. aciphylum* invading carpet-forming mosses on moist, level ground (Fig. 3a.8) followed by the establishment of *P. alpestre* on the higher, drier areas of the resulting *Chorisodontium* turf. Individual stands may attain areas in excess of 2500 m$^2$, probably through coalescence of originally discrete turfs. The surface of the larger banks generally becomes undulating, terraced or fissured as a result of wind, solifluction, frost-induced phenomena and other factors. Eventually, growth of the banks results in their upper surface being blown clear of snow in winter, allowing colonization by epiphytic crustose and fruticose lichens, followed by plant death and erosion and, in some instances, recolonization by mosses. Wind-induced ripple systems in *C. aciphyllum* turf may also be the scene of succession involving mosses alternating with epiphytic lichens and algae. Thus the formation of the banks appears to involve directional succession, and their maintenance cyclic succession.

Despite periodic surface erosion the *Polytrichum/Chorisodontium* banks have accumulated peat deposits up to 2m deep (exceptionally to 3m on Elephant Island), and their potential stability is emphasized by radiocarbon dates for basal peat extending back 5000 years. Peat formation is unusual under aerobic, non-waterlogged conditions and its occurrence here is attributable both to the development of permafrost below depths of 20 − 30 cm, and to low rates of decomposition in the active layer resulting in part from intrinsic characteristics of the mosses concerned. Approximately half the original primary production becomes incorporated into the permanently frozen peat (Fenton, 1980). Similarly, upward growth of the banks, averaging 0.9 − 1.1 mm yr, is approximately half the annual shoot extension, the difference being accounted for by compression following partial decay of the plant material. It is striking that many of the banks have a vertical face along the downslope edge (Fig. 3a.7), often with an overhang at the base. This feature appears to be due not to erosion, but to a combination of other factors including downslope movement of the upper part of the banks over the permanently frozen, rigid core, and inhibition of growth at the base of the downslope edge by persistent snow banks or permanent ice (Fig. 3a.17).

## 3a.6. ENVIRONMENTAL RELATIONSHIPS OF ANTARCTIC VEGETATION

### 3a.6.1. At the community level

A much narrower range of flowering plant species and communities occurs in southern polar regions than at comparable Arctic latitudes. How far does this reflect climatic severity as opposed to post-Pleistocene dispersal barriers? The following points are pertinent to this issue:

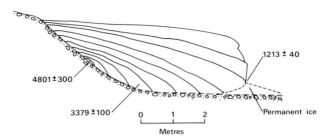

Fig. 3a.17. Cross-section of a moss bank on Signy Island formed principally by *Chorisodontium aciphyllum*, showing radiocarbon dates (yr BP) and hypothetical moss bank surfaces at 500 yr intervals. Reproduced from Fenton (1982) by courtesy of the Regents of the University of Colorado.

1. Mean summer air temperatures in the maritime region are lower, and mean winter air temperatures higher, than those at most Arctic localities supporting extensive phanerogamic vegetation.
2. The maritime Antarctic offers a wide range of habitats with respect to water availability, winter snow cover, and edaphic factors.
3. Mineral nutrient availability is unlikely to be seriously limiting in most maritime Antarctic habitats, but few soils are highly saline.
4. The two native flowering plants, particularly *Deschampsia antarctica*, are among those which extend to the highest altitudes supporting phanerogams on the sub-Antarctic island of South Georgia.
5. Within the Antarctic zone these two vascular species are locally abundant only at sites with a particularly favourable temperature regime.
6. Few species have shown a potential to become established in experimental introductions of other flowering plants into the maritime Antarctic.

Points 1 — 3 suggest that mean summer temperature is the principal factor making the maritime Antarctic environment more severe than that of many arctic regions with a substantial vascular flora. Points 4 — 6 suggest that phanerogamic vegetation would fail to develop extensively under maritime Antarctic conditions even in the absence of dispersal barriers. Thus, although the argument is not fully resolved, I consider low summer temperatures to be the most potent single factor restricting the extent and diversity of phanerogamic vegetation in the maritime Antarctic.

The situation concerning cryptogams is rather different and here the South Sandwich Islands are of particular interest. The oldest known rocks on these isolated volcanic islands (Fig. 3a.1) are only 3 million years old and thus long postdate the break-up of Gondwana. The general vegetation is sparse and species poor, but local areas influenced by volcanic heat and moisture support a richer flora (Fig. 3a.14), including at least a dozen bryophytes so far unrecorded elsewhere in the Antarctic zone. This suggests that several species are capable of crossing ocean barriers, but are only able to become established under these atypical Antarctic conditions. Similarly, viable air-borne spores of micro-organisms not known to occur in the soil biota have been trapped at Antarctic sites (Corte and Daglio, 1964; Heal *et al.*, 1967), and several mosses otherwise unknown from the Antarctic have already become established by fumaroles on Deception Island originating from the 1967 and 1969 eruptions. Thus, while removal of dispersal barriers might well result in additional species becoming established, it seems unlikely that the overall character of maritime Antarctic vegetation would change significantly under present climatic conditions.

Temperature effects on community distribution are evident elsewhere in the maritime region, as carpet-forming mosses, banks of *Polytrichum alpestre* and *Chorisodontium aciphyllum* and other cryptogamic communities, as well as flowering plants, are most extensive in low altitude sites of northerly aspect. However, water availability has a more profound effect on distribution of the major cryptogamic community types. Relationships between topography and growth form at a well vegetated locality such as Signy Island are summarized in Fig. 3a.18. The highest, most windswept cliffs and scree are largely devoid of macrophytes. Lichens, particularly crustaceous forms, tend to occupy drier, more exposed habitats than mosses. The bryophyte growth forms can be regarded as a series from compact short turfs and small cushions in dry habitats, through tall turfs and carpets, to large, often loose hummocks by running water. Shoot arrangement in bryophyte colonies of different growth forms may influence water relations in a manner that rationalizes the distribution shown in Fig. 3a.18, and could explain the relative abundance of short turfs and small cushions in the vegetation of porous volcanic substrata (Gimingham and Smith, 1971). In experiments, the short, compact cushions and turfs showed a greater capacity for both absorption and retention of capillary water, than did the taller, looser carpets and hummocks.

The floristic composition and distribution of cryptogamic communities thus appears to be dependent chiefly on water availability, and to a lesser extent on exposure, at the sub-formation level, whereas a wide range of other factors may control the distribution of species sharing a common growth form. Minor variation in water availability is likely to be involved in some cases, but the chemical nature of the substrate has an important influence, exemplified by the distribution of calcicole and calcifuge mosses in relation to marble outcrops. Other significant factors include variation in snow cover, substrate stability and degree of nutrient enrichment by marine organisms. The principal sources of mineral nutrients in Antarctic substrata are physical weathering of rock (particularly K, Ca), precipitation and sea spray (Na, Mg and small amounts of N and P), excrement of marine animals (N, P) and limited atmospheric nitrogen fixation by lichens and blue-green algae (Horne, 1972). None of the major nutrients appears to be seriously limiting in the maritime Antarctic, with the possible, local exception of nitrogen (Collins *et al.*, 1975), but shortage of available nitrogen is an important factor at many inland sites in continental Antarctica.

Apart from the local abundance of algae, the growth form types in continental Antarctic macrophytic vegetation are those which occupy the driest, most exposed of the vegetated sites in the maritime region. Water availability is generally regarded as the most important single factor leading to general vegetational impoverishment in continental Antarctica, which is thus regarded as a cold desert. Bryophytes, in particular, become abundant only on the margins of lakes and melt water streams, and

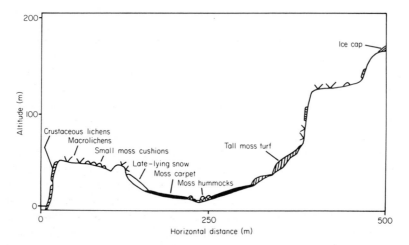

Fig. 3a.18. Diagram indicating the distribution of the major cryptogamic growth from types across an idealized valley in the maritime Antarctic. Reproduced from Longton (1982) by courtesy of Chapman and Hall Ltd.

around snow banks which melt gradually through much of the summer. Less extensive moss cover is associated with local areas of water seepage, and hollows where snow accumulates due to wind influence during light summer flurries. Water may also be available in summer from melting ground ice. However, this source alone does not appear to be capable of supporting well-developed stands of mosses, though it may be important in maintaining micro-organisms and occasional moss cushions. Nakanishi (1977) has shown that variation in the species composition of bryophyte vegetation on Prince Olav Coast is correlated with habitat water supply. Lichens are also most prominent in sites irrigated by melt water during at least part of the summer. Some species are capable of absorbing moisture from saturated air, and the greatest abundance of *Buellia frigida* and other lichens in some of the dry valleys occurs at altitudes above 600 m where cloud cover is most frequent (Schofield, 1972).

While extensive areas of rock and scree are undoubtedly too dry to support macrophytes, sites that are at least seasonally wet but non-vegetated occur in both coastal and montane regions, thus emphasizing that additional factors are involved in restricting the development of vegetation in continental Antarctica. Wind is of great significance, influencing evaporation, temperature, erosion by abrasive soil and ice particles, and weathering of rock surfaces. Snow, with its protective and irrigating properties, tends to accumulate in sheltered situations, as do the deposits of finely divided mineral soil favoured by continental Antarctic mosses. Wind may be of more significance than insolation, at least locally, as lichens, mosses and endolithic algae occur principally on sheltered south-west facing slopes in some localities (Broady, 1981; Nakanishi, 1977). However, life in general appears to be more abundant in north-south orientated valleys than those aligned east-west, as the latter receive little direct solar radiation due to the low elevation of the sun even at mid-day.

By analogy with the Arctic, it is unlikely that availability of light energy for photosynthesis severely restricts the development of vegetation in continental Antarctica, with its summer regime of continuous solar radiation at relatively low intensities. Thus any positive correlation between insolation and the occurrence of vegetation is likely to result from indirect effects such as the influence of radiation on temperature. Lichen abundance is associated with rock colour in some localities, being greatest on dark rocks with low albedo. Rock surface temperatures can increase from $-15°C$ to $28°C$ in 3 hr (Llano, 1965). However, wide diurnal temperature fluctuations, and high day-time values, on dry, well insolated rock surfaces may be detrimental to lichens, and thus a preference by lichens for dark coloured rocks may be due to factors other than temperature. One possibility is that water from light summer snowfall is made available most effectively on dark, relatively warm rock by rapid melting. A preference by lichens for dolerite rather than for paler coloured sedimentary rocks in the Theron Mountains has been attributed to the greater stability of the dolerite, with its low susceptibility to frost shattering (Lindsay and Brook, 1971). Substrate stability may also be important in controlling lichen distribution on talus slopes. In some areas lichens are most abundant in habitats covered by snow throughout much of the summer, and the resulting combination of low light intensity, low but relatively stable temperature, and availability of melt water could be particularly favourable for their metabolism (Gannutz, 1971).

Endolithic lichens in the dry valleys are most common on the warmer, northern face of a given rock, where temperatures in the microphytic zone are higher than at the rock surface during periods of sunshine. The contrast between the barren nature of most rock surfaces in this region and the extensive lichen colonization of sub-surface layers is related to insulation of the latter habitat against the rapid temperature fluctuations through $0°C$ which are common at the surface (Friedmann, 1982). Endolithic microphytes commonly show a preference for pale coloured substrata which enhance light penetration.

Schofield (1972) has stressed the importance of salinity in restricting macrophytic vegetation over wide areas of gravelly soil, noting that lichens and mosses are less tolerant than algae. As available nitrogen is limiting in many continental Antarctic habitats it is not surprising that an association between the occurrence of lichens and nesting colonies of birds such as snow petrels (*Pagodroma nivea*) has been noted on several inland nunataks. However, the lack of such a relationship has been reported in other localities, while in Marie Byrd Land the abundance but not the diversity of lichens was greatest on outcrops influenced by birds. As the degree of nitrophily varies among Antarctic lichens the range of species established in a particular mountain system may determine the extent to which the local vegetation is associated with nesting sites. *Lecanora tephroeceta*, a nitrophilous species collected from a snow petrel colony, grows best in pure culture where supplied with ammonia (a product of uric acid in bird urine) as a nitrogen source. *Prasiola crispa* shows optimum growth when supplied with uric acid, while the non-nitrophilous *Bryum algens* utilizes a wider range of nitrogen compounds (Schofield and Ahmadjian, 1972).

At risk of oversimplification, it is suggested firstly that low summer temperature is primarily responsible for the insignificant role of phanerogams in maritime Antarctic vegetation, secondly that water availability is of major importance in determining both the distribution of cryptogamic growth form types in relation to topography at a given site, and the contrast in plant cover between the maritime and continental regions, and thirdly that a variety of other factors influence local plant distribution. However, this view is based largely on observations of plant cover in relation to apparent environmental variation. Only rarely have the correlations been quantified, and conclusions based on such correlations tested by experiment.

### 3a.6.2. At the species level

A full understanding of the factors controlling distribution of plant communities requires detailed information, for individual species, on micro-climate and on plant responses to environmental variation. Conditions affecting low-growing Antarctic plants differ profoundly from the climate indicated by standard meteorological records, particularly during periods of strong insolation and under snow. This point is illustrated in Fig. 3a.19 by temperature data for *Bryum argenteum* on Ross Island, South Victoria Land, covering five summer days. Mean temperatures near midday were 0°C 2 m above the ground compared with 10 − 11°C among the moss shoots where individual readings reached 19°C. It is also evident from these, and from maritime Antarctic data, that moss temperatures are commonly a few degrees above air temperature at night, thus reducing the frequency and severity of diurnal freeze − thaw cycles in summer (Longton, 1972; Walton, 1982). This may be due to the water content of moss colonies reducing rates of temperature change. Undue emphasis on daily maxima can give a misleading impression, as indicated in Fig. 3a.20 which shows that, despite high maxima, the greatest frequency of temperature readings in *B. argenteum* on Ross Island was within 2.5 degrees of 0°C both at night and by day during a 24-day mid-summer period; night here refers to a 9 hr period in each 24 hr when the east-facing site was shaded by a rock outcrop leading to consistently low insolation (Fig. 3a.19). The data in Fig. 3a.20 also indicate wide differences between the summer temperature regimes affecting Antarctic and boreal populations of this cosmopolitan species. Ground temperatures during winter have been shown to remain relatively stable at temperatures between 0°C and − 20°C in snow-covered habitats favoured by mosses, and several Antarctic bryophytes have been shown to survive at − 15°C for 3 yr in the dark.

Vegetative reproduction must predominate in Antarctic bryophytes as sporophytes are rarely, or sporadically, produced except in *Pottia austro-georgica* and a few other principally epilithic species in the maritime region. Many maritime Antarctic lichens produce spores, but fertility decreases with exposure and climatic severity (Smith, 1972). The available evidence suggests that extension growth of Antarctic mosses and lichens occurs primarily during the summer snow-free period, although in some species stem elongation may begin before snow melt (Collins, 1977; Hooker, 1980a,b). *Bryum argenteum* from Ross Island showed maximum growth at a day/night temperature regime of 22/15°C in long-term experiments, growth being very slow under simulated Antarctic conditions: no consistent differences

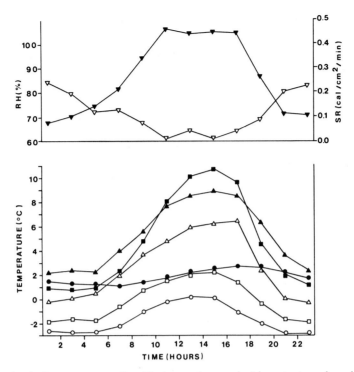

Fig. 3a.19. Microclimatic data for *Bryum argenteum* on Ross Island averaged over twelve 2-hour time intervals per day during 5 days from 28 December 1971. Solar radiation (▼); relative humidity (▽); air temperature at 200 cm (○), 30 cm (□) and 2 cm (△) above the surface; soil temperature at 2 cm (▲) and 20 cm (●) below the surface; temperature among moss shoots (■). Readings at 15-, 7.5- and 30-minute intervals for temperature, solar radiation and relative humidity respectively. Reproduced from Longton (1974) by courtesy of the American Bryological and Lichenological Society.

were detected between plants of polar, temperate and tropical provenances in frost resistance or growth/temperature relationships, both an Antarctic and a tropical strain being capable of slow growth at 5/ − 5°C (Longton, 1981). However, inherent clinal variation in annual stem elongation and leaf length occurs in the bipolar moss *Polytrichum alpestre*, resulting in the formation of progressively shorter, more compact turf with increase in latitude (Longton, 1979b).

The growth response to temperature in *B. argenteum* agrees with Rastorfer's (1970) indication of an optimum temperature for net photosynthesis of c. 25°C in continental Antarctic material under laboratory conditions. His data suggest that net photosynthesis in *B. argenteum* should be possible 24 hr per day under the light and temperature regime on Ross Island in summer as has been shown for some Arctic mosses by field experiments (Oechel and Sveinbjornsson, 1978). Nevertheless microclimatic conditions on Ross Island appear to be decidedly suboptimal for metabolism in *B. argenteum* and also for the endemic *B. antarcticum*.

Laboratory studies on several maritime Antarctic mosses indicated that water availability on Signy Island is generally adequate to permit assimilation at substantial, and at times optimal rates. *Polytrichum alpestre* showed a temperature optimum for net photosynthesis of 5 − 10°C in material previously maintained in a simulated Antarctic temperature regime, this value rising to 15°C in plants pretreated under warmer conditions. *Drepanocladus uncinatus* demonstrated less capacity for acclimation, but showed a broad response curve with maximum net assimilation at 15°C, and 40% of the maximum rate at 0°C. It was predicted that maximum daily net photosynthesis in both species occurs in mid-summer, though with the rate then depressed at mid-day by high temperature which reaches 20 − 30°C among

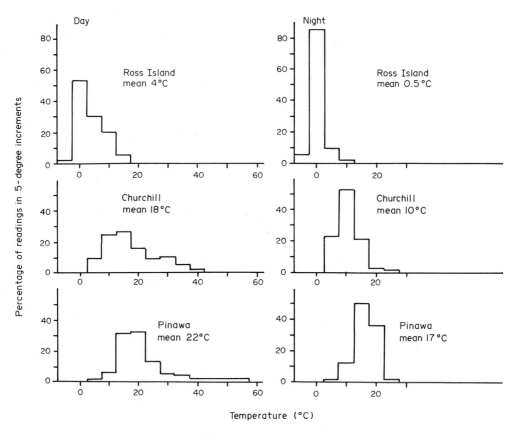

Fig. 3a.20. Mean temperatures and frequency distribution of temperature readings during the day (left) and night (right) in stands of *Bryum argenteum* during summer on Ross Island, and at sub-Arctic (Churchill) and boreal forest (Pinawa) sites in Manitoba, Canada. Ross Island data based on 24 days in December 1971 and January, 1972, day length 15 hours; Churchill data based on 17 days in August 1973, day length 15 hours; Pinawa data based on 21 days in June and July 1973, day length 17 hours. Readings from thermistors placed 2 – 3 mm below the surface of the moss turfs, recorded at intervals of 15 minutes on Ross Island and 30 minutes at Churchill and Pinawa. Re-drawn from Longton and MacIver (1977).

the moss leaves. Significant carbon assimilation is also likely beneath light snow cover in spring. There has been little work on drought resistance in Antarctic mosses but investigations elsewhere suggest that many mosses are able to survive severe cytoplasmic dehydration such as may be anticipated in the continental region.

Metabolic studies on Antarctic lichens in the laboratory have shown that optimal temperatures for net assimilation are generally less than 15°C, and not uncommonly near 0°C, with positive values down to − 12 to − 18°C in frozen thalli of some species (Lange and Kappen, 1972). This capacity for net photosynthesis at low temperature is apparently due more to reduced respiration rates under cold conditions than to adaptation in gross photosynthesis. Immediate return to normal values of both photosynthesis and respiration has been recorded in some continental Antarctic species on rewarming after rapid cooling to − 196°C, although in other species metabolism recovered slowly after a year at − 20°C. Desiccated lichen thalli have been shown to resume photosynthesis through absorption of water condensing on their surfaces, or even of water vapour, thus confirming a potential to benefit from conditions in cloud, or in air beneath snow. Light compensation points are low under cold conditions, a significant point in terms of photosynthesis under snow, and in endolithic or sublithic habitats. Metabolism in continental Antarctic lichens may be most active in spring when liquid water is most freely available, but thallus growth is concentrated in late summer under the relatively moist conditions prevailing in the maritime Antarctic. Little lichen growth occurs during winter on Signy Island, and the slow recovery of growth under apparently favourable conditions in spring results from loss of respirable substrate during winter and desiccation effects associated with freezing stress (Hooker, 1980b,c). Radial growth rates in crustose lichens on Signy Island average 0.4 mm per year in *Caloplaca elegans*, but only 4 mm per century in *Rhizocarpon geographicum*. The latter figure indicates that some individual thalli may be older than 500 − 700 years, and thalli 300 − 600 years old have been recorded in fruticose species of *Usnea*. Studies in the Arctic suggest that potential longevity in *Rhizocarpon* is at least 4000 years (Beschel, 1961).

The two native angiosperms flower regularly throughout the maritime Antarctic but viable seed develops only during climatically favourable years on Signy Island, where seedling mortality is also high. Experiments on South Georgian material suggested that both species are well adapted to growth and reproduction under long days and low but fluctuating summer temperatures, although the experimental conditions available were warmer than field micro-climate in the maritime Antarctic. Optimum temperature for net photosynthesis in *Deschampsia antarctica* varies with latitude of origin, being 13°C for Signy Island and 5°C for more southerly provenances. Positive net assimilation in both species occurs from − 5°C to 35°C, thus spanning the normal field temperature range in summer (Smith, 1984).

It is thus clear that many Antarctic plants are well adapted to the severe environment in terms of resistance to frost and desiccation and relationships between metabolism and environment. The ability to survive for protracted periods in a desiccated, frozen, inactive state, and to resume metabolism and growth rapidly during brief periods of favourable conditions, must be a prerequisite for many organisms in the continental region. High rates of photosynthesis and growth are not necessarily advantageous though for cryptogams with opportunistic growth patterns in habitats where low stature is of adaptive value and where competition is often minimal. It must also be remembered that most of the Antarctic metabolic studies considered here have been carried out in the laboratory rather than *in situ*, leading to problems in interpreting results (Davis and Harrisson, 1981). Moreover, conflicting results have in some cases been reported for the same species, e.g. *Polytrichum alpestre* (Collins, 1977; Rastorfer, 1972), thus emphasizing the need for careful standardization of methodology. Work on Antarctic plant ecophysiology is extensively reviewed by Smith (1984), and the bryological studies are considered in their bipolar context by Longton (1980, 1982).

## 3a.7. PLANTS IN ANTARCTIC ECOSYSTEMS

Primary production may locally reach surprisingly high levels in closed stands of maritime Antarctic vegetation. Results summarized by Smith (1984), indicate net annual primary production (dry wt) of up to *c.* 400 g m$^{-2}$ yr$^{-1}$ n continuous turfs of *Deschampsia antarctica,* 400 − 600 g m$^{-2}$ yr$^{-1}$ in tall moss turfs and 500 − 900 g m$^{-2}$ yr$^{-1}$ in moss carpets. Minor additions to these values may be anticipated from associated algae and lichens. Lichen production on Signy Island has been estimated as 34 − 150 mg g$^{-1}$ yr$^{-1}$ in four fruticose species, but no data are available on an area basis. The above values for mosses are of the same order as maximum bryophyte production so far reported throughout the world, and are comparable with community production in Arctic tundra. They have been confirmed by several workers using both harvest techniques and prediction from metabolic and microclimatic data. Even in continental Antarctica, late season green biomass of 200 g m$^{-2}$ has been reported for an open stand of

*Bryum algens* and *Grimmia lawiana* at Mawson. A production estimate of *c.* 250 g m$^{-2}$ yr$^{-1}$ for closed turf was obtained for *Bryum argenteum* on Ross Island from physiological and microclimatic data. In the dry valleys of South Victoria Land, Friedmann (1980) has estimated that total biomass of endopedic micro-organisms may reach 32-177 g m$^{-2}$ rock surface, based on determination of Kjeldahl nitrogen, but this figure is likely to include dead material. Production values on a land area basis in continental Antarctica must be infinitesimal due to the low proportion of vegetated terrain.

Ecosystem processes have been studied most intensively at two sites on Signy Island occupied respectively by tall turfs of *Polytrichum alpestre* and *Chorisodontium aciphyllum*, and by carpets of *Calliergon sarmentosum*, *Drepanocladus uncinatus* and other hydric mosses (Davis, 1981). Net primary production in each site averaged *c.* 400 g m$^{-2}$ yr$^{-1}$. In neither case was there evidence of significant direct grazing of bryophytes and lichens, although micro-algae formed an important dietary component of arthropods, tardigrades and rotifers. Thus utilization of primary production is mainly through the detritus pathway in these, as in other Antarctic communities. The greater build up of peat beneath the turf-forming rather than carpet-forming mosses apparently resulted from more rapid decomposition of the latter (14 − 25% yr compared with less than 2% in the tall turfs). This difference exists despite conditions being more waterlogged and anaerobic in the carpets. It was attributed by Davis to the stem structure of the turf-formers rendering them less susceptible to decay, to the occurrence of permafrost in the deeper peats, and to greater nitrogen limitation on the microflora in the semi-ombrogenous moss turfs than in the soligenous carpets. Plant death and erosion may also restrict peat formation beneath the moss carpets. However, others (Collins *et al.*, 1975) suggest that there is a surplus of available nitrogen in some moss turfs on Signy Island, emphasizing that the control of key ecosystem processes remains imperfectly understood, in even the most intensively studied Antarctic communities.

Antarctic vegetation has few direct counterparts in Arctic and alpine regions. In terms of cover and growth form representation, that in continental Antarctica may be regarded as an impoverished extension of the polar desert of inland regions of the high Arctic, whereas the closest approach to maritime Antarctic vegetation is seen on Arctic islands such as Svalbard where there are extensive bryophyte and lichen dominated communities. In both cases, however, flowering plants are more strongly represented in the Arctic, at least at low altitudes, reflecting the higher summer temperatures experienced there than in the Antarctic. Its distinctive nature renders Antarctic terrestrial vegetation of considerable scientific importance, both as an object of study in its own right, and in view of its potential for enhancing our understanding of lower latitude ecosystems. Interactions between plants and their environment is unusually amenable to investigation, since grazing and competition are generally minimal. Studies of adaptation to the severe environmental conditions are of particular interest, as many of the indigenous Antarctic species also range widely in climatically more favourable areas. An understanding of responses to stress may become vital as environmental degradation places lower latitude vegetation under increasing pressure, while information on the functioning of relatively simple communities in the Antarctic could improve our concepts of the more complex systems elsewhere. It may be noted that the age of certain peat banks and lichen thalli demonstrate that instability is not an inevitable feature of simple ecosystems.

It is therefore imperative that significant areas of the major terrestrial vegetation types should be shielded from man-induced modification. This is a challenging task because Antarctic vegetation is vulnerable, as a result of its limited extent, the susceptibility, particularly of bryophyte and macrolichen communities, to trampling and other physical disturbance, low or insignificant rates of recovery, and other factors discussed by Cameron (1972) and Schofield (1972). For example, careless housekeeping at a station on Goudier Island, notably indiscriminate scattering of coal ash and garbage, has resulted in the destruction of lichen communities, and probably in the loss of bryophyte species from the island. More subtle responses of biotic communities to human interference may also be anticipated. On Candlemas Island the distribution of tall turf- and cushion-forming mosses is severely restricted, such vegetation occurring principally on plains along the north-west coast. In 1964 it was noted that extensive areas of *Drepanocladus uncinatus* and *Polytrichum alpinum* in this area were buried beneath a thin layer of mud comprising trampled soil and excrement around the margin of a penguin colony. Though the relationship is not proven, the destruction of vegetation in this instance may quite possibly be linked with reduction of whale stocks, through krill surplus and consequent increase in the penguin population. Extreme vigilance and care in matters of conservation are thus required of all engaged in research and other activities in the Antarctic. More specifically it is suggested that the Agreed Measures under the Antarctic Treaty should be revised so that the provisions for conservation of plants become as stringent as those for animals.

# ACKNOWLEDGEMENTS

This account is an outgrowth of work undertaken with the British Antarctic Survey and United States Antarctic Research Program. I am indebted to Drs. R.I. Lewis Smith and D.W.H. Walton for helpful comments on the manuscript and for information on current research by members of the British Antarctic Survey.

# REFERENCES

Allen, S.E. and Heal, O.W. (1970) Soils of the maritime Antarctic zone. In *Antarctic Ecology*, Vol.2. Ed. M.W. Holdgate, pp. 693 – 6. Academic Press, London.

Allison, J.S. and Smith, R.I. Lewis (1973) The vegetation of Elephant Island, South Shetland Islands. *Bull. Br. Antarct. Surv.* 33 and 34, 185 – 212.

Beschel, R.E. (1961) Dating rock surfaces by lichen growth and its application to glaciology and physiography (lichenometry). In *Geology of the Arctic*, Vol.2, Ed. G.O. Raasch, pp. 1044 – 62. University of Toronto Press, Toronto.

Broady, P.A. (1981) The ecology of chasmolithic algae at coastal locations of Antarctica. *Phycologia* 20, 259 – 72.

Cameron, R.E. (1972) Pollution and conservation of the Antarctic terrestrial ecosystem. In *Conservation Problems in Antarctica*, Ed. B.C. Parker, pp. 267 – 305. Virginia Polytechnic Institute and State University, Blacksburg.

Cameron, R.E., Lacy, G.H., Morelli, F.A. and Marsh, J.B. (1971) Farthest south soil microbial and ecological investigations. *Antarct. J. U.S.* 6, 105 – 6.

Claridge, G.G.C., Campbell, I.B., Stout, J.D. and Dutch, M.E. (1971) The occurrence of soil organisms in the Scott Glacier Region, Queen Maud Range, Antarctica. *N.Z. J. Sci.* 14, 306 – 12.

Collins, N.J. (1977) The growth of mosses from two contrasting communities in the maritime Antarctic: measurement and prediction of net annual production. In *Adaptations within Antarctic Ecosystems*, Ed. G.A. Llano, pp. 921 – 33. Smithsonian Institution, Washington.

Collins, N.J., Baker, J.H. and Tilbrook, P.J. (1975) Signy Island, maritime Antarctic. *Ecol. Bull. (Stockholm)* 20, 345 – 74.

Corte, A. and Daglio, C.A.N. (1964) A mycological study of the Antarctic air. In *Biologie Antarctique*, Eds. R. Carrick, M.W. Holdgate and J. Prévost, pp. 115 – 20. Hermann, Paris.

Davis, R.C. (1981) Structure and function of two Antarctic terrestrial moss communities. *Ecol. Monogr.* 51, 125 – 43.

Davis, R.C. and Harrison, P.M. (1981) Prediction of photosynthesis in maritime Antarctic mosses. *CNFRA* 51, 241 – 7.

Dodge, C.W. (1965) Lichens. *Monogr. Biol.* 15, 194 – 200.

Dodge, C.W. (1973) *Lichen Flora of the Antarctic Continent and Adjacent Islands*. Phoenix Publishing, Canaan, New Hampshire.

Fenton, J.H.C. (1980) The rate of peat accumulation in Antarctic moss banks. *J. Ecol.* 68, 211 – 28.

Fenton, J.H.C. (1982) The formation of vertical edges on Antarctic moss peat banks. *Arct. Alp. Res.* 14, 21 – 26.

Fenton, J.H.C. and Smith, R.I. Lewis (1981) Distribution, composition and general characteristics of the moss banks of the maritime Antarctic. *Bull. Br. Antarct. Surv.* 51, 215 – 36.

Filson, R.B. (1966) The lichens and mosses of MacRobertson Land. *A.N.A.R.E. Sci. Rep. Ser. B (II), Bot.* 82, 169 pp.

Fogg, G.E. (1967) Observations on the snow algae of the South Orkney Islands, *Phil. Trans. R. Soc.* B252, 279 – 87.

Follman, G. (1965) Una asociacion nitrofila de liquenes epipetricos de la Anartica occidental con *Ramalina terebrata* (Tayl. et Hook.) como especie caracterizante. *Inst. Antart. Chileno Publ.* 4, 18 pp.

Friedmann, E.I. (1980) Endolithic microbial life in hot and cold deserts. *Origins of Life* 10, 223 – 35.

Friedmann, E.I. (1982) Endolithic micro-organisms in the Antarctic cold desert. *Science* 215, 1045 – 53.

Gannutz, T.P. (1971) Ecodynamics of lichen communities in Antarctica. In *Research in the Antarctic*, Ed. L.A. Quam, pp. 213 – 26. American Association for Advancement in Science, Washington D.C.

Gimingham, C.H. and Smith, R.I. Lewis (1970) Bryophyte and lichen communities in the maritime Antarctic. In *Antarctic Ecology*, Vol. 2, Ed. M.W. Holdgate, pp. 752 – 85. Academic Press, London.

Gimingham, C.H. and Smith, R.I. Lewis (1971) Growth form and water relations of mosses in the maritime Antarctic. *Bull. Br. Antarct. Surv.* 25, 1 – 21.

Heal, O.W., Bailey, A.D. and Latter, P.M. (1967) Bacteria, fungi and protozoa in Signy Island soils compared with those from a temperate moorland. *Phil. Trans. R. Soc.* B252, 191 – 7.

Holdgate, M.W. (1964) Terrestrial ecology in the maritime Antarctic. In *Biologie Antarctique*, Eds. R. Carrick, M.W. Holdgate and J. Prévost, pp. 181 – 94. Hermann, Paris

Hooker, T.N. (1980a) Growth and production of *Cladonia rangiferina* and *Sphaeropharus globosus* on Signy Island, South Orkney Islands. *Bull. Br. Antarct. Surv.* 50, 27 – 34.

Hooker, T.N. (1980b) Factors affecting the growth of Antarctic crustose lichens. *Bull. Br. Antarct. Surv.* 50, 1 – 19.

Hooker, T.N. (1980c) Growth and production of *Usnea antarctica* and *U. fasciata* on Signy Island, South Orkney Islands. *Bull. Br. Antarct. Surv.* 50, 35 – 49.

Horne, A.J. (1972) The ecology of nitrogen fixation on Signy Island, South Orkney Islands. *Bull. Br. Antarct. Surv.* 27, 1 – 18.

Janetschek, H. (1970) Environments and ecology of terrestrial arthropods in the high Antarctic. In *Antarctic Ecology*, Vol.2, Ed. M.W. Holdgate, pp. 871 – 85. Academic Press, London.

Kanda, H. (1981) Flora and vegetation of mosses in ice-free areas of Soya Coast and Prince Olav Coast, East Antarctica. *Hikobia* Supplement 1, 91 – 100.

Koob, D. (1967) Algae distribution. *Antarct. Map Folio Ser.* 5, 13 – 15.

Lange, O.C. and Kappen, L. (1972) Photosynthesis of lichens from Antarctica. *Antarct. Res. Ser.* 20, 83 – 95.

Lindsay, D.C. (1971) Vegetation of the South Shetland Islands. *Bull. Br. Antarct. Surv.* 25, 59 – 83.

Lindsay, D.C. (1977) Lichens of cold deserts. In *Lichen Ecology*, Ed. M.R.D. Seaward, pp. 183 – 209. Academic Press, London.

Lindsay, D.C. (1978) The role of lichens in Antarctic ecosystems. *Bryologist* 81, 268 – 76.

Lindsay, D.C. and Brook, D. (1971) Lichens from the Theron Mountains. *Bull. Br. Antarct. Surv.* 25, 95 – 8.

Llano, G.A. (1965) The flora of Antarctica. In *Antarctica*, Ed. T. Hatherton, pp. 331 – 50. Methuen, London.

Longton, R.E. (1967) Vegetation in the maritime Antarctic. *Phil. Trans. R. Soc.* B 252, 213 – 35.

Longton, R.E. (1972) Reproduction of Antarctic mosses in the genera *Polytrichum* and *Psilopilum* with particular reference to temperature. *Bull. Br. Antarct. Surv.* 27, 51 – 96.

Longton, R.E. (1973a) The occurrence of radial infection patterns in colonies of polar bryophytes. *Bull. Br. Antarct. Surv.* 32, 41 – 9.

Longton, R.E. (1973b) A classification of terrestrial vegetation near McMurdo Sound, continental Antarctica. *Can. J. Bot.* 51, 2339 – 46.

Longton, R.E. (1974) Microclimate and biomass in communities of the *Bryum* association on Ross Island, Continental Antarctica. *Bryologist* 77, 109 – 27.

Longton, R.E. (1979a) Vegetation ecology and classification in the Antarctic zone. *Can. J. Bot.* 57, 2264 – 78

Longton, R.E. (1979b) Studies on growth, reproduction and population ecology in relation to microclimate in the bipolar moss *Polytrichum alpestre* Hoppe. *Bryologist* 82, 325 – 67.

Longton, R.E. (1980) Physiological ecology of mosses. In *The Mosses of North America*, Eds. R.J. Taylor and A.E. Leviton, pp. 77 – 113. Pacific Division AAAS, San Francisco.

Longton, R.E. (1981) Inter-population variation in morphology and physiology in the cosmopolitan moss *Bryum argenteum* Hedw. *J. Bryol.* 11, 501 – 20.

Longton, R.E. (1982) Bryophyte vegetation in polar regions. In *Bryophyte Ecology*, Ed. A.J.E. Smith, pp. 123 – 65. Chapman and Hall, London.

Longton, R.E. and Holdgate, M.W. (1979) The South Sandwich Islands. IV. Botany. *Sci. Rep. Br. Antarct. Surv.* 94, 53 pp.

Longton, R.E. and MacIver, M.A. (1977) Climatic relationships in Antarctic and Northern Hemisphere populations of a cosmopolitan moss, *Bryum argenteum* Hedw. In *Adaptations within Antarctic Ecosystems*, Ed. G.A. Llano, pp. 899 – 919. Smithsonian Institution, Washington.

Nakanishi, S. (1977) Ecological studies of the moss and lichen communities on the ice-free areas near Syowa Station, Antarctica. *Antarct. Rec.* 59, 68 – 96.

Oechel, W.C. and Sveinbjornsson, B. (1978) Primary production processes in Arctic bryophytes at Barrow, Alaska. In *Vegetation and Production Ecology of an Alaskan Arctic Tundra*, Ed. L. Tieszen, pp. 269 – 98. Springer-Verlag, New York.

Pickard, J. (1982) Vegetation of the Vestfolds Hills, Antarctica: a progress report. *J. Hattori Bot. Lab.* 52, 295 – 7.

Rastorfer, J.R. (1970) Effects of light intensity and temperature on photosynthesis and respiration of two East Antarctic mosses, *Bryum argenteum* and *Bryum antarcticum*. *Bryologist* 73, 544 – 56.

Rastorfer, J.R. (1972) Comparative physiology of four west Antarctic mosses. *Antarct. Res. Ser.* 20, 143 – 61.

Robinson, H.E. (1972) Observations on the origin and taxonomy of the Antarctic moss flora. *Antarct. Res. Ser.* 20, 163 – 77.

Rudolph, E.D. (1963) Vegetation of the Hallet Station area, Victoria Land, Antarctica. *Ecology* 44, 585 – 6.

Schofield, E. (1972) Preserving the scientific value of cold desert ecosystems: past and present practices and a rationale for the future. In *Conservation Problems in Antarctica*, Ed. B.C. Parker, pp. 193 – 225. Virginia Polytechnic Institute and State

University, Blacksburg.

Schofield, E. and Ahmadjian, V. (1972) Field observations and laboratory studies of some Antarctic cold desert cryptogams. *Antarct. Res. Ser.* 20, 97 – 142.

Seppelt, R.D. and Ashton, D.H. (1978) Studies on the ecology of the vegetation at Mawson Station, Antarctica. *Aust. J. Ecol.* 3, 373 – 88.

Smith, R.I. Lewis (1972) Vegetation of the South Orkney Islands with particular reference to Signy Island. *Sci. Rep. Br. Antarct. Surv.* 68, 124 pp.

Smith, R.I. Lewis (1982) Farthest south and highest occurrences of vascular plants in the Antarctic. *Polar Rec* 21, 170 – 3.

Smith, R.I. Lewis (1984) Terrestrial plant ecology of the sub-Antarctic and Antarctic. In *Antarctic Ecology*, Ed. R.M. Laws. Academic Press, London.

Smith, R.I. Lewis and Corner, R.W.M. (1973) Vegetation of the Arthur Harbour — Argentine Islands region of the Antarctic Peninsula. *Bull. Br. Antarct. Surv.* 33 and 34, 89 — 122.

Smith, R.I. Lewis and Gimingham, C.H. (1976) Classification of cryptogamic communities in the maritime Antarctic. *Bull. Br. Antarct. Surv.* 43, 25 – 47.

Tilbrook, P.J. (1970) The terrestrial environment and invertebrate fauna of the maritime Antarctic. In *Antarctic Ecology*, Vol. 2, Ed. M.W. Holdgate, pp. 886 – 96. Academic Press, London.

Ugolini, F.C. (1977) The protoranker soils and the evolution of an ecosystem at Kar Plateau, Antarctica. In *Adaptations within Antarctic Ecosystems*, Ed. G.A. Llano, pp. 1091 – 110. Smithsonian Institution, Washington.

Vishniac, H.S. and Hempfling, W.P. (1979) Evidence of an indigenous microbiota (yeast) in the dry valleys of Antarctica. *J. Gen. Microbiol.* 112, 301 – 14.

Walton, D.W.H. (1982) The Signy Island terrestrial reference sites: XV. Micro-climate monitoring, 1972 – 74. *Bull. Br. Antarct. Surv.* 55, 111 – 26.

Wise, K.J.A. and Gressitt, J.L. (1965) Far southern animals and plants. *Nature* 207, 101 – 2.

# CHAPTER 3b

# Terrestrial Habitats — Invertebrates

LAURITZ SØMME

Zoological Institute, University of Oslo, P.O. Box 1050, Blindern, Oslo 3, Norway

## CONTENTS

## 3b.1. INTRODUCTION

The terrestrial animal life of the Antarctic is dominated by the invertebrates, represented by several groups from protozoans to insects and mites (Fig. 3b.1). Other groups important to warmer geographical regions are missing or poorly represented. Examples of these are earthworms and land snails, as well as several families and orders of insects and arachnids. The number of invertebrate species in the Antarctic is low compared to other regions, but the diversity of species and higher taxa increases northwards through the Maritime Antarctic to the sub-Antarctic islands.

In contrast to the invertebrates almost all of the vertebrates in the Antarctic depend on the ocean, and can be regarded as belonging to the marine ecosystem. Exceptions are a few species of birds that find their food mainly on land. The Kerguelen and South Georgia pintails partly feed in freshwater. The only

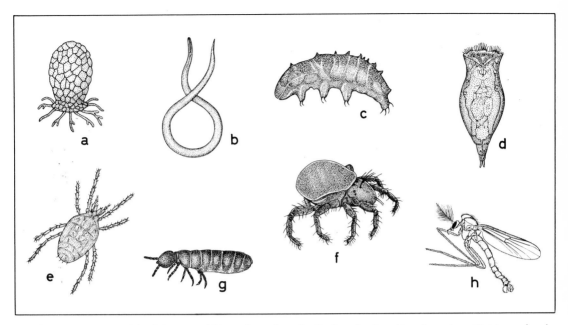

Fig. 3b.1. Representatives of free-living terrestrial invertebrates from the Continental and maritime Antarctic. a. Testate amoebae, b. Nematode, c. Tardigrade, d. Rotifer, e. Prostigmatid mite, f. Cryptostigmatid mite, g. Collembola, h. Midge.

songbird of the Antarctic, the South Georgia pipit, lives on seeds and insects, while the sheathbill is an omnivorous scavenger. Apart from cats, rats and mice introduced by man to several of the sub-Antarctic islands, and the reindeer and other herbivores introduced to South Georgia and Iles Kerguelen (see Ch. 6b) there are no true terrestrial mammals.

## 3b.2. SURVEY OF THE INVERTEBRATE FAUNA

### 3b.2.1. Protozoans

Numerous species of unicellular organisms belonging to the Protozoa are found in the Antarctic. Some of them live in freshwater, while others prefer moist habitats on land. They occur in moss carpets and ground covered by higher plants, as well as in peat, mineral soils and animal guano. Four main groups of Protozoa are represented. They comprise the flagellates, which have one or more filamentous flagella, the naked amoebae and the testate amoebae (Fig. 3b.1a) that move about and capture food by means of pseudopods, and finally the ciliates, with bands of short cilia on their bodies. The number of species decreases with increasing latitude, but more than 65 species from the Antarctic Peninsula, Signy Island and South Georgia are considered as definitely established (Smith, 1978). In East Antarctica several species of all groups are represented in moss communities and in barren sand without visible vegetation (Sudzuki, 1979).

### 3b.2.2. Nematodes

Nematodes (Fig. 3b.1b) are an extremely successful group of small animals, occurring almost everywhere in soils, plants, animals and dead organic materials. Most of the species live in terrestrial or freshwater habitats, but some are found in the ocean as well. Antarctic nematodes were first discovered in the Maritime zone about eighty years ago. At present approximately 70 species are known from the Continental, Maritime and sub-Antarctic zones (Maslen, 1979).

At Signy Island most of the nematodes are microbial feeders, some live on fungi and some are omnivorous (Spaull, 1973). No species feeding on vascular plants are found, but a single carnivorous species is present in moss communities. The nematode biomass estimates for the Signy Island habitats are within the range found for habitats in other parts of the world. Their numbers in soil and vegetation may reach several millions per $m^2$.

From a biogeographical point of view the nematodes offer an interesting parallel to Antarctic terrestrial arthropods, as the distribution of most species is restricted to certain areas. Forty species of nematodes are recorded from the maritime Antarctic, and thirty-four are known only from this zone. Whilst three species are common to the Continental and Maritime zones, seven species are known from the Continent only. About half of the sub-Antarctic species are restricted to this zone, while the remaining ones are also found in other parts of the world.

### 3b.2.3. Tardigrades and rotifers

The tardigrades are minute animals, less than one millimetre in length, related to the arthropods. So far twenty-three species of tardigrades have been reported from the Antarctic (Jennings, 1976). Some of them are found as far south as 77°S in Victoria Land. Most of the species have a world-wide distribution, and only a few of them are restricted to the Antarctic. The terrestrial tardigrades (Fig. 3b.1c) live mainly in mosses, mats of foliose algae and dead plant material. Densities vary from less than twenty thousand to more than fourteen million animals per $m^2$. Relatively little is known about the physiology of these animals, but the high respiration rates of *Macrobiotus furciger* suggest a physiological adaptation to the extreme environment of the maritime Antarctic. Tardigrades are known to tolerate severe desiccation, surviving in a state of anhydrobiosis, and this may account for the widespread distribution of many species.

Rotifers (Fig. 3b.1d) are often found associated with the tardigrades. Like them they live partly in fresh water, but several species inhabit moist terrestrial habitats. At Signy Island these tiny animals, less than half a millimetre long, are found in mosses at densities of more than one hundred thousand per $m^2$. The number of rotifer species in Antarctic mosses appear to be just as great or greater than in temperate areas (Jennings, 1979).

### 3b.2.4. Enchytraeids

The enchytraeids are a family of small, white worms related to earthworms. They occur along the shoreline of the Maritime and sub-Antarctic islands, but are also found in more typical terrestrial habitats, such as moss banks and soil. More research on their distribution and ecology is certainly needed.

### 3b.2.5. Insects

Compared to other geographic regions the number of insect species in the Antarctic is very limited (Gressitt, 1967) (Table 3b.1). From the Continental and Maritime zones 67 species have been recorded, but only twenty-two of them are free-living, the remaining ones being parasites on warm-blooded

TABLE 3b.1. Numbers of Insect Species Present in the Continental and Maritime Antarctic and on the Sub-Antarctic Island of South Georgia.

| | Continental zone | Maritime zone | South Georgia |
|---|---|---|---|
| Collembola (springtails) | 12 | 8 | 16 |
| Mallophaga (biting lice) | 18 | 32 | 35 |
| Anoplura (sucking lice) | 3 | 3 | 1 |
| Thysanoptera (thrips) | — | — | 1 |
| Hemiptera (bugs, etc.) | — | — | 2 |
| Coleoptera (beetles) | — | — | 7 |
| Siphonaptera (fleas) | 1 | 1 | 1 |
| Diptera (flies and midges) | — | 2 | 13 |
| Hymenoptera (parasitic wasps) | — | — | 1 |

animals. Collembola (or springtails) (Fig. 3b.1g) are the only free-living insects on the Continent. They are primitive, wingless species, up to one or two millimetres in length. In addition to Collembola, two species of midges (Fig. 3b.1h) are found in the Maritime zone. One of them has lost its ability to fly through wing reduction, a condition frequently encountered in insects from cold and windy areas. The fauna of the sub-Antarctic islands is much richer comprising a larger variety of insect groups (Gressitt, 1970). At South Georgia representatives of flies, beetles, aphids, thrips and parasitoid wasps are also present (Table 3b.1).

Among the parasitic insects the biting lice are the most numerous, living among the bases of feathers. The few species of sucking lice are restricted to phocine seals. The Antarctic flea (*Glaciopsyllus antarcticus*) is found on the Continent in nests of fulmars and petrels. The close contact with their warmblooded hosts protects the parasitic insects from the effects of the Antarctic climate.

What they lack in number of species some of the Antarctic insects make up in numbers of individuals. Dense clusters of the Collembola *Cryptopygus antarcticus* are frequently observed under foliose algae or floating on small pools of melt water. At Signy Island the population of this species in a moss carpet varied from 36,000 to 95,000 specimens per $m^2$ (Tilbrook, 1977).

## 3b.2.6. Mites and spiders

Approximately seventy species of mites are known from the Continental and Maritime zones (Gressitt, 1967), and a similar number from South Georgia (Gressitt, 1970). Commonly found species are feather mites that are parasites on birds, and nasal mites living in the noses of seals. Free-living Antarctic mites mostly belong to the soft-bodied prostigmatid mites (Fig. 3b.1e) and the hard-shelled cryptostigmatid beetle mites (Fig. 3b.1f). The first group includes some of the toughest terrestrial animals of the world inhabiting nunataks on the Antarctic continent. One of them, *Nanorchestes antarcticus*, which is only a third of a millimetre long, is found as far south as 85°S.

Cryptostigmatid mites frequently form clusters, and the Maritime species *Alaskozetes antarcticus* is often found in dense aggregations comprising hundreds or thousands of immature and adult specimens. No convincing explanation for this behaviour has so far been suggested.

Spiders are not found in the Continental and Maritime zones. The most southerly spiders of the world live on South Georgia, where four species are present.

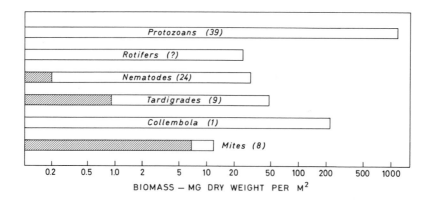

Fig. 3b.2. Estimates of the biomass of different groups of terrestrial invertebrates in a moss community at Signy Island in the maritime
Antarctic. Figures in parenthesis indicate number of species within each group; hatched areas are the biomass of carnivorous species.

## 3b.3. FOOD WEBS

Relatively few details of the feeding ecology of terrestrial Antarctic invertebrates are available. Generally most of them feed mainly on bacteria, algae and fungi. Although abundant, very little moss is consumed, while micro-algae growing in the moss carpets may be the favourite food of many species. Under experimental conditions the Collembola *Cryptopygus antarcticus* has a preference for micro-algae, but also eats fungi and in the laboratory will feed on homogenized moss turf. Examination of gut contents and faecal pellets reveals that filamentous fungi and micro-algae are the most important food sources in the field. The mite *Alaskozetes antarcticus* is both a herbivore and a detritivore, feeding on lichens, foliose algae and organic debris mainly of vertebrate origin.

Very few predators are present in the terrestrial ecosystems of the Continental and Maritime zones. Some of the mites, however, prey on other invertebrates. At Signy Island *Gamasellus racovitzai*, a fast-moving, yellow species, feeds mainly on *Cryptopygus antarcticus*, but also consumes other mites.

As a consequence of the low number of species the food webs of Antarctic terrestrial ecosystems are simple compared to those of other regions. In the moss communities at Signy Island only two trophic levels of invertebrates exist (Davis, 1981). The primary consumers, feeding mainly on the microflora (i.e. bacteria, yeast, algae and filamentous fungi) consist of protozoans, rotifers, tardigrades, nematodes, mites and Collembola. They are themselves the prey of a small group of carnivores, including one species of nematode, one species of tardigrade and the mite *Gamasellus racovitzai*. An indication of the relative abundance and importance of the different invertebrates in this community is seen from their biomass (Fig. 3b.2). Compared to the plant feeders the biomass of the carnivores is very small. Among the primary consumers the biomass of protozoans exceeds all other groups, reflecting the presence of tremendous numbers. Their ecology is poorly known, but it is believed that bacteria constitute their main food source.

A striking feature of these moss communities is that only a small part of the green plants is consumed. Combined with a slow breakdown of dead plant material, this causes thick layers of peat to accumulate beneath the moss surface. In contrast the primary consumers of arctic and temperate ecosystems are more numerous and diversified, consuming a larger part of the plants. The secondary consumers are also more numerous, and of greater importance for the regulation of plant feeding populations than in the Antarctic.

The terrestrial food webs of the sub-Antarctic islands are more complicated, but so far poorly investigated. At South Georgia the beetles *Perimylops antarcticus* and *Hydromedion sparsutum* are partly moss feeders, having an almost unlimited supply of food at their disposal. The litter and soil formed underneath the large tussock grass along the coast harbours a complex community of terrestrial arthropods. Among the primary consumers are mites, Collembola, beetles and fly larvae, preyed upon by spiders, carnivorous beetles and mites.

## 3b.4. BIOGEOGRAPHY

During the breakup of Gondwanaland the theories of continental drift (page 40) suggest that 65 million years ago the Antarctic continent was still in contact with South America and Australia. At that time climatic conditions were more favourable, and the Continent had a rich fauna and flora. These historical connections between the land masses are reflected today in the biogeography of the circum-antarctic flora and fauna. Within insects close relationships are found between species from different continents, such as the chironomid midges from South America, Australia and New Zealand (Brundin, 1967).

Because of the climatic changes that took place, very little of the original fauna of the Antarctic remains. There is evidence to suggest, however, that the present fauna of the South Polar region contains both an important relic element and a later immigrant element, which is particularly obvious in the outer areas.

From a terrestrial point of view it is convenient to divide the south polar region into three zones on climatic and vegetational grounds (Holdgate, 1964; Wallwork, 1973). An outer sub-Antarctic zone comprises the islands lying in the region of the Convergence between 46°S and 55°S, while the Maritime zone comprises the Antarctic Peninsula, some of the islands on the Scotia Ridge and Bouvetøya. Finally, the Continental zone covers the main Antarctic land masses.

The fauna of the different zones includes a high proportion of endemic species, i.e. species restricted to limited geographical areas. Of twenty-three species of prostigmatid mites from the Continental zone, only *Nanorchestes antarcticus* is known to occur elsewhere. The fauna of cryptostigmatid mites comprises three endemic species, and ten out of twelve species of Collembola are found only on the Continent. The two exceptions are *Friesea grisea* from Victoria land and East Antarctic and *Cryptopygus antarcticus* that recently has been recorded from Dronning Maud Land.

Four out of sixteen species of cryptostigmatid mites and four out of ten species of prostigmatid mites are endemic to the Maritime zone. The other mite species and all of the eight Collembola (Table 3b.1) are also found on the sub-Antarctic islands.

The sub-Antarctic zone has a much richer fauna. At present 56 species of cryptostigmatid mites (Travé, 1981) and about 70 species of Collembola (Deharveng, 1981; Wallwork, 1973) have been recorded. Great variations in faunal composition are found when different areas are compared. In Fig. 3b.3 Marion, Crozet, Kerguelen and Heard Islands are considered as one area, Macquarie Island as another area, while South Georgia and the Maritime zone are treated together as the third area.

While ten species of cryptostigmatid mites are endemic to South Georgia and the Maritime zone, this fauna is also influenced by its close proximity to South America, having nine species in common with that continent (Fig. 3b.3A). Similarly, some of the species from Macquarie also occur in New Zealand. The fauna of Marion, Crozet, Kerguelen and Heard have nine species in common with Macquarie, but are very different from South Georgia and the Maritime zone. Twenty-four out of thirty-seven species

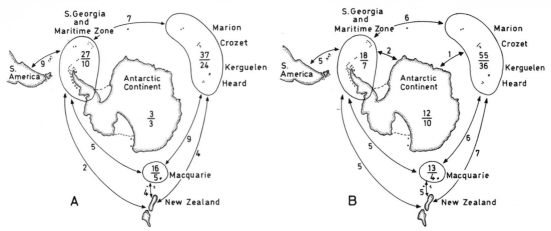

Fig. 3b.3. Distribution of species and subspecies of cryptostigmatid mites (A) and Collembola (B) in different areas of the Antarctic. The proportion of species recorded to the number of endemic species are indicated from each area. Number of species common to two areas are shown by figures and arrows.

have so far only been recorded from this eastern group of islands, which may be considered as a separate, small biogeographical province (Travé, 1981).

The Collembola of the sub-Antarctic zone comprise a western group, an eastern group and a circum-Antarctic group (Wallwork, 1973). In the west, seven species are endemic to South Georgia and the Maritime zone, and so far thirty-six out of fifty-five species have only been recorded from Marion, Crozet, Kerguelen and Heard (Fig. 3b.3B). As for the cryptostigmatid mites, the western and eastern fauna of Collembola are quite different. Some of the sub-Antarctic Collembola also occur in South America and New Zealand, and a few have an even larger, world-wide distribution.

How species from one area are spread to new localities in the Antarctic is not known. Most arthropods, and in particular prostigmatid mites and Collembola, would be subject to desiccation if blown over long distances by the wind. A more likely course is survival under more sheltered conditions during transport by birds. That dispersal of some sort does take place is seen from the fact that terrestrial arthropods are found on the younger volcanic islands of the Antarctic. Bouvetøya is notable in this respect, situated about 1700 km from its closest neighbouring island, and more than 2000 km from South Africa, the nearest continent. The island lies on the mid-Atlantic ridge, and is made of volcanic rocks from 600,000 to three million years old. The fauna of terrestrial arthropods comprises three species of Collembola and six species of mites, which must have been transported over vast distances before reaching this remote island.

## 3b.5. PROBLEMS OF SURVIVAL

Survival under the harsh conditions of the polar regions is not simple. The organisms must tolerate both extreme low winter temperatures, as well as short, cold summers, and their adaptations must involve several kinds of ecological and physiological mechanisms. Low temperature tolerance, increased metabolic rates, life cycle strategies, anaerobiosis, choice of microhabitats and morphological adaptations are among the factors involved.

The choice of habitats in the Antarctic is greatly restricted by the availability of water. The occurrence of mites and Collembola on nunataks and coastal land areas of the Continent is mainly confined to the sparse vegetation of moss and lichens, where some moisture is retained. Since there is no soil, melt water rapidly disappears between rocks and coarse gravel on most of the snow-free areas.

Large parts of the mountain slopes in the Maritime zone and on South Georgia are also covered by loose scree, where no plants can grow, and no invertebrates may live. Thus terrestrial arthropods colonizing land areas of the Antarctic must not only tolerate low temperatures, but must also find habitats with sufficient moisture content.

On inland Continental nunataks apparently barren areas exist, where some moisture is received by melting of neighbouring ice during the summer. Although moss and lichen do not grow, the gravel contains micro-organisms such as bacteria, algae and fungi (Janetschek, 1967). They form the basis of a 'chalikosystem', which represents the outermost border of life and supports a few species of Collembola and mites.

## 3b.5.1. Cold-hardiness

Although it has been shown that some insects are able to survive the actual formation of ice in their tissues (freezing-tolerant), the majority of insects and other terrestrial arthropods are killed by ice formation (freezing-susceptible). In the same way that small volumes of highly purified water will supercool in the range of $-30°C$ to $-40°C$ before ice crystals form, some freezing-susceptible insects may be considered as liquid containers in which the conditions for supercooling are extremely favourable. They depend on supercooling to survive temperatures below the freezing point of their body fluids. Supercooling to $-40°C$ and $-50°C$ has been recorded in insect eggs (Sømme, 1982), and below $-60°C$ in some Arctic insect larvae.

The limit of supercooling in terrestrial arthropods is measured as their supercooling point, the temperature at which spontaneous freezing occurs during gradual cooling. It is marked by a rapid temperature rise, which accompanies the latent heat emission during freezing of water in a supercooled state. In freezing-susceptible species the supercooling point represents the lower limit for survival. Freezing may also take place at higher temperatures, as the chances of spontaneous freezing also increases with prolonged time intervals in the supercooled state.

Among the Collembola and free-living mites of the Antarctic no freezing-tolerant species has been found (Sømme, 1981). All species studied so far depend entirely on supercooling for winter survival, and are instantly killed if freezing occurs. The ability to supercool, however, is influenced by several factors. One factor of great importance, which has been well established in insects from temperate areas, is the presence of nucleators in the gut contents, which will initiate freezing in the alimentary canal. These may be mineral dust particles consumed with the food, or coarse food particles themselves.

In the cryptostigmatid mite *Alaskozetes antarcticus* and the Collembola *Cryptopygus antarcticus* and *Parisotoma octooculata* the ability to supercool is related to their feeding activity (Sømme and Block, 1982; Young and Block, 1980). Supercooling point distribution of all three species shows division into a low group and a high group of freezing temperatures. In *Alaskozetes* about half of the specimens with access to food have supercooling points in the low group, and the other half in the high group. Following starvation for three weeks more than 80% of the mites fall into the low group. A similar shift in the direction of the low group is observed in the Collembola as a result of starvation.

Another factor of great importance for supercooling in terrestrial arthropods is the presence of cryoprotectant substances. In species from temperate areas it has frequently been observed that the supercooling points are depressed during the winter by the accumulation of glycerol or other polyhydric alcohols and sugars (Sømme, 1982). The effect of glycerol on supercooling is purely physical. Because of the hydrophilic properties of this substance the amount of free water is reduced, and the chances of ice nucleation are decreased.

In *Alaskozetes* accumulation of glycerol takes place at $0°C$, and is enhanced in specimens kept at $-5°C$

and $-10°C$ (Young and Block, 1980). An inverse relationship exists between the concentration of glycerol and the supercooling point. The mean supercooling point of *Alaskozetes* without glycerol is $-26.5°C$, but the mites can supercool to $-30°C$ with a concentration of 50 g glycerol $mg^{-1}$ water. The protection offered by glycerol is thus of considerable importance for *Alaskozetes* to survive severe winters in the maritime Antarctic.

Cryoprotectant substances are also important for the cold-hardiness of *Cryptopygus antarcticus*. Glycerol was found in species acclimated at low temperatures in the laboratory (Sømme and Block, 1982), and fairly high levels of mannitol, another polyhydric alcohol, appeared in some samples. The concentrations of the sugars fructose and trehalose also increased, suggesting a combined cryoprotectant effect of all four substances.

Contrary to the condition in Collembola and mites, larvae of the Antarctic midge *Belgica antarctica* are able to survive a certain degree of freezing (Baust, 1980). Overwintering takes place in the larval stage, and the larvae survive freezing at $-15°C$, but do not tolerate exposures at $-20°C$.

In the Anvers Island area these larvae are found in the thin organic layers on rocks close to the shore, and on small, low islands. Because of their limited degree of freezing tolerance they are not capable of surviving the low ambient temperatures of the area. However, their microhabitat temperatures are nearly constant around $0°C$ throughout the year. The basis of this thermostability is the close contact with the sea. Both bedrock and the thin organic layer remain in thermal equilibrium with sea water, eliminating the risk of large temperature fluctuations.

### 3b.5.2. Metabolic rates

Some invertebrates have the ability to undergo instantaneous and seasonal metabolic compensation for temperature changes. The increased metabolic rate following cold acclimation, i.e. a period of exposure to low temperatures, increases the activity and performance of the animals compared to warm acclimated specimens. In some marine invertebrates compensation is complete within the temperature range of their environment, making it possible for them to maintain the same level of activity under all conditions.

Some sort of compensation would also be expected in terrestrial invertebrates that are active during the cold summers of the polar regions. Increased metabolic rates may be required to maintain development, reproduction and other activities under extreme temperature conditions, and terrestrial arthropods of the Antarctic may live in a state of permanent cold acclimation.

Compared to temperate species, the metabolic rate of the Antarctic mite *Alaskozetes antarcticus* in the range of $0°C$ to $10°C$ is twice as high. Similarly, the predatory Antarctic mite *Gamasellus racovitzai*, has a metabolic rate three to four times higher than its temperate relatives. These elevations bring the metabolic rate at Antarctic temperatures to levels comparable to temperate species in their normal temperature range. In this way the Antarctic mites may be able partly to avoid the depressant effect of low temperatures on activity, feeding, growth and reproduction. At Signy Island where both species are commonly found, they maintain their biological functions in the temperature range of $-4°C$ to $+15°C$, which is typical of summer daytime there. In fact, Antarctic terrestrial arthropods may be active at still lower temperatures. Under experimental conditions six species of mites and two species of Collembola were able to walk slowly even at temperatures in the range of $-4°C$ to $-8°C$.

Elevated respiration has also been demonstrated in the Collembola *Cryptopygus antarcticus* and *Parisotoma octooculata* from the maritime Antarctic and in *Isotoma klovstadi* from the Continental zone (Block, 1979). Their metabolic rates are comparable to that of species from the Arctic and alpine regions, but in general are higher than in temperate lowland species.

### 3b.5.3. Life cycles

Extending the life cycle is an important adaptation in many species of Arctic and alpine invertebrates. While temperate species have one or more generations per year, species living under extreme conditions may need two or more years to complete their life cycles. Extended life cycles are often connected with the ability to postpone part of the development when conditions in one year are unfavourable. In this way the species is not dependent upon success in any one year, but has the possibility of utilizing two or more successive years. As a result many species over-winter in more than one stage.

The Antarctic midges resemble their northern relatives in having a short adult life, and a flexible life cycle. Several larval stages are present all the year around. The full life cycle of the Collembola *Gomphiocephalus hodgsoni* is extended over two to three seasons at McMurdo Sound (Janetschek, 1967), and all stages are found during the winter.

This is also the case for *Cryptopygus antarcticus*, where the proportions of specimens of various sizes remains remarkably constant throughout the year (Tilbrook, 1977). Initial growth appears to be relatively rapid, causing a build up of specimens in the following size group. It seems likely that each generation spans more than one year with sporadic and opportunistic growth when climatic conditions are favourable. These Collembola may become quite old, and are known to have lived for more than three years in laboratory cultures. In comparison temperate species normally have one or more generations per year.

Several species of prostigmatid mites and the predator *Gamasellus racovitzai* over-winter both as adults and juveniles. *Alaskozetes* has a life-cycle length of at least one year (Block, 1980), and all individuals probably over-winter twice during their life span. A mixed population of immature and mature individuals is found in all seasons. To survive the winter all stages must be sufficiently cold-hardy. In fact, the smaller juvenile instars appear to be slightly better adapted to low temperatures than the adults (Block and Sømme, 1982).

### 3b.5.4. Anaerobiosis

During the early part of the winter, because of repeated freezing and thawing, the surface of the ground is frequently covered by a layer of ice. These conditions develop in particular in places where much of the snow is swept away by the wind. Terrestrial arthropods living in such surface habitats are

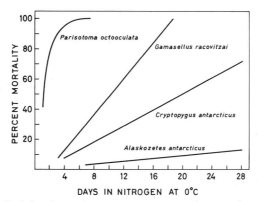

Fig. 3b.4. Mortality of Antarctic Collembola and mites during storage in an oxygen-free atmosphere at 0°C for increasing time intervals.

not only exposed to low winter temperatures, but may also become entirely enclosed by ice. Collembola and mites probably survive in porous parts of the substrate, in crevices and micro-hollows. Due to their own respiration, and also that of plants and other organisms, an oxygen deficiency is likely to develop. In extreme situations the animals may exist in complete anaerobiosis — life without oxygen. Some insects from temperate mountainous areas are able to survive several months in an atmosphere of pure nitrogen at 0°C.

In the Antarctic the ability to survive without oxygen has been studied in two species of Collembola and two species of mites from Signy Island (Block and Sømme, 1982; Sømme and Block, 1982). The reactions of these to anaerobic conditions were quite different (Fig. 3b.4). *Alaskozetes antarcticus* was the most tolerant, and more than 80% survived twenty-eight days of anaerobiosis at 0°C. *Cryptopygus antarcticus* also survived relatively long periods, while *Parisotoma octooculata* lived only for a few days. Mortality rates of *Gamasellus racovitzai* lay between those of the two Collembola species. The reason for these differences is not known, but they may reflect differences in microhabitats of the different species during over-wintering.

## 3b.6. CONCLUDING REMARKS

Our knowledge of Antarctic terrestrial invertebrates is still sporadic, and more work on their taxonomy and biogeography is needed. Population and life-cycle studies have been conducted on a few species in the most accessible areas, but little is known about species on remote nunataks and islands. Physiological data have mainly been obtained from summer animals and laboratory studies, while more studies should be performed on specimens directly from their over-wintering habitats. Some of the most detailed results on the ecology of Antarctic terrestrial invertebrates are derived from the studies on somewhat atypical moss communities at Signy Island. Due to their simplicity the understanding of Antarctic ecosystems is of great importance for basic ecological research. Studies on the interaction between species and the energy flow through these systems reveal principles that are useful for studies of more complex systems in other parts of the world.

As long as our knowledge is relatively incomplete it is difficult to assess the effects of human activities in the Antarctic on terrestrial invertebrates. Presumably many of these organisms, living on the border of their existence, may be very sensitive to any change and pollution of their environments. For this reason alone more studies on Antarctic terrestrial invertebrates should be conducted. In themselves these animals, living in the most remote and unfavourable areas of the world, are fascinating objects of research.

## REFERENCES

Baust, J.G. (1980) Low temperature tolerance in an Antarctic insect: a relict adaptation. *Cryo Letters* 1, 360 – 71.
Block, W. (1979) Oxygen consumption of the Antarctic springtail *Parisotoma octooculata* (Willem) (Isotomidae). *Rev. Ecol. Biol. Sol.* 16, 227 – 33.
Block, W. (1980) Survival strategies in polar terrestrial arthropods. *Biol. J. Linn. Soc.* 14, 29 – 38.
Block, W. and Sømme, L. (1982) Cold-hardiness of terrestrial mites at Signy Island, maritime Antarctic. *Oikos* 38, 157 – 67.
Block, W. and Young, S.R. (1978) Metabolic adaptations of Antarctic terrestrial micro-arthropods. *Comp. Biochem. Physiol.* 61A, 363 – 8.
Brundin, L. (1967) Insects and the problem of austral disjunctive distribution. *Ann. Rev. Entomol.* 12, 149 – 68.
Davis, R.A. (1981) Structure and function of two Antarctic terrestrial moss communities. *Ecol. Monogr.* 51, 125-43.
Deharveng, L. (1981) Collemboles des iles subantarctiques de l'Ocean indien mission J. Travé 1972 → 1973. *C.N.F.R.A.* No. 48, 33 – 108.
Gressitt, J.L. (1967) Entomology of Antarctica, Introduction. *Antarctic Res. Ser.* 10, 1 – 33.

Gressitt, J.L. (1970) Subantarctic entomology and biogeography. *Pacific Insects Monogr.* 23, 295 – 374.

Holdgate, M.V. (1964) Terrestrial ecology in the Maritime Antarctic. In *Biologie antarctique*, Eds. R. Carrick, M.W. Holdgate and J. Prévost, pp. 181 – 94. Hermann, Paris.

Janetschek, H. (1967) Arthropod ecology of South Victoria Land. *Antarctic Res. Ser.* 10, 205 – 93.

Jennings, P.G. (1976) Tardigrada from the Antarctic Peninsula and Scotia Ridge region. *Bull. Br. Antarct. Surv.* 44, 77 – 95.

Jennings, P.G. (1979) The Signy Island terrestrial reference sites. X. Population dynamics of Tardigrada and Rotifera. *Bull. Br. Antarct. Surv.* 47, 89 – 105.

Masleen, N.R. (1979) Additions to the nematode fauna of the Antarctic region with keys to taxa. *Bull. Br. Antarct. Surv.* 49, 207 – 29.

Smith, H.G. (1978) The distribution and ecology of terrestrial protozoa of Sub-Antarctic and Maritime Antarctic islands. *Sci. Rep. Br. Antarct. Surv.* 95, 1 – 104.

Sømme, L. (1981) Cold tolerance of alpine, Arctic and Antarctic Collembola and mites. *Cryobiol.* 18, 212 – 20.

Sømme, L. (1982) Supercooling and winter survival in terrestrial arthropods. *Comp. Biochem. Physiol. A*, in press.

Sømme, L. and Block, W. (1982) Cold hardiness of Collembola at Signy Island, maritime Antarctic. *Oikos* 38, 168 – 76.

Spaull, V.W. (1973) Distribution of Nematode feeding groups at Signy Island, South Orkney Island, with an estimate of their biomass and oxygen consumption. *Bull. Br. Antarct. Surv.* 37, 21 – 32.

Sudzuki, M. (1979) On the microfauna of the Antarctic region, 3. Microbiota of the terrestrial interstices. *Mem. Nat. Inst. Pol. Res.* No. 11, 104 – 26.

Tilbrook, P.J. (1977) Energy flow through a population of the Collembolan *Cryptopygus antarcticus* in *Adaptations within Antarctic Ecosystems*, Ed. G.A. Llano, pp. 935 – 46. Gulf Publ. Co., Houston.

Travé, J. (1981) Biogeographie des oribates (Acariens) subantarctique. *C.N.F.R.A.* No. 48, 149 – 58.

Wallwork, J.A. (1973) Zoogeography of some terrestrial micro-arthropoda in Antarctica. *Biol. Rev.* 48, 233-59.

Young, S.R. and Block, W. (1980) Experimental studies on the cold tolerance of *Alaskozetes antarcticus*. *J. Insect Physiol.* 26, 189 – 200.

# CHAPTER 3c

# Terrestrial Habitats — Inland Waters

## JULIAN PRIDDLE

British Antarctic Survey, Natural Environment Research Council, High Cross, Madingley Road, Cambridge CB3 0ET, U.K.

## CONTENTS

## 3c.1. INTRODUCTION

Lakes and pools are attractive subjects for ecological investigations. An early concept of the 'trophic-dynamic' ecosystem model (Lindemann, 1942) was based largely on lakes, which have the advantage that it is possible to recognize a natural boundary for the ecosystem. The cycling of material or energy within the system and exchange with the outside world can then be studied easily. A lake also provides a 'summary' of the surrounding terrestrial environment in that its character will be influenced strongly by the quality of inflow water.

Early investigations of Antarctic lakes concentrated on the simplicity of the ecosystems, suggested by the paucity of the fauna and flora. We now know, however, that some processes occurring in Antarctic lakes are as complex as those in temperate or tropical lakes. Other features of the lakes are unique, resulting from the combined effects of extreme climate and biogeographical isolation. The growth of animals and plants at uniformly low temperatures, often with long periods of darkness, and the cycling

Fig. 3c.1. Freshwater lakes on Signy Island, South Orkney Islands, in late January. The lakes range from oligotrophic Tranquil Lake in the foreground to mesotrophic Light Lake in the background. (Photo: C. Ellis-Evans)

of biologically important elements in the enclosed environments of frozen lakes are topics currently under investigation (Fig. 3c.1).

Inland water-bodies are sparsely distributed in Antarctica. The polar ice-cap contains an estimated 90% of the world's surface freshwater but the presence of liquid water as lakes and pools is generally restricted to coastal areas and offshore islands (Fig. 3c.2). The range of water-bodies is wide, from small melt water pools scattered around the edge of the Continent to oligotrophic and eutrophic lakes in the maritime Antarctic and on the sub-Antarctic islands. Some saline and stratified lakes are also known.

## 3c.2. THE ANTARCTIC LAKE ENVIRONMENT

Two major features of the Antarctic region have a profound influence on the character and development of lakes. The cold climate results in the freezing of the upper layer of many lakes for most or all of the year. During this time exchange with the catchment is severely curtailed and the part of the lake which remains liquid is effectively a closed system. In this respect there is an interesting parallel with temperate lakes. These may become thermally stratified in summer with the warm upper layer (epilimnion) being mixed and providing the site for planktonic plant growth. The cooler lower layer (hypolimnion) is separated from the epilimnion by a zone of rapid density change associated with a sharp temperature change. The hypolimnion is self-contained and provides a 'sink' for the organic production of the epilimnion. When an Antarctic lake is frozen it is equivalent to an isolated hypolimnion. Solar

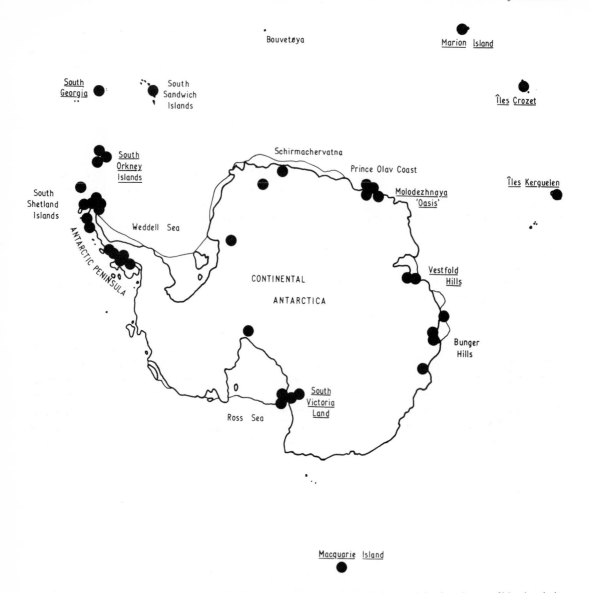

Fig. 3c.2. Map of Antarctica showing the locations of inland water-bodies. Those names which are underlined are the sites of lakes described in the text. Modified from Priddle and Heywood (1980).

radiation supplying the energy essential for plant growth is also in short supply then. The period of freezing coincides with the winter months when day-length is short or when, at higher latitudes, the sun is continuously below the horizon for part of the year. A snow layer accumulates on the ice. It has a high reflectivity (albedo) and may also absorb more than 90% of the solar radiation which enters it. Photoautotrophic organisms, which use light energy to convert inorganic carbon in the water to organic compounds, are the major endogenous producers in a lake ecosystem and these will obviously be strictly limited in their growth and distribution by the extreme seasonality of the light climate.

Secondly, the barren nature of recently deglaciated catchments usually implies that the water entering a lake will contain little dissolved material. Biologically-useful nutrients such as nitrogen and phosphorus may be in short supply. This will further inhibit the growth of organisms in the lakes and dramatic differences may be observed between otherwise similar lakes, one of which may receive only glacial melt water whereas the other may be close to a penguin rookery or other source of natural nutrient enrichment.

### 3c.3. FRESHWATER LAKES

Antarctic lakes follow an evolutionary series as their catchment areas become deglaciated (Priddle and Heywood, 1980). Nutrient content of the water usually increases as a catchment changes from being ice-covered to exposed rock and scree, and later develops vegetation cover. Further nutrient input commonly occurs in lakes close to the sea, where lakes receive wind-blown sea-spray and effluent from seal or seabird colonies.

Lakes and pools may form in basins eroded in ice. These are usually shallow and freeze solid during the winter. Their lifespan is short as their formation represents a part of a rapid melting event. The presence of the algae which cause coloured snow indicates that amounts of inorganic nutrients sufficient for some plant growth may be derived from snow with its minute traces of rock dust.

The lakes which form next in the series occur at the edge of the ice-sheet. These proglacial lakes in rock basins are typically found in lateral moraines and are partially dammed by ice. Although they are less transitory than lakes in ice basins, they are still susceptible to major fluctuations in size as the ice dam continues to melt. Like that of proglacial lakes on ice, inflow to these lakes is very dilute.

As the ice-sheet recedes further, lake basins in rock or drift may be exposed or formed by morainic deposits. Lakes occupying these basins are not exposed to changes in their basins caused by ice-melt. As the catchment becomes vegetated, the residence time of groundwater is increased and inflow to the lake contains a higher content of dissolved ions. Nutrient levels are still low and the lakes would be described as 'oligotrophic'. Biological enrichment may however produce a profound alteration in the nature of these lakes.

The three lakes described below have been selected as representative of the main stages in this evolutionary series.

TABLE 3c.1. Physical, Chemical and Biological Properties of Three Freshwater Lakes

| Location | Lake Glubokoye Molodezhnaya (67°40′S, 45°51′E) | Moss Lake Signy Island (60°43′S, 45°38′W) | Heywood Lake |
|---|---|---|---|
| Max. depth (m) | 38.0* | 10.4* | 6.4 |
| Volume ($m^3 \times 10^{-3}$) | — | 53.4 | 96.2 |
| Summer pH | 5.5 | 7.0 | 7.1 |
| Summer $Cl^-$ (g $m^{-3}$) | 14.6 | 22.7 | 34.6 |
| Summer $PO_4 - P$ (mg $m^{-3}$) | — | 0.5 | 15.9 |
| Summer $NO_3$-N (mg $m^{-3}$) | — | 3.8 | 7.7 |
| Summer chlorophyll $a$ (mg $m^{-3}$) | — | 1.9 | 7.9 |
| Max.C-fixation by phytoplankton (mg $m^{-3}$ $d^{-1}$) | c.250[+] | 31 | 800 |

*See text regarding fluctuations of maximum depth.
[+]Carbon fixation estimated from oxygen changes under ice.
Data from: Ellis-Evans, 1981a; Heywood et al., 1980; Kaup, 1975; Kriss et al., 1968; Light et al., 1981.

### 3c.3.1. Lake Glubokoye — a proglacial lake

This proglacial lake is situated in the Molodezhnaya ice-free area ('oasis') near the coast of Enderby Land (45°E). It is ice-free in some summers but freezes to a depth of 3.5 m during the winter. It is ice-dammed and has changed in size as the outflow has melted. A 21 m drop in water level in the lake was recorded when a major breach occurred in the ice-dam near the outflow (Klokov, 1970). The present level appears to be approximately 13 m maximum depth, compared with the earlier value of 38 m (Kriss et al., 1968).

The ionic content of the lake water is low (Table 3c.1) as the 9 km² oasis area where the lake is situated has very little vegetation and the lake is fed mainly by melt water. The lake is clear and water temperatures rise to nearly 7°C in summer (MacNamara, 1970).

There are few planktonic organisms. Seven diatom species and no animals have been found in plankton net samples (Opaliński, 1972). The presence of smaller algal cells, which would be too small to be caught by the net, may however be indicated by the values of primary productivity in the water column. Measurements of carbon fixation gave figures of $200 - 300$ mg C m$^{-3}$ day$^{-1}$ under late winter ice-cover (Kaup, 1975) and this implies a larger, more active population than that found by Opaliński. The low overall metabolism in the water column is amply demonstrated by the very small bacterial biomass (Kriss et al., 1968).

The majority of the biomass accumulation in the lake is attributable to the benthos. Here, the bottom silt is overlain by a 'bacterio-algal layer ranging from 1 cm to 1 m in thickness' (Opaliński, 1972). Growing amongst this layer are moss plants which are overgrown with algae. Mosses have been dredged from the deepest part of the lake (then 32 m). The biota of the algal layer consists largely of blue-green algae and diatoms, although the ubiquitous filamentous green alga *Oedogonium* has also been noted. The fauna associated with this layer is dominated by testate amoebae but also includes tardigrades, rotifers and nematodes. Copepods are the only animals which are capable of moving into the water column. Many of the animals are present as resting stages (Korotkevich, 1964).

### 3c.3.2. Moss Lake — a nutrient-poor freshwater lake

The location of this lake on Signy Island in the South Orkney Islands (61°S, 45°W) exposes it to less extreme climatic conditions than those experienced by Lake Glubokoye. The island group lies on the Scotia Arc and has a maritime climate. Signy is more deglaciated than other members of the group and a little over half of its area becomes free of snow and ice during the summer. Moss Lake is situated on the east side of the island, 48 m above sea-level. Its catchment is a cirque dammed by a moraine and lies at the head of a small valley containing three lakes. The catchment is small (9 ha) and consists mainly of bare scree and an ice-field. Marine influence is very limited, as the lake is 800 m from the coast, and ionic content of the lakewater is low (Table 3c.1).

The lake is small with a maximum depth of 10.4 m. Subterranean drainage through a porous part of the moraine causes a drop in water level during the winter. The surface water freezes to a maximum depth of about 1 m and the lake is ice-covered for about ten months. Ice forming on the surface of lake water consists of clear columnar crystals with their long axes parallel to each other. The vertical orientation of these crystals allows them to act as light guides, enhancing the transmission of solar radiation to the water beneath. Bands of frozen slush accumulate through the winter as freezing and thawing alternate and this material absorbs much of the light incident upon it. Freshly fallen snow comprises the top layer.

INLAND WATERS

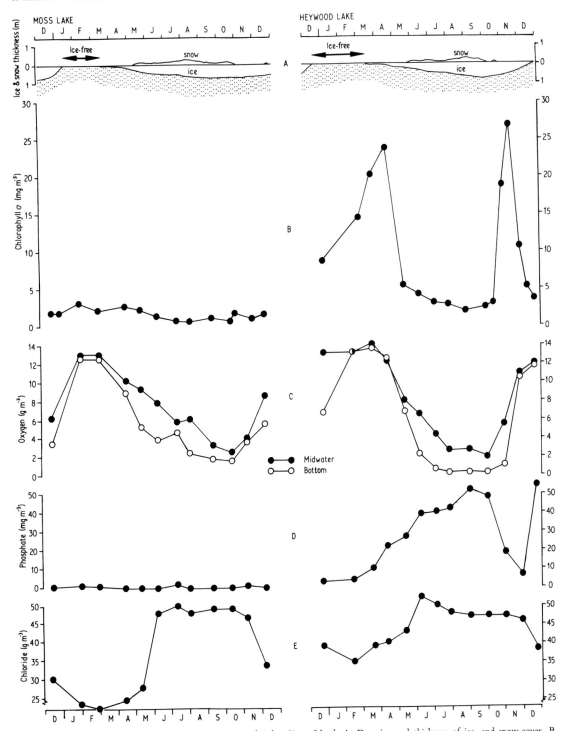

Fig. 3c.3. The seasonal cycles of Moss Lake and Heywood Lake, Signy Island. A, Duration and thickness of ice- and snow-cover. B, phytoplankton biomass as indicated by the concentration of particulate chlorophyll *a*. C, midwater and deepwater oxygen concentrations. D, concentration of dissolved phosphate, an important biological nutrient the solubility of which increases under anoxic conditions. E, concentration of chloride ion, a substance little affected by biological activity and which reflects the concentration of ions in winter due to removal of pure water as ice.

The chemical composition of the lake-water changes during winter (Fig. 3c.3). In the closed system, oxygen is gradually consumed and the proportion of reduced ions, e.g. nitrite, increases. The lake cools to about 0°C. When the ice melts in December or January, high winds rapidly mix the lake water throughout the water-column and oxygen saturation is restored. The water temperature rises to about 5°C.

Moss Lake derives its name from the luxuriant growths of aquatic mosses which cover approximately 40% of the lake area below the 5 m depth contour. Above this zone they are very sparsely distributed as this part of the lake is subjected to ice-scour during the winter decrease in water level. Two species are represented and one of these, *Calliergon sarmentosum*, is also present terrestrially on Signy Island. The other, a species of *Drepanocladus*, appears to be exclusively aquatic. Both exhibit unusual growth forms with large leaves and stems up to 0.4 m long. Remaining areas of the lake trough benthos are covered by a thin, perennial 'felt' consisting mainly of blue-green algae.

The benthic plant communities are perennial, suggesting that growth is possible for much of the year. Ice- and snow-cover in winter dramatically reduce underwater irradiance such that the midwinter minimum is 0.03% of the maximum value recorded under summer open-water conditions (Priddle, 1980). Field experiments with both the aquatic mosses and the blue-green algal community indicate that they are incapable of growing during the period around midwinter but can maintain net production during August and then grow successfully for the rest of the year. Laboratory experiments confirm the very low light requirements of these plants. Their respiration rate is reduced at low temperatures, allowing the accumulation of photosynthetically fixed carbon even at low light levels. It also seems likely that the large leaves of the mosses are advantageous for photosynthesis in low light. Growth rate throughout the rest of the year is, however, slow and annual production by the benthos is low.

Planktonic algae, mainly unicellular forms, are scarce and their population maximum occurs in spring and summer. The microbial population in the water column is also sparse (Ellis-Evans, 1981a). Activity is low in winter but there is a rapid response to the spring thaw. Bacteria utilize the majority of phytoplankton production and total annual bacterial biomass production is $18-35\%$ of total phytoplankton production. Bacterial metabolism is very slow in Moss Lake, with turnover times for glucose and acetate exceeding 1000 h in winter.

In common with other Antarctic lakes, the fauna is poorly developed and several groups important in temperate lakes are not represented. Fish, molluscs and aquatic insects are absent — presumably reflecting the biogeographical isolation of Antarctica. Rotifers account for the majority of species (24 species) and the remaining taxa are crustaceans (7 species), nematodes, a platyhelminth worm, tardigrades, an annelid worm and protozoa. The most conspicuous animals are two species of copepod — *Pseudoboeckella poppei* and the predatory *Parabroteas sarsi*.

Invertebrates are present throughout the year but maximum abundance occurs during the summer. Oxygen concentration under the winter ice-cover does not drop so far as to preclude animal survival (Fig. 3c.3). The great diversity of rotifer species is generally associated with the benthos as planktonic food supply is scarce and seasonal. Many species are known from other localities where they would be found in the plankton. In Moss Lake they tend to show certain modifications of the appendages and size changes which may be associated with their bottom-dwelling habit. Reproduction of the copepods and most other species coincides with the spring phytoplankton peak. Early larval stages feed on suspended particles and feed on benthic organisms only as later stages and adults (Heywood, 1970). The trophic structure of the ecosystem is apparently simple and only seven species have been identified as predators, of which the most significant is *Parabroteas* (Priddle and Dartnall, 1978).

The structure of the aquatic moss provides a very diverse habitat in comparison with the simple algal felts (Priddle and Dartnall, 1978). The moss stems provide a variety of niches and the stems can be divided into zones like the different floors of a skyscraper (Fig. 3c.4). The moss is first colonized by algae adhering to the leaves of the apical bud. As the leaves expand, material becomes trapped in the leaf

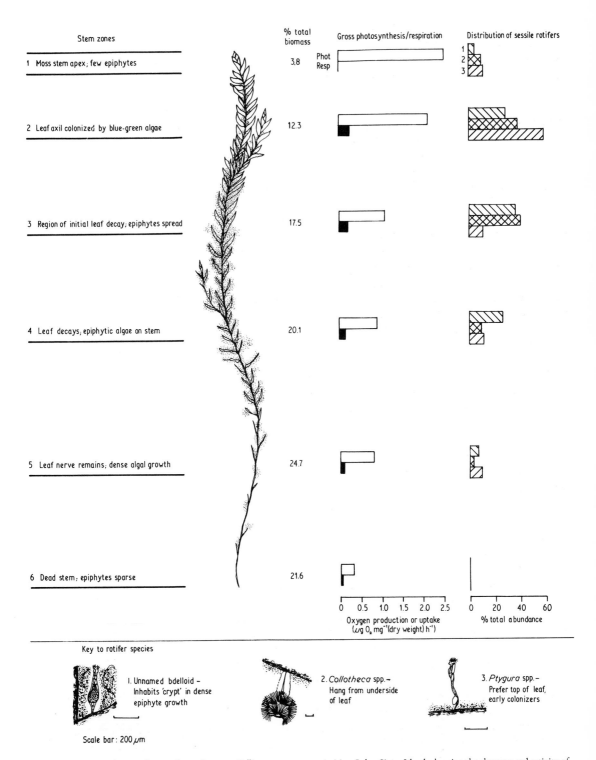

Fig. 3c.4. Zonation of a typical stem of aquatic moss, *Calliergon sarmentosum*, in Moss Lake, Signy Island, showing the character and activity of the epiphytic community and the settlement preferences of three genera of rotifers. Data from Priddle and Dartnall (1978).

axil and algae continue to spread over the leaf surface. Animals feed on the epiphytic algae, and some species, including six rotifers, become attached. Further down the stem, the moss leaves die and decay. Here algal production predominates and swimming animals graze the attached cells. At the bottom of the moss plant only the stem remains and this decays once it becomes buried in the sediment.

### 3c.3.3. Heywood Lake — a naturally enriched freshwater lake

This lake is also situated on Signy Island and provides a contrast with Moss Lake. It is closer to the coast and therefore receives a more significant input of nutrients from the rich marine ecosystem. This input is both direct as wind-blown spray and indirect enrichment by seals, especially elephant seals (*Mirounga leonina*), which congregate around the lake and swim in it. The position of Heywood Lake in a vegetated coastal catchment probably implies that it would have a higher nutrient status than Moss Lake despite enrichment by seals, but recent increases in seal numbers during the mid-1960s have accentuated the differences between the lakes. The productivity of Antarctic lakes is characteristically small as low nutrient availability combines with the distinct seasonality of the irradiance regime to restrict plant growth. Although nutrient enrichment in Heywood Lake has been relatively small, the effect has been more dramatic than would be expected in a temperate lake (Table 3c.1, Figs. 3c.3 and 3c.5).

Heywood Lake is small, although its surface area (4.5 ha) is the largest of the Signy Island lakes, and its maximum depth is only 6.4 m. Several inflow streams feed the lake from a catchment dominated by moss banks and seal wallows, but much of the inflow volume derives from a neighbouring lake. Raised nutrient content encourages the development of a large phytoplankton community in summer. Such algae may exploit the entire water-column in shallow, well-mixed lakes and attain high biomass. The water however becomes turbid and vertical transmission of solar radiation may fall below 20% m$^{-1}$. Growth of phytoplankton cells therefore takes place mainly in the upper layers of Heywood Lake and depends on conditions of complete wind-mixing and is limited to the ice-free season. Two distinct phytoplankton communities develop, differing in species composition (Light *et al.*, 1981). One appears under summer conditions and is characterized by flagellate Cryptophyta. These algae are responsible for a very high production rate (up to 4 g C m$^{-2}$ day$^{-1}$) during the short summer season. At this time daily

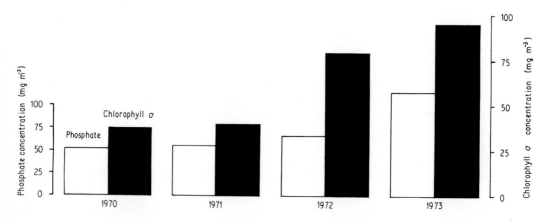

Fig. 3c.5. Increases of maximum values of dissolved reactive phosphate and phytoplankton biomass in Heywood Lake, Signy Island, during 1970 – 73 when the elephant seal population around the lake increased. Data from Light (1977).

irradiance input may be higher than in temperate regions, as long day-lengths and high solar radiation coincide, but the growing season is short. In contrast phytoplankton cells are virtually absent from the water-column in winter. Ice-cover halts the circulation of water within the lake and also drastically reduces irradiance. Planktonic algae are unable to grow under these conditions and probably sediment to the lake bottom. An increase of underwater irradiance coincides with the melting of snow on the lake ice and melting also occurs at the lake edge. Green algae are washed into the body of the lake and grow rapidly after an initial 'lag' phase. These cells differ from the summer, open-water community in having a very high cellular proportion of chlorophyll pigment. This probably indicates a response to the low irradiance conditions under ice. Carbon fixation by this community is low but growth rates are rapid in spite of low temperatures.

Winter freezing remains the most important influence on the lake ecosystem. The period of ice-cover is usually slightly shorter than that of Moss Lake but ice- and snow-thicknesses are similar. The increased trophic status of the lake is shown both in the chemical composition of the lake water and the seasonal behaviour of its constituents (Fig. 3c.3). Increased biological activity within the lake, particularly of phytoplankton, results in depletion of phosphorus and nitrogen in summer.

Microbial numbers and activity reflect the population dynamics of the phytoplankton, as in Moss Lake (Ellis-Evans, 1981a,b). Maximum bacterial heterotrophic activity coincides with the open-water phytoplankton maximum. There are large differences between turnover times for glucose and acetate in summer (shortest time 39 h for glucose and 27 h for acetate) and in winter (c. 500 h for both compounds). The annual bacterial production is 7 – 46% of phytoplankton production (Ellis-Evans, 1981b). Microbial decomposition proceeds at significant levels in the presence of large amounts of organic substrate and in winter, when the lake is sealed by ice, dissolved oxygen becomes depleted, especially close to the lake bottom (Fig. 3c.3).

The high primary production by phytoplankton contrasts with the behaviour of the benthic plant population. In the shallow-water zone which freezes in winter, the species composition of the benthic algal community resembles that found in Moss Lake and is probably similarly productive. Benthic plants are, however, very much less abundant in deeper water and mosses are now absent, and this decrease in biomass may be related to summer turbidity associated with phytoplankton growth. It appears that the benthic plant community has decreased further within recent years, as Light and Heywood (1973) reported benthic moss from Heywood Lake and collections from the 1960s suggest a more flourishing benthic community than at present.

The planktonic nature of plant productivity influences the animals. Zooplankton biomass is high for an Antarctic lake but has low species diversity and low activity in winter. Few of the zooplankton taxa found in Heywood Lake are absent from Moss Lake. The biology of those which are present is strongly linked to the more marked seasonality of plant production. *Branchinecta gainii*, a freshwater fairy-shrimp is present in Heywood Lake, where it over-winters in the egg stage. The nauplius larvae feed on phytoplankton under ice in September – October. The common copepod, *Pseudoboeckella poppei*, also shows a seasonality of breeding, although some ovigerous females may be found in the population throughout the year (Heywood, 1970). Low oxygen tensions may have an effect on some other animals. Only six species of rotifer are found in Heywood Lake, whereas Moss Lake has twenty-four.

## 3c.3.4. Freshwater lakes on sub-Antarctic islands

Most of the sub-Antarctic islands have lakes and pools. As with the truly Antarctic freshwater lakes their character reflects both climatic influences and their geographic isolation. The sub-Antarctic islands have a climate which is milder than that of the Antarctic islands and coast, and are more deglaciated.

However, many are very isolated and long distances separate them from sources of colonizing organisms (Fig. 3c.2). In this context South Georgia is unique in that it lies on the Scotia Arc which forms a series of stepping stones from Tierra del Fuego and the Falkland Islands to the Antarctic Peninsula. The other islands for which freshwater biological data are available — Marion Island, Macquarie Island, Îles Crozet and Îles Kerguelen — are truly oceanic.

These oceanic islands have mild climates characterized by a small annual temperature range and high precipitation. Average air temperatures vary from 2°C to 5°C. Prolonged winter ice-cover on lakes is therefore rare. Water temperature in shallow water bodies may rise as high as 14°C. High precipitation (greater than 2.5 m per annum on Marion Island) is the most important influence on the freshwater ecosystem, resulting in very dilute inflow to lakes, with nitrogen and phosphorus concentrations often below detectable limits using conventional analyses (Grobbelaar, 1975). Vascular terrestrial vegetation in the catchments will retain ions more effectively than the cryptogamic vegetation of the maritime Antarctic catchments.

Sub-Antarctic lakes are often more complex ecosystems than those at higher latitudes. The most conspicuous feature of many of these lakes is the presence of rooted and floating fringing vegetation, reflecting the absence of continuous winter ice-cover and the better-developed terrestrial vegetation of the catchments (see Chapter 5a). The lake basin may also contain submerged vascular plants such as the water-weeds *Myriophyllum* (Macquarie Island) and *Potamogeton* (Marion Island), and larvae of terrestrial insects (South Georgia, Macquarie Island). Molluscs and fish are apparently absent from freshwater habitats and these groups would be expected to suffer significant dispersal problems. Both are present on the Falkland Islands (Weller, 1975).

As with truly Antarctic lakes, biological enrichment plays an important rôle in the development of some lakes and pools on subAntarctic islands. Large shallow pools on Marion Island have low production by phytoplankton and the majority of plant growth takes place in the benthos (Grobbelaar, 1975). Some of these pools have however been enriched by inflow derived from penguin rookeries and here phytoplankton dominates the freshwater plant community (cf. Heywood Lake) and carbon fixation may reach $0.8$ g C m$^{-1}$ day$^{-1}$ in summer, with accompanying high turbidity (only 20% light penetration at 10 cm depth) (Grobbelaar, 1974).

The influence of climate on biological distribution is demonstrated by the numbers of freshwater invertebrates recorded in collections from the islands of the Scotia Arc and the Antarctic Peninsula (Dartnall and Heywood, 1980). There is a decrease in the number of species, which is correlated with increasing latitude. South Georgia has ten species of freshwater crustaceans, only one less than the Falkland Islands. The lakes of Signy Island have eight species, whilst sampling a few lakes on the Antarctic Peninsula has yielded four species and only one — the ubiquitous copepod *Pseudoboeckella poppei* — is recorded from Alexander Island (68°S).

## 3c.4. SALINE LAKES

A number of saline and hypersaline water bodies is known from the Antarctic (Burton, 1981). They have formed as the result of two contrasting processes. Some lakes can be shown to have originated as sea water trapped in bays or fjords after the retreat of a marine transgression. Others are situated in ablation areas where evaporation loss exceeds precipitation. Here the groundwater becomes concentrated by evaporation or sublimation and lakes may form which are many times more saline than sea water. Lakes of marine origin may also become more concentrated by this process. In both cases, subsequent alteration of the local hydrological regime may result in melt water flowing into the catchment. Fresh water entering the lake basin will not mix with the dense saline water and the lake will become stratified (meromictic) with an ice-covered freshwater layer overlying the original saline water-body. A different form of meromictic lake is uniquely polar. Epishelf lakes form between the land coast and the attached floating ice-shelf — the upper layer is fresh water but beneath it is sea water which communicates with the sea beneath the ice-shelf (Priddle and Heywood, 1980). Such lakes have a tidal rise and fall, and marine fish and invertebrates may live in the lower layer, a long distance from the open sea.

Saline lakes and pools present more extreme environments than those of the freshwater lakes. The high ionic content of the water depresses its freezing point and water-bodies are known whose water would be expected to freeze at temperatures as low as $-55°C$. Whereas the ice-cover on freshwater lakes helps to isolate the ecosystem from the rigorous winter climate, saline water-bodies remain exposed and their winter temperatures approach ambient air temperatures.

## 3c.4.1. Don Juan Pond — a hypersaline pool

Don Juan Pond is situated in Wright Valley, South Victoria Land (Fig. 3c.2), an area ice-free owing to ablation. Beach lines around the lake suggest that it is a relic of a much larger, probably freshwater lake with a depth of approximately 10 m. Its present depth is only 10 cm Total dissolved solids content of the water is very high ($541$ kg m$^{-3}$) and is thirteen times as concentrated as sea water. This high concentration of solutes (Table 3c.2) results in a theoretical freezing point of $-48°C$. The lake does not freeze in winter and probably has a temperature similar to that of the surrounding air. The high solute concentration is maintained by the inflow of groundwater concentrated by sublimation at low temperatures.

Few, if any, organisms can survive in such a hostile environment. A dried mat of algae, flagellates and bacteria has been reported from salt flats at the edge of the pool. It seems unlikely however that these microbes can survive in the rugged environment of Don Juan Pond and perhaps they grow actively only when moistened by freshwater streams originating from melting permafrost (Wright and Burton, 1981). Three strains of bacteria and a yeast have been isolated from the pool. Their salt tolerance was unusually high but their abilities to function at the very low temperatures obtaining in the water has been questioned.

Don Juan Pond may represent the fate of several freshwater lakes in ablation areas where the net loss of water from the catchment causes elevated concentrations of salts both in the lake and in groundwater inflow.

## 3c.4.2. Deep Lake — a hypersaline lake of marine origin

Several saline lakes in the coastal areas of Continental Antarctica have been formed after sea water has been trapped in fjords and bays following isostatic rebound of the land after ice-retreat (Burton, 1981; Priddle and Heywood, 1980). Deep Lake is one of the larger members of a group of such lakes in the Vestfold Hills area (Fig. 3c.2). The lake-water is approximately 5.5 times as concentrated as sea water with a total dissolved solids content of $280$ kg m$^{-3}$ (Table 3c.2). The theoretical freezing point of the water is $-28°C$ and the lake remains open through the winter, becoming isothermal at approximately $-14°C$.

As with Don Juan Pond, the high salinity and low temperature combine to create an environment incapable of supporting a diverse ecosystem. Only one algal species has been recorded in the plankton and there are apparently no herbivores. Photoautrophic carbon fixation in the water-column is undetectable but some carbon uptake by the benthic unicellular algae has been recorded. The presence of

TABLE 3c.2. Physical and Chemical Properties of Saline and Meromictic Water-bodies

| Location | Don Juan Pond Wright Valley, S. Victoria Land (77°32′S, 161°33′E) | Lake Vanda | | Deep Lake Vestfold Hills (68°33′S, 78°15′E) |
|---|---|---|---|---|
| Max. depth (m) | — | 67.5 | | 36.0 |
| Mean depth (m) | 0.11 | — | | 14.2 |
| Volume ($m^3 \times 10^{-3}$) | $c.0.7^*$ | $c.860\,000^*$ | | 9100 |
| Sample depth (m) | $A^+$ | 11(fresh) | 66(saline) | $A^+$ |
| $Cl^-$ (kg $m^{-3}$) | 212 | 0.19 | 82 | 159 |
| $SO_4^{2-}$ (kg $m^{-3}$) | 0.11 | 0.006 | 0.82 | 2.5 |
| $Na^+$ (kg $m^{-3}$) | 1.15 | 0.047 | 2.36 | 72 |
| $Ca^{2+}$ (kg $m^{-3}$) | 114 | 0.069 | 26 | 2.2 |
| $PO_4 - P$ (mg $m^{-3}$) | — | 0.62 | 162 | 10.0 |

*Approximate calculation by present author.
$^+$A = average value for whole water-column.
Note that most chemical concentrations are expressed in different units from those in Table 3c.1.
Data from: Armitage and House, 1962; Kerry *et al.*, 1977; McLeod, 1964; Meyer *et al.*, 1962; Vincent *et al.*, 1981.

organisms in the lake-water does not necessarily imply that they are capable of growth in such a medium. Laboratory tests with bacteria isolated from Deep Lake show that they were unable to grow on media made with more than 25% lake-water.

### 3c.4.3. Lake Vanda — a stratified lake

Lake Vanda occupies an undrained basin in Wright Valley, South Victoria Land. It is a deep lake (Table 3c.2) but evidence of the previous maximum lake level is found 57 m higher than the present lake surface — suggesting that the lake was once nearly twice as deep as at present.

The lake consists of two ecosystems. The upper layer is nutrient-poor fresh water covered by 3.5 m of ice even in summer. This layer extends down to *c.* 50 m depth, below which salinity gradually increases to approximately four times that of sea water. This saline layer is slightly less than 18 m deep in the deepest part of the lake and has become warmed to an average temperature of 25°C. The cause of this high temperature is not completely understood and it is known to occur to a lesser degree in some other meromictic lakes in South Victoria Land. Both geothermal heating and solar warming by a 'greenhouse effect' have been proposed.

The freshwater upper layer resembles other oligotrophic Antarctic lake ecosystems. Lake Vanda is further south than the freshwater lakes discussed here and so experiences a longer period of continuous darkness in winter (*c.* 30 days). This and the persistent thick ice-cover severely limits algal growth. Even so, phytoplankton densities of up to 4000 cells per litre have been recorded and algae belonging to eight genera identified. Particulate carbon fixation is low even in summer (7.2 mg C $m^{-3}$ $h^{-1}$ for the top 25 m of the water-column). Single species of each of the ciliates, tardigrades and rotifers have been found in the zooplankton. A diverse benthic community is also found in the freshwater layer. The filamentous blue-green algae *Schizothrix* and *Lyngbya* form a mat community and a specimen of an aquatic moss has been collected. In summer the ice at the edge of the lake melts forming a shallow moat 1 – 10 m wide. Protozoa and invertebrates such as nematodes, rotifers and tardigrades are found here.

The boundary between the freshwater and saline layers provides an interesting instance of the spatial partitioning of microbiological processes (Vincent *et al.*, 1981). The transition from freshwater to saline conditions is quite sharp and occurs between 50 and 54 m depth, at which the salinity is 1.5 times that of sea water. Ionic concentration then increases gradually below this depth and below 60 m the water is anoxic. Biological nitrogen cycling in which inorganic nitrogen is converted to soluble compounds involves pathways of oxidation (nitrification) and reduction (denitrification) which result from the metabolism of specific micro-organisms. Conditions in the transition zone in Lake Vanda appear to be ideal for nitrifying bacteria which produce high levels of dissolved nitrous oxide, reaching 20,000 times atmospheric saturation at the bottom of the boundary layer. Slightly deeper than this nitrogen-rich layer is a pronounced peak in algal carbon fixation. Denitrifying bacteria require different conditions for growth and these are found at the top of the anoxic layer, where much of the nitrous oxide produced in the boundary layer is broken down.

## 3c.5. CONCLUSIONS

Antarctic lakes are extreme environments in many respects, especially in the severity of their climate, and their extreme seasonality and biogeographical isolation. Some of the saline water-bodies are virtually sterile. In contrast, some freshwater lakes and pools may represent localized regions of high productivity compared with otherwise barren terrestrial areas (Light and Heywood, 1975). The small size of some lakes make them very susceptible to relatively minor differences in catchment character. Natural processes such as enrichment by seals and seabirds may cause extreme biological changes when imposed upon the very low nutrient regimes of proglacial lakes. Investigation of the range of Antarctic lake ecosystems, with their simple biota and pronounced seasonal changes, may provide insights into ecological processes of world-wide interest.

## ACKNOWLEDGEMENTS

I would like to thank Drs. R.B. Heywood and H.J.G. Dartnall (British Antarctic Survey) for many helpful discussions and permission to use unpublished data. Dr. D.W.H. Walton and my mother kindly criticized early drafts of this chapter.

## REFERENCES

Armitage, K.B. and House, H.B. (1962) A limnological reconnaissance in the area of McMurdo Sound, Antarctica. *Limnol. Oceanogr.* 7, 36 – 41.

Burton, H.R. (1981) Chemistry, physics and evolution of Antarctic saline lakes: a review. *Hydrobiologia* 82, 339 – 62.

Dartnall, H.J.G. and Heywood, R.B. (1980) The freshwater fauna of South Georgia. *Bull. Br. Antarct. Surv.* 50, 115 – 18.

Ellis-Evans, J.C. (1981b) Freshwater microbiology in the Antarctic. 1, Microbial numbers and activity in oligotrophic Moss Lake, Signy Island. *Bull. Br. Antarct. Surv.* 54, 85 – 104.

Ellis-Evans, J.C. (1981a) Freshwater microbiology in the Antarctic. 2, Microbial numbers and activity in nutrient-enriched Heywood Lake, Signy Island. *Bull. Br. Antarct. Surv.* 54, 105 – 21.

Grobbelaar, J.U. (1974) Primary production in freshwater bodies of the sub-Antarctic island Marion. *South Afr. J. Antarct. Res.* 4, 40 – 45.

Grobbelaar, J.U. (1975) The lentic and lotic freshwater types of Marion Island (sub-Antarctic): a limnological study. *Verh. int. Verein. theor. angew. Limnol.* 19, 1442 – 9.

Heywood, R.B. (1970) Ecology of the freshwater lakes of Signy Island, South Orkney Islands. 3, Biology of the copepod *Pseudoboeckella silvestri* Daday (Calanoidea, Centropagidae). *Bull. Br. Antarct. Surv.* 23, 1 – 17.

Heywood, R.B., Dartnall, H.J.G. and Priddle, J. (1980) Characteristics and classification of the lakes of Signy Island, South Orkney Islands, Antarctica. *Freshwat. Biol.* 10, 47 – 59.

Kaup, E.B. (1975) O perbichnoy produktsii ozer oazica Molodezhnogo. (On the primary productivity of the lakes of the Molodezhnaya oasis). *Trudy sov. antarkt. Eksped.* 65, 149 – 152.

Kerry, K.R., Grace, D.R., Williams, R. and Burton, H.R. (1977) Studies on some saline lakes of the Vestfold Hills, Antarctica. In *Adaptations within Antarctic Ecosystems*, Ed. G.A. Llano, pp. 839 – 58. Gulf Publishing, Houston.

Klokov, V.D. (1970) Proryv vod iz ozera Glubokogo na stantsii Molodezhnoi. (Water discharge from Lake Glubokoye at Molodezhnaya Station). *Inf. Byull. sov. antarkt. Eksped.* 77, 96 − 9 (English translation: 1972, 7, 533-5).

Korotkevich, V.S. (1964) Concerning the population of water-bodies in the oases of East Antarctica. *Soviet Antarctic Research* 1, 154 − 61.

Kriss, A. Ye., Aleksandrov, A.M., Kozlovski, A.M., Ledenyeva, K.V. and Leflat, O.N. (1968) Mikrobiologicheskie issledovaniya ozera Glybokogo vblizi stantsii Molodezhnoi. (Microbiological research in Lake Glubokoye near Molodezhnaya Station). *Inf. Byull. sov. antarkt. Eksped.* 70, 44 − 8 (English translation: 1969, 7, 159 − 62).

Light, J.J. (1977) Production and periodicity of Antarctic freshwater phytoplankton. In *Adaptations within Antarctic Ecosystems*, Ed. G.A. Llano, pp. 829 − 37. Gulf Publishing, Houston.

Light, J.J., Ellis-Evans, J.C. and Priddle, J.(1981) Phytoplankton ecology in an Antarctic lake. *Freshwat. Biol.* 11, 11 − 26.

Light, J.J. and Heywood, R.B. (1973) Deepwater mosses in Antarctic lakes. *Nature, Lond.* 242, 535 − 6.

Light, J.J. and Heywood, R.B. (1975) Is the vegetation of Continental Antarctica predominantly aquatic? *Nature, Lond.* 256, 199 − 200.

Lindemann, R.L. (1942) The trophic − dynamic aspect of ecology. *Ecology* 23, 399 − 418.

McLeod, I.R. (1964) The saline lakes of the Vestfold Hills, Princess Elizabeth Land. In *Antarctic Geology*, Ed. R.J. Adie, pp. 65 − 72. New-Holland Publishing, Amsterdam.

MacNamara, E.E. (1970) Some limnological observations from Enderby Land, Antarctia. *Limnol. Oceanogr.* 15, 768 − 75.

Meyer, G.H., Morrow, M.B., Wyss, O., Berg, T.E. and Littlepage, J.L. (1962) Antarctica: the microbiology of an unfrozen saline pond. *Science, N.Y.* 138, 1103 − 4.

Opaliński, K.W. (1972) Flora and fauna of freshwater bodies of the Thala Hills oasis (Enderby Land, Eastern Antarctica). *Pol. Arch. Hydrobiol.* 19, 383 − 98.

Priddle, J. (1980) The production ecology of benthic plants in some Antarctic lakes. 1, *In situ* production studies. *J. Ecol.* 68, 141 − 53.

Priddle, J. and Dartnall, H.J.G. (1978) The biology of an Antarctic aquatic moss community. *Freshwat. Biol.* 8, 469 − 80.

Priddle, J. and Heywood, R.B. (1980) The evolution of Antarctic lake ecosystems. *Biol. J. Linn. Soc.* 14, 51 − 66.

Vincent, W.F., Downes, M.T. and Vincent, C.L. (1981) Nitrous oxide cycling in Lake Vanda, Antarctica. *Nature, Lond.* 292, 618 − 20.

Weller, M.W. (195) Notes on formation and life of ponds of the Falkland Islands and South Georgia. *Bull. Br. Antarct. Surv.* 40, 37 − 47.

Wright, S.W. and Burton, H.R. (1981) Biology of Antarctic saline lakes. *Hydrobiologia* 82, 319 − 38.

# CHAPTER 4

# Marine Habitats — Introduction

## W.N. BONNER and D.W.H. WALTON

British Antarctic Survey, Natural Environment Research Council, High Cross, Madingley Road, Cambridge CB3 0ET, U.K.

Within the Antarctic, if it is defined as the region limited by the Antarctic Convergence, the sea covers more than twice the area of the land. Nowhere in the world is there a greater contrast between the ecosystems supported by the sea and the land than in the Antarctic. The terrestrial ecosystem (Chapter 3), restricted as it is to the few rock exposures and coasts of the land mass, is largely barren and supports a sparse flora and fauna made up of very few species. Fluctuations in climate, and particularly of temperature, are extreme and production is low in both relative and absolute terms.

The marine ecosystem presents a striking contrast. Areas of the Southern Ocean are very productive and in absolute terms overall production is high. This high production sustains rich species assemblies of both plants and animals, reviewed by Sayed El Sayed. The region is characterized by a unique short food chain, where most of the primary production from the phytoplankton is channelled through a single herbivore, Antarctic krill, to the higher predators, the seals, whales and birds. Krill is a unique organism with a startlingly high biomass. It represents the key species in the marine ecosystem.

Although the short food chain with krill as its central number is dominant in the ecology of the area, it is only part of a much more complex food web. There are other herbivores besides krill and many other carnivores, including fish and squid. Part of the food web is made up by a very rich and diverse benthic fauna, supported by a seasonal rain of detritus from the water-column above. Microbes occur both in the water-column and in the sediments, but their significance in the Antarctic marine ecosystem is unknown.

Seasonality is one of the characteristics of the Antarctic marine ecosystem. This is caused mainly by variation in the amount of light entering the water in summer and winter. The contrast between summer and winter light levels increases as one moves poleward, and reaches an extreme in the pack ice zone. Snow-covered pack allows very little light to enter the sea. On the other hand, the amount of temperature variation in the sea is very small, and so the marine environment, while very seasonal, is also predictable. Gordon Picken describes the Antarctic benthos and its specializations.

Antarctic fish, described by Karl-Hermann Kock, are an interesting group. Nearly all species and many higher groups are confined to the region. Many Antarctic fish have developed specialized physiological adaptations to the extreme conditions under which they live. Some produce antifreezes in the body fluids, while others dispense with the oxygen-carrying pigment, haemoglobin, in their blood. Their role in the ecosystem as krill consumers is locally important, but they are in competition for this resource with other vertebrates, namely penguins, seals and whales.

Another group of potential krill consumers are the squids. Little is known of the biology of squid in the Antarctic, but from their importance in the diet of other animals it is clear that they are a very abundant group. Information on the role of squid in the Antarctic marine environment is beginning to accumulate, as Malcolm Clarke relates, but puzzling differences exist between squid caught in research nets and those found in the stomachs of predators.

This chapter deals with the primary producers, the zooplankton of the pelagic zone, and the wide variety of organisms that make up the benthos. It looks in greater detail at the two primarily marine large predator groups, the fish and the squid. The secondarily marine higher predators, the seals, whales and sea birds, are dealt with in the next chapter.

The extensive and diverse marine ecosystem has controlling influence on all other life in the Antarctic. The interactions of the marine system with life on land are described in Chapter 7 by Andrew Clarke.

The Antarctic has historically provided a variety of marine harvests. Currently krill and various species of fish are exploited. Because of the great importance of the first of these to the entire system, conservation is of vital importance. In Chapter 8 Bryan Sage describes the present position of conservation of Antarctic marine living resources.

# CHAPTER 4a

# Plankton of the Antarctic Seas

## SAYED Z. EL-SAYED

Department of Oceanography, Texas A&M University, College Station, Texas 77843, U.S.A.

## CONTENTS

## 4a.1 INTRODUCTION

The study of Antarctic plankton holds more than a casual interest and more than academic importance. With the burgeoning human population and its increased demand for animal protein, several nations in recent years have shown a growing interest in the exploitation of the living resources of the Southern Ocean. Notable among these resources is the shrimp-like organism, Antarctic krill, *Euphausia superba*. The potential catch of krill has been estimated to be well in excess of the total annual harvest of fish and shellfish from all the world's oceans. Krill is the key organism in the Antarctic food web upon which all higher species, whales, seals, penguins, winged birds, fish and squid depend, directly or indirectly, for their food. If these resources are to be exploited wisely, an urgent need exists for a better understanding of the complex ecosystem of the Southern Ocean. Since these resources, most notably krill, and their food base, the phytoplankton, are integral components of that ecosystem, they need especially careful study. This knowledge of Antarctic plankton has practical application for the management and harvesting of Antarctic marine living resources.

Before examining Antarctic plankton it would be instructive to discuss the unique physical/chemical settings of the Antarctic seas since these strongly affect the biology, distribution and abundance of the planktonic organisms. The unique features of the waters surrounding Antarctica are characterized by: (a) the presence of pack-ice around the Antarctic continent and the seasonal waxing and waning of the pack-ice zone; (b) the variability of the light regime which alternates between perpetual darkness in the winter and continuous daylight during summer; (c) the extensive cloud cover; and (d) the Circumpolar Current which serves to connect circum-Antarctic waters into a single system, the Southern Ocean, through the transport of nutrients and plankton. It is in response to the physical/chemical environments that the Antarctic planktonic organisms have developed their characteristic features. For example, the marked variations in incoming solar radiation have a pronounced effect on plant and animal life; the rise and fall of primary production in the Southern Ocean appear to be a direct response to the amount of energy received, as will be discussed later.

## 4a.2 THE PHYSICAL/CHEMICAL ENVIRONMENT

The Antarctic Convergence (Chapter 2a) forms one of the major boundary zones of the world's oceans, dividing the southern waters into the Antarctic and sub-Antarctic regions. The main transport of water between 40° and 60°S is from west to east in what is known as the West Wind Drift or Circumpolar Current. Close to the continent easterly or southeasterly winds prevail, which cause transport in a westerly direction. Superimposed upon this eastward circumpolar movement are north – south components (Fig. 4a.1). At the northern limit of the circumpolar waters the Antarctic Surface Water sinks beneath the less dense southward-flowing sub-Antarctic water. The region in which this phenomenon occurs is called the Antarctic Convergence or the Polar Front. This natural boundary has a great influence on the distribution of phytoplankton, zooplankton, fish and birds. It is a region of sharp temperature gradients and increased nutrient salts and concentrations. Its position is not stationary; it fluctuates with changing temperature, pack-ice and currents, moving between 54°S and 62°S. As the Antarctic Surface Water sinks below the surface at the Convergence, and continues northward, it forms the Antarctic Intermediate Water, which is characterized by a salinity minimum and a relatively high oxygen maximum. At the edge of the continent (in the Weddell Sea in particular), very cold water sinks and moves northward along the shelf and into the deep, forming the Antarctic Bottom Water. Immediately above this water mass, there is an exceptionally thick, relatively warmer, more saline and nutrient-rich water mass called the Circumpolar Deep Water. The upward movement of this southward-moving water mass creates a zone of upwelling, which supplies to the surface water vast quantities of nutrients, contributing to the luxuriant phytoplankton growth close to the Antarctic coasts.

Because the region lies in relatively high latitudes, the extreme variations in incoming solar radiation between summer and winter have a pronounced effect on marine plant life. Figure 4a.2 illustrates these

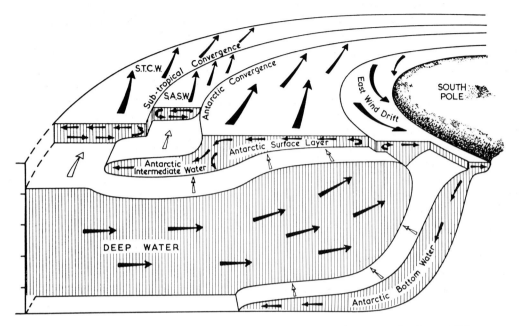

Fig. 4a.1. Schematic representation of the currents and water masses of the Southern Ocean (after Foxton, 1956).

effects by plotting the mean monthly values of radiant energy against primary production and the standing crop (biomass per unit volume) of phytoplankton. As previously mentioned, the increase and later decline of the phytoplankton biomass and primary productivity mirror the amount of energy received.

Light penetration in Antarctic waters is determined not only by its intensity, angle of incidence, surface reflectance, absorption by suspended particles, and so forth, but also by the presence of thick fast-ice (near the continent) and pack-ice, which appreciably reduce the amount of submarine illumination. Winter pack-ice halves the area of open water south of the Antarctic Convergence, expanding from 4 million square kilometres in February (austral summer) to 22 million square kilometres in September, the end of winter in the Southern Hemisphere. At its greatest extent, the sea ice amounts to about half of Antarctica (Fig. 4a.3). Throughout its cycle of expansion and contraction, the circum-Antarctic ice belt generally moves from east to west except for a local clockwise circulation of ice in the Weddell Sea. The biological implication of the extent of the pack-ice is highly significant since the pack-ice is the habitat of special assemblages of microflora and microfauna. The formation of winter pack-ice and its break-up in summer are important stages in the annual cycle of the marine ecosystem. At break-up, diatoms and other phytoplankton are released into the water column, and the speed of that break-up may affect the total seasonal production.

As a result of upwelling, the rich nutrient salts of the Antarctic waters do not appear to limit phytoplankton growth. Even at the peak of phytoplankton growth, the concentration of nutrients remains well above the limiting values; for example, no discernible seasonal variations in the nitrate values have been observed in Antarctic waters regardless of the periodic growth rate of phytoplankton. Phosphate distribution, on the other hand, shows a marked decline during late spring and early summer. However, the low values encountered in the latter period are still higher than some of the maximum values reported in the temperate regions.

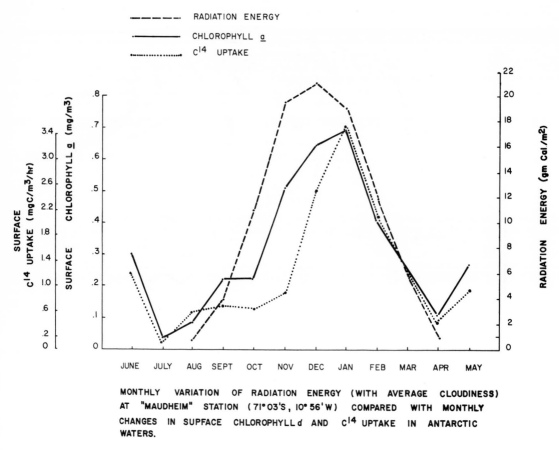

RADIATION ENERGY

CHLOROPHYLL a

c$^{14}$ UPTAKE

MONTHLY VARIATION OF RADIATION ENERGY (WITH AVERAGE CLOUDINESS) AT "MAUDHEIM" STATION (71°03'S, 10°56'W) COMPARED WITH MONTHLY CHANGES IN SUPFACE CHLOROPHYLL d AND c$^{14}$ UPTAKE IN ANTARCTIC WATERS.

Fig. 4a.2. Monthly variation of radiation energy (with average cloudiness) at Maudheim station (71°03'S, 10°56'W) compared with monthly changes in surface chlorophyll a and $^{14}$C uptake in Antarctic waters (after El-Sayed, 1971).

## 4a.3 PHYTOPLANKTON

### 4a.3.1. Diatoms

The Antarctic phytoplankton population, in general, is dominated by diatoms with smaller numbers of dinoflagellates and silicoflagellates. Nearly 100 species of diatoms have been found in Antarctic waters, and the exceptional richness of the diatom flora is certainly related to the great abundance of phytoplankton in southern waters. Although many of the diatom species are considered cosmopolitan, Hasle (1969) lists three species as having distribution limited to the Antarctic Zone and 20 diatom taxa restricted to the southernmost part of the Antarctic Zone. This endemism seems to be fairly high among Antarctic diatoms, but the general distribution pattern is circumpolar. Very few species are thought to be bipolar in distribution (with truly disjunct biogeographical populations) but examples are

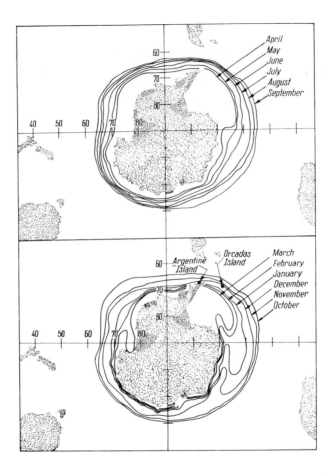

Fig. 4a.3. Monthly variations in the boundary of the pack-ice (after J.O. Fletcher, adapted from *Atlas of Antarctic*, 1960).

*Thalassiosira antarctica*, *Porosira glacialis*, and *Nitzschia cylindrus*. The diatoms in the pack-ice may prove to have more endemic populations.

Although the biology of Antarctic species of phytoplankton needs more study, only a few species are known to produce resting spores. Production of these resting spores has been triggered in growth studies by low nutrients or high salinities, but not by low light or temperature. One would expect that the survival rate of resting spores would be high during the long austral night when cells freeze in the ice, even though experimental evidence suggests other triggering mechanisms. More work needs to be done both in the laboratory and in the field if scientists are to understand the survival of seed populations during unfavourable conditions.

There are both centric (often drum-shaped cells) and pennate (boat-shaped cells) diatoms in the Antarctic waters, but the pennate diatom colonies of the genus *Nitzschia* (Fig. 4a.4) are striking under the microscope: individual, thin, flat cells connect end to end and face to face to form gently spiralling, ribbon-like colonies. Although there are many single cells, colony formation is common in austral waters.

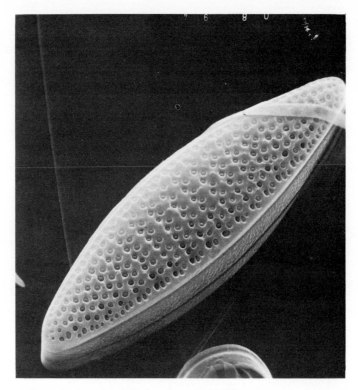

Fig. 4a.4. The pennate diatom, *Nitzschia kerguelensis* (SEMX3000), (Photo credit: Dr. G.A. Fryxell).

## 4a.3.2. Dinoflagellates

Because Antarctic phytoplankton studies have largely been devoted to diatoms, and because Antarctic dinoflagellates are rather small and some are encased in weak and easily deformed plates, there is a paucity of knowledge about the distribution and abundance of these organisms in the Antarctic waters. Furthermore, since the number of dinoflagellates is significantly smaller than the number of diatoms, the dinoflagellates' biomass in the Antarctic has probably been underestimated and their study overlooked. However, in some water columns, the dinoflagellates dominate, and they do contribute an important share of the biomass.

Antarctic dinoflagellates actually belong to only a few thecate genera (those encased in cellulose plates), chiefly *Protoperidinium* and *Dinophysis*. Of these *Protoperidinium* (Fig. 4a.5) is by far the more abundant in species and in individuals. The genus, *Ceratium*, which is so prevalent in other seas, is generally absent from Antarctic waters. In fact the Antarctic Convergence forms a striking boundary between the presence and absence of *Ceratium*.

Naked, or thin-walled, dinoflagellates are probably not as rare in Antarctic waters as they seem. The assumption that they are rare is due, in part, to the preservation and fixing processes, which render them unrecognizable. However, eight naked species of *Gymnodium*, four of *Gyrodinium* and two of *Amphidinium* have been identified; and *Pyrocystis pseudonoctiluca* has also been reported (Balech, 1975).

As far as we know, the Southern and Northern polar waters have only a very few cosmopolitan dinoflagellate species in common. Three bipolar species are known: *Protoperidinium antarcticum, P. saltans* and *P. thulescence.* The latter has also been reported in sub-Antarctic and Arctic waters. Endemism is high among Antarctic dinoflagellates, although some occasionally are found in sub-Antarctic waters, probably because of the isolated 'rings' or meanders of Antarctic waters in the region of the Convergence. No less than 35 species, however (many belonging to the genus *Protoperidinium*), are limited to the Antarctic region. In no other oceanic region do the dinoflagellates attain such a high degree of endemisms (80 − 85%) as in the Antarctic.

### 4a.3.3. Other algae

*Phaeocystis poucheti* is one of the few non-diatom algae which may be exceedingly plentiful in southern Antarctic waters. This unicellular motile and colonial brown − yellow algae belongs to the order Prymnesiales. It has a colonial palmelloid stage which can be large enough to be visible to the naked eye. *Phaeocystis* is very widely distributed in warm and temperate waters as well as in Antarctic and Arctic seas. It can produce gelatinous blooms in great abundance sometimes known as 'Dutchman's baccy juice'. These gelatinous colonies contain great numbers of cells, each with two chloroplasts.

Silicoflagellates are unicellular chrysophycean algae which include both motile and non-motile cells. Occasionally the cells may be linked together to form small, rather simple colonies. The silicoflagellates have an internal skeleton with conspicuous siliceous hollow tubes, forming a spectacular symmetrical structure containing many chloroplasts. Only one species of silicoflagellate, *Dictyocha (Distephanus) speculum*, is generally abundant in Antarctic waters, and sometimes outnumbers any species of diatom. It is larger than many diatoms, and it should not be overlooked as a contributor to phytoplankton biomass and primary production in Antarctic waters. It is a good indicator of cold water.

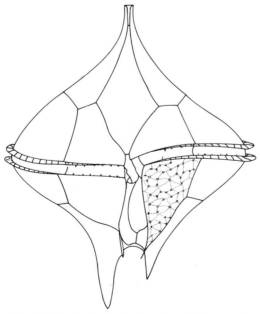

Fig. 4a.5. The thecate dinoflagellate, *Protoperidinium antarcticum.*

### 4a.3.4. Species composition and water masses

Research in distribution and abundance of the phytoplankton has indicated that species composition and abundance change in relation to the distribution of the water masses. These changes are most evident through analysis of the phytoplankton samples collected along north – south crossings of the Antarctic Convergence. North of the Convergence, the composition of the phytoplankton is of subtropical character, with the dinoflagellates rather than the diatoms being by far the predominant element. Dinoflagellates are not so dominant, however, in the Pacific and Indian sectors of the Antarctic, where Russian investigators have reported that 99% of the total number of cells are composed of diatoms.

### 4a.3.5. Distribution and abundance

Measurement of the amount of plant pigment, chlorophyll $a$, in a given volume of water, is used to estimate the phytoplankton standing crop (biomass per unit area or volume) and has been extensively used to study the phytoplankton distribution and abundance. Large phytoplankton standing crops are found in the Scotia Sea, in the Bransfield and Gerlache Straits, the waters adjacent to the Antarctic Peninsula, in the southern Ross Sea and in the southwestern Weddell Sea. Coastal waters off the Antarctic continent and off the sub-Antarctic Islands are, in general, characterized by much higher standing crops than are open oceanic regions. Small phytoplankton standing crops are found in the Drake Passage, in the Bellingshausen Sea, and in the open oceanic regions of the Antarctic seas, except for isolated pockets of dense phytoplankton biomass.

The vertical distribution of chlorophyll $a$ exhibits maximum values at the depths to which 25 – 50% of surface light intensity penetrates. Below this, there is a gradual decrease in phytoplankton biomass to the depth of the photic zone (about 100 m) with little beyond this point.

Near the coastal regions and in the vicinity of islands, diatom blooms are rather common. A dense bloom, extending for more than 100 km, of the diatom *Thalassiothrix antarctica* was reported by Hardy and Gunther (1935). Another extensive bloom of the diatom *Thalassiosira tumida* (Fig. 4a.6) was encountered in the southwestern Weddell Sea (El-Sayed, 1971). It was estimated that the area affected by that bloom was at least 15,500 km$^2$ (6,100 mi$^2$) in extent. Dense patches, like thick soup, of a distinctly yellow-brownish colour, were visible for miles on end (Fig. 4a.7). The area of the bloom was noticeably teeming with Adélie penguins, crab-eater seals, and other marine vertebrate life.

Extensive blooms of *Phaeocystis* have been reported in the Ross and Weddell Seas. Massive blooms of this algae form jelly-like colonies which clog plankton nets. In the Ross Sea, a *Phaeocystis* bloom from the surface to depths of 100 – 150 m has been observed (El-Sayed *et al.*, 1979). Distinct discolouration of water (greenish/grey to greenish/brown), accompanied this bloom, resulting in considerable submarine light attenuation.

Primary production, the fundamental level in the food web of the Antarctic Ocean, has generally correlated very well with the data on the distribution of the standing crop of phytoplankton. As with variations in the standing crop, primary productivity shows wide point-to-point variation. Low values have been reported in the Drake Passage, the Bellingshausen Sea, and in oceanic waters. Higher values are found in coastal regions and in the vicinity of the Antarctic and sub-Antarctic islands (e.g. the South Orkney Islands and in the vicinity of Crozet and Kerguelen Islands) than in adjacent areas. Values as high as 3.2 and 3.6 grams of carbon fixed per square metre per day were reported in the Gerlache Strait and near Deception Island, respectively, during the austral summer. These values, which are comparable to any of the highly productive areas off Peru, southeast Arabia, Somalia, southwest Africa, etc. have,

Fig. 4a.6. Photograph of the bloom of the diatom *Thalassiosira tumida* observed on 10 February 1968 (Photo: S.Z. El-Sayed).

no doubt, perpetuated a belief in the richness of the Antarctic waters. Despite these high values, however, the Antarctic Ocean, taken as a whole, does not show an outstandingly high rate of primary production.

Based on *in situ* primary productivity experiments, using the radioactive carbon isotope ($^{14}C$) uptake method, a mean primary productivity value of 0.134 gram carbon fixed per square metre of the water column per day was calculated (El-Sayed, 1978). This production value is similar to those in other open oceanic waters. However, in the Antarctic, practically all of the production is compressed into the 120 – 150 day austral light regime rather than spread throughout the year.

In marked contrast to the generally low standing crop and primary production values reported for oceanic regions, investigators have encountered several deep oceanic stations where rich phytoplankton populations have been noted (El-Sayed and Weber, 1982). For example, at a station located about 700 km from the head of the Weddell Sea an immensely rich assemblage of the diatom *Corethron criophilum*, reaching more than 1.4 million cells per litre was found at a depth of 5 m (Fryxell and Hasle, 1971). Similar rich stations were recorded in the Pacific and Indian sectors; however, these oceanic stations are exceptional. The factors contributing to their high productivity are not clear.

Several attempts have been made to explain the variability of productivity values found in the Southern Ocean in terms of the physical, chemical and biological factors, which control phytoplankton production there. Investigators have found it difficult, if not impossible, to deduce rate-limiting factors by examining their relationship to any one of the parameters (e.g. temperature), as many factors vary simultaneously, including the species composition of the phytoplankton, where significant changes take place across the Antarctic Convergence (Holm-Hansen *et al.*, 1977). It can be concluded, however, that the large seasonal variations of light, sea ice cover, stability of the water column, and grazing by the zooplankton may be more significant in predicting the factors controlling phytoplankton productivity than are upwelling and the regeneration of nutrients.

### 4a.3.6. Seasonal variation

Antarctic phytoplankton has the ability to survive the protracted darkness of the austral winter and to resume growth with the advent of spring. The primary productivity cycle is initiated by early seasonal blooming along the ice-edge, as the shade-adapted microalgal communities (which persist underneath

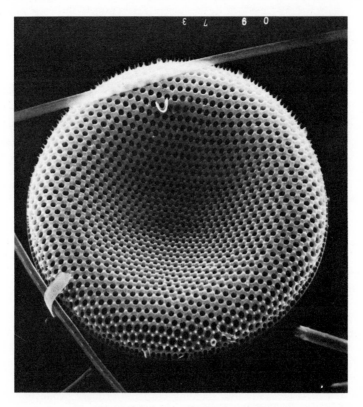

Fig. 4a.7. The centric diatom, *Thalassiosira tumida* (SEMX1500) (Photo: Dr. G.A. Fryxell).

the pack-ice in winter) are released into the water. With increasing sunlight, pelagic algae (phytoplankton) appear in the upper water layers, and these, coupled with the ice algae and slow disintegration of the pack-ice, reach a production climax in late spring and early summer. Due to the persistence of the ice at high latitudes, production is delayed by nearly a month or two. During the growing season, the phytoplankton growth rate has been estimated to nearly double once each day. Such doubling matches the maximum rate exhibited by phytoplankton in temperate water.

### 4a.3.7. Ice algae

Since the early observations made of the microalgae in the fast ice of the Davis Sea by Vangefen aboard the *Gauss* (1901 − 02), much interest has been shown in recent years in studying the taxonomy, physiology and ecology of the micro-flora and micro-fauna of the pack-ice. The groups of micro-flora known to be present in the pack-ice include diatoms, chrysophyceans, dinoflagellates and green flagellates. Of these, the diatoms are ubiquitous and generally dominant. Dinoflagellates are generally rare, although occasionally they may achieve dominance. *Phaeocystis*, on the other hand, may be present in some areas in enormous numbers at certain times of the year. The wide distribution of ice diatoms seems to give credence to the opinion that ice micro-flora may be a major factor in determining the annual production in Antarctic waters. The ice regime is perhaps of even greater importance in determining regional differences in phytoplankton standing crop and primary production. Not only may the ice layer at its break-up release a significant number of organisms into the water column, but also the period during which the seas are open may be of critical importance in controlling the rate of build-up of the standing crop and hence of the total seasonal production. Thus a short season may permit only a partial build-up of the standing crop; a long season, on the other hand, allows adequate time for the phytoplankton to build up and to continue production at a high level. Under 'long season' conditions, grazing by the zooplankton probably becomes the most important factor regulating phytoplankton populations.

Taken together, the more accurate measurements of primary productivity made in recent years underscore three significant findings: (a) that there is much greater regional variability in the productivity parameters than had been previously thought, (b) that the high productivity in the Antarctic waters occurs, in fact, only in inshore waters, and not in oceanic regions; and (c) that the productivity of the Southern Ocean as a whole, is perhaps not as high as we had been led to believe.

## 4a.4. ZOOPLANKTON

The extensive studies made by the *Discovery* Investigations have added a wealth of knowledge about the composition, distribution and life histories of the most dominant zooplankton species. One of the crowning achievements of scientific research in the Antarctic is the classic *Discovery* Report published in 1962, on the natural history and distribution of Antarctic krill by the late J.W.S. Marr.

### 4a.4.1. Composition

The dominant organism among Antarctic zooplankton is the shrimp-like Antarctic krill, *Euphausia superba*, which in the past supported enormous stocks of whales. Although there are many species of Antarctic zooplankton which vary considerably in size, density, distribution and behaviour, discussion

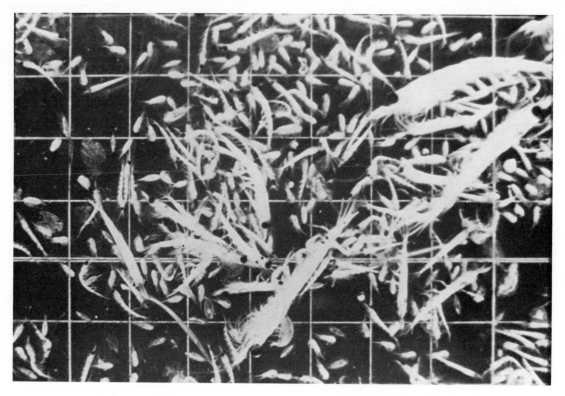

Fig. 4a.8. Antarctic zooplankton (Photo: Dr. J. Dearborn).

in this section will be limited to those components that constitute the bulk of the zooplankton biomass. Special attention must be given to krill since it forms the most important group biologically and economically in the Antarctic marine ecosystem.

Copepod crustaceans constitute the bulk of Antarctic mesoplankton (middle-sized zooplankton), accounting, on an average, for about 73% of its total biomass. The three most prevalent species of *Calanoides acutus, Calanus propinquus*, and *Rhincalanus gigas*. The chaetognaths (arrow worms) are an important ecological group because of their abundance among zooplankton, second only to the copepods, and because of their role as carnivorous predators. In both abundance of individuals and frequency of occurrence, *Eukronia hameta* is the most common of the eighteen species of chaetognaths which are found in both the Antarctic and sub-Antarctic regions. Also because different species of chaetognaths are characteristic of different water masses, the geographic, bathymetric and density distribution of these species reflect features of ocean dynamics, and several of the chaetognaths are useful as indicators of movement of water masses.

According to T.L. Hopkins (1971), who made an extensive study of Antarctic zooplankton in the late 1960s, the principal contributors to zooplankton biomass in the samples taken in the Pacific sector of the Antarctic were the copepods, chaetognaths, and euphausiids, in that order. He found that the copepods constituted 68% and 60%, respectively, of the standing crop in the Antarctic and sub-Antarctic regions sampled. Russian investigators (Voronina, 1966) found similar percentages (73% and 62%) for the Antarctic and sub-Antarctic copepods in the Indian Ocean sector. The paucity of the euphausiids in Hopkins' plankton samples can be attributed to the fact that his collections were made further north in the West Wind Drift, that is, in areas not frequented by the krill. Russian investigators found similarly low numbers of krill because they used a type of net which is poorly suited to capturing adult krill.

## 4a.4.2. Distribution

Although most, if not all, Antarctic zooplankton species are circumpolar in distribution, they frequently show centres of concentration at different depths or in different latitudes. As a result, there are both quantitative and qualitative differences among species. Zooplankton may become concentrated in the oceanic areas of convergence and divergence, and the total zooplankton biomass at the Antarctic Convergence was found to be 3 to 3.5 times greater than in the adjacent areas (Voronina, 1968).

Despite the extensive investigations of zooplankton carried out in the Southern Ocean, it is difficult to arrive at reasonable estimates of the zooplankton standing crop in Antarctic waters because of problems related to sampling techniques, patchy distribution and those associated with seasonal vertical migration. It is usually assumed, however, that the biomass of zooplankton, in general, is the same as that of krill.

Over one-half of the zooplankton biomass of the upper 1000 m of sampled waters occurs in the top 250 m, in the late spring, summer, and early autumn months (November to April); while during the remainder of the year (May to October) most of the biomass is below 250 m (Hopkins, 1971).

The annual total biomass of zooplankton in the upper 1000 m in the Pacific sector averages 3.5, 2.55 and 2.76 g dry weight/$m^2$ in the Antarctic, sub-Antarctic and Antarctic Convergence, respectively. In the Atlantic sector, on the other hand, there is an average zooplankton/micro- nekton abundance of 5 to 9 g dry weight/$m^2$ in the upper 1000 m (Hopkins, 1966).

No reliable data estimating the annual production of zooplankton are available at present. However, if we assume that the life span of most zooplankton organisms exceeds one year, the total annual

Fig. 4a.9. Krill (*Euphausia superba*) (Photo: U. Kils).

production might be the same as the average standing crop (about 10 g m$^{-2}$). If we take this value as representative of the Antarctic and multiply it by the mean surface area of open water south of the Convergence, we obtain a biomass (dry weight) value which is equivalent to 178 million tonnes. This figure should be regarded as the low estimation of zooplankton biomass for several reasons: (a) there is some uncertainty about the amount of open water in the pack ice; (b) we know very little about the distribution and abundance of zooplankton in the area covered permanently by pack-ice; and (c) there is evidence that in sampling, zooplankton can often avoid nets.

### 4a.4.3. Seasonal variations

Southern Ocean zooplankton exhibits seasonal variations: in summer, it is much more plentiful in coastal than in open-ocean waters; in winter, it is found at great depths.

The seasonal changes of the zooplankton biomass are characterized by two peaks. The first peak seems to result from the formation of pre-spawning concentrations of organisms following the spring migration upward in water column. This change represents merely a vertical re-distribution of plankton rather than an actual increase in abundance, as can be seen from the fact that the biomass in the upper 1000 m layer as a whole remains practically unchanged. The second (summer) peak depends on the development of the new generation until the annual peak is reached. Later the downward migration of older organisms marks a more or less sharp decline in the biomass to a level which persists throughout the long austral winter. The summer peak of zooplankton first appears in northern latitudes and then develops further to the south. In summer, in the two northern zones, the zooplankton biomass peaks two months later than the phytoplankton does. In the southern zone, on the other hand, one month elapses between the peaks of phyto- and zooplankton (Hart, 1942).

### 4a.4.4. The phytoplankton/zooplankton relationship

In the waters around South Georgia an inverse correlation was found between phytoplankton density and the numbers of herbivorous zooplankton (Hardy and Gunther, 1935). A comparable inverse relationship between zooplankton and phytoplankton exists in the Weddell Sea (El-Sayed and Mandelli, 1965). Grazing by zooplankton is the main factor controlling Antarctic phytoplankton; in the northern region of the Antarctic this might be 3 to 5 times as intense as in the English Channel; whereas, near South Georgia, it is comparable to the level in the English Channel (Hart, 1942). Zooplankton populations increase following the phytoplankton peak, and herbivorous organisms can be shown to increase first, with carnivores building up at a later stage (Currie, 1964; Foxton, 1964). Little is known of the detailed food chain sequence involved, but this broad pattern is likely to be of general significance.

### 4a.5. KRILL

### 4a.5.1. The organism

Unlike other zooplankton krill are relatively large and occur in swarms of great abundance. Of the eleven species of krill found in the Antarctic, the most important are *Euphausia superba* (Fig. 4a.9), *E. frigida*, *E. crystallarophias*, *Thysanoessa macrura* and *E. valentini*. These euphausiids are distributed as follows: in waters north of the Antarctic Convergence, *E. valentini*; in the pack-ice, *E. crystallarophias*, and in the open waters south of the convergence, dense patches of the larger species of *E. frigida* and *E. superba*.

The largest concentrations of *E. superba* are in the Atlantic sector, particularly in the northern Weddell Sea, the Scotia Sea, north of the South Orkney Islands, around the South Shetlands Islands, and west of the South Sandwich Islands (Fig. 4a.10). Breeding takes place in at least four areas: the Bellingshausen Sea, the Bransfield Strait, the Davis Strait, and in the vicinity of South Georgia. The number of breeding stocks is unknown; however, two separate ones may exist. It is believed that krill spawn once or twice a year (for one, two or maybe more years); fecundity is between 2000 to 3000 eggs per spawning. Longevity is still a matter of controversy; however, there are indications that krill may live for more than four years.

Krill occur in large swarms, dense enough to discolour surface water. Underwater photography reveals a swarm structure of increasing density from the periphery to the centre, with the krill in parallel alignment, all heads in the same direction. These swarms are amorphous in shape, may be up to several hundred metres across, and can reach concentrations of thousands of individuals per cubic metre. According to Soviet scientists, krill swarms can extend to depths of 40 to 50 m; although concentrations most suitable for commercial catch are found between 1 and 10 m. In 1975/76, the West German trawler, *Weser*, averaged 8 to 12 tons of krill per hour of towing time, with a maximum catch of 35 tons in 8 minutes. A super-swarm of krill estimated at more than 2.5 million tons was observed near Elephant Island in February, 1981 (Fig. 4a.11).

The swarming habits of krill make it easy for baleen whales to feed on these animals, with blue whales preferring adolescent krill and fin whales favouring adults. Research has indicated that adult and juvenile krill are often found in separate layers, with juveniles being closer to the surface. The fact that different age groups of *Euphausia superba* swarm separately may ultimately be useful for commercial harvesting and fishery management.

## 4a.5.2. Krill biology

While most aspects of krill biology are under intensive study, definitive answers are needed to questions on growth rates, rate of advance to maturity, longevity, fecundity, and spawning potential. Moreover, as harvesting is already under way at several widely spaced sites, it is necessary to know the numbers of spawning events, temporal and spatial differences within the populations throughout their circumpolar range, and the factors inducing their spawning behaviour.

Another question involves the relationship between the distribution of krill and their physical/chemical environment. The Southern circumpolar seas, although seasonally variable, are considered relatively uniform in thermal regime and radiant energy. Inter-seasonal and inter-annual variations in krill populations are also unknown factors. British scientists reported shortages of krill in the 1977 – 78 season around South Georgia, a region known to be one of the major feeding grounds of baleen whales. During that year, young Antarctic fur seals showed a corresponding poor growth rate and, compared with earlier years, there was a low survival rate among chicks of black-browed albatrosses and Gentoo penguins.

## 4a.5.3. Krill stocks

The existence of large stocks of krill in the Antarctic has been known for many years, but interest in their commercial exploitation arose in the early 1960s with the decline of whaling in the Southern Ocean and with the recent establishment by coastal states in other parts of the world of 200-mile wide exclusive economic zones. This change has forced long-distance fishing fleets to hunt for harvests in such waters as the Southern Ocean.

Estimates of both the standing stocks of krill, and the extent of dependence of many species on them as a food source vary widely, if not wildly. Great discontinuities in the distribution and swarming behaviour of krill populations themselves, as well as diverse methods of estimating stocks are responsible for the variety of opinion on the matter. Estimates of krill stocks range from 125 million tonnes to 5000

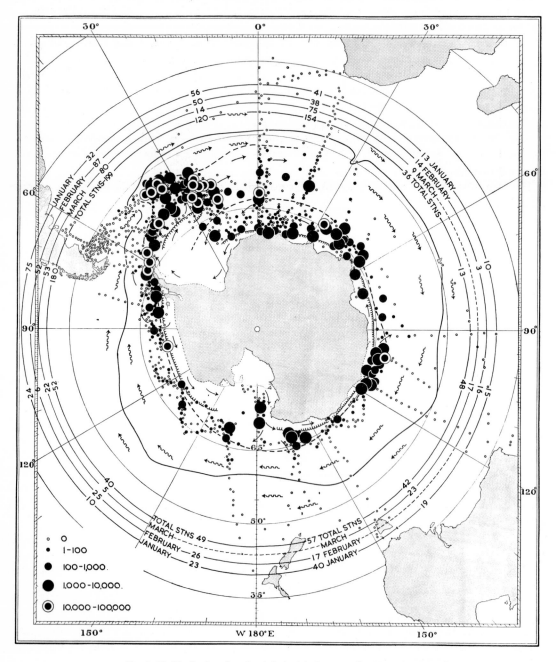

Fig. 4a.10. Distribution of staple whale food in Summer (after Marr, 1962).

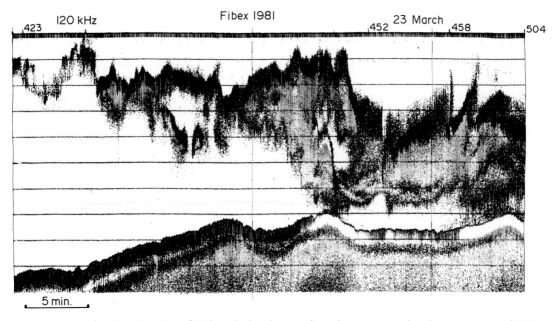

Fig. 4a.11. A patch of krill located northeast of Elephant Island, as shown on the 120 kHz system. Depth marks are at 20 m intervals (Photo: M.C. Macaulay).

million tonnes. The bases from which these estimates were made, coupled with the relatively small areas in which sampling was conducted, render these estimates tenuous and less than useful. However, krill stocks are clearly very large. Although the seasonal and annual variations are known, they have generally not been taken into account when making the computations. Nonetheless, inter-annual differences are known to occur (cold versus warm years), since annual harvesting success has been inconsistent (poor, for instance in 1977 – 78).

Annual production of krill is even more obscure than the other parameters of the Antarctic ecosystem. Few have attempted computations because data are so incomplete. However, estimates of these values range from 13 to 400 million metric tons. Thus it might pose a considerable threat to the Antarctic ecosystem if large-scale commercial fishing were undertaken in the face of estimates varying 20 to 30 fold.

## 4a.6. CONCLUSION

Krill exploitation is a good example of the Antarctic resource-use problem. The krill harvest is increasing and more nations are investigating techniques for harvesting and marketing it. Scientific questions abound, including harvesting effects on krill itself and on other species such as whales, seals, and birds, all of which depend on it for food. It is in response to these problems and in an effort to provide answers to these questions that an international scientific body has developed a ten-year programme known as BIOMASS (Biological Investigations of Marine Antarctic Systems and Stocks).*

*This has been sponsored by four international scientific bodies:
   Scientific Committee on Antarctic Research
   Scientific Committee on Ocean Research
   International Association for Biological Oceanography
   Advisory Committee on Marine Resources Research of FAO (Food and Agriculture Organization of the United Nations).

The principal objective of BIOMASS is to gain a better understanding of the structure and dynamics of the Antarctic marine ecosystem as a basis for future management of the harvest of living resources.

In order to obtain the information upon which these management decisions must be based, there is a need for a greater understanding of the complex ecosystem of which these resources are integral components. In an effort to provide hard-core data, therefore, and to arrive at more accurate estimates of the trophic levels of the Antarctic Ocean than are currently available, studies of krill and its food base (phytoplankton) and other components of Antarctic plankton, are essential. Such investigations are, in fact, the central focus of the BIOMASS programme, and are presently being undertaken by several nations interested in the wise management of living resources in the Southern Ocean.

## ACKNOWLEDGEMENT

The author is grateful to Dr. E. Balech, Dr. T.L. Hopkins and Dr. G.A. Fryxell for reading and commenting on parts of the paper. Grateful acknowledgement is expressed to the Division of Polar Programmes/National Science Foundation for the financial support during the past several years that enabled the author to collect the data discussed in the paper.

## REFERENCES

Balech, E. (1975) *Clave Ilustrada de Dinoflagelados antarcticos*. Instituto Antarctico Argentino Direccion Nacional del Antartico. Pub. IAA No.11. Buenos Aires, 99 pages.

Currie, R.I. (1964) Environmental features in the ecology of the Antarctic Seas. In *Biologie Antarctique*. Eds R. Carrick, M.W. Holdgate and J. Prevost, pp. 81 – 86. Hermann, Paris.

El-Sayed, S.Z. (1971) Observations on phytoplankton bloom in the Weddell Sea. In *Biology of the Antarctic Sea IV*, Eds G.A. Llano and I.E. Wallen, pp.301 – 12. American Geophysical Union, Washington D.C.

El-Sayed, S.Z. (1978) Primary productivity and estimates of potential yields of the Southern Ocean. In *Polar Research: to the present and the future*, Ed. M.A. McWhinnie, pp. 141 – 60. Westview Press for the American Association for the Advancement of Science, Boulder, Colorado.

El-Sayed, S.Z., Biggs, D.C., Stockwell, D.A., Warner, R.A. and Meyer, M.A. (1979) Biogeographic and metabolic studies of the phytoplankton and zooplankton populations of the Ross Sea, Antarctica. *Ant. J. U.S.* 13(4), 131 – 33.

El-Sayed, S.Z. and Mandelli, E.F. (1965) Primary production and standing crop of phytoplankton in the Weddell Sea and Drake Passage. In *Biology of the Antarctic Sea II*, Ed. G.A. Llano, pp.87 – 106. American Geophysical Union, Washington, D.C.

El-Sayed, S.Z. and Weber, L.H. (1982) Spatial and temporal variations in phytoplankton biomass and primary productivity in the southwest Atlantic and the Scotia Sea. *Polar Biol.* 1, 83 – 90.

Foxton, P. (1956) The distribution of the standing crop zooplankton in the Southern Ocean. *Discovery Rep.* 28, 193 – 235.

Foxton, P. (1964) Seasonal variations in the plankton of Antarctic waters. In *Biologie Antarctique*, Eds R. Carrick, M.W. Holdgate and J. Prevost, pp.311 – 18. Hermann, Paris.

Fryxell, G.A. and Hasle, G.R. (1971) *Corethron criophilum* Castracane: its distribution and structure. In *Biology of the Antarctic Sea IV*, Eds G.A. Llano and I.E. Wallen, pp.335-46. American Geophysical Union, Washington D.C.

Hardy, A.C. and Gunther, E.R. (1935) The plankton of the South Georgia whaling grounds and adjacent waters, 1926 – 1972. *Discovery Rep.* 11, 1 – 456.

Hart, T. (1942) Phytoplankton periodicity in Antarctic surface waters. *Discovery Rep.* 21, 263 – 348.

Hasle, G.R. (1969) An analysis of the phytoplankton of the Pacific Southern Ocean: abundance, composition, and distribution during the Brattegg Expedition, 1947 – 48. Hvalråad. Skr. No.52, 168pp.

Holm-Hansen, O., El-Sayed, S.Z., Franceschini, G.A. and Cuhel, R. (1977) Primary production and the factors controlling phytoplankton growth in the Antarctic Seas. In *Adaptations within Antarctic Ecosystems*, Ed. G.A. Llano, pp.11 – 50. Smithsonian Institute, Washington.

Hopkins, T.L. (1966) A volumetric analysis of the catch of the Isaacs-Kidd midwater trawl and two types of plankton nets in the Antarctic. *Aust. J. mar. Freshwat. Res.* 17, 147 – 54.

Hopkins, T.L. (1971) Zooplankton standing crop in the Pacific sector of the Antarctic. In *Biology of the Antarctic Sea IV*, Eds G.A. Llano and I.E. Wallen, pp.347 – 62. American Geophysical Union, Washington D.C.

Marr, J.W.S. (1962) The natural history and geography of the Antarctic krill (*Euphausia superba* Dana). *Discovery Rep.* 32, 33 – 464.

Voronina, N.M. (1966) On the distribution of zooplankton biomass in the Southern Ocean. *Okeanologia, Akaud. Nauk, USSR* 6(6), 1041 – 54.

Voronina, N.M. (1968) The distribution of zooplankton in the Southern Ocean and its dependence on the circulation of water. *SARSIA* Vol.34.

# CHAPTER 4b

# Marine Habitats — Benthos

## GORDON B. PICKEN

Department of Zoology, University of Aberdeen, Aberdeen AB9 2TN, Scotland, U.K.

## CONTENTS

## 4b.1. INTRODUCTION

The chaotic expanses of pack-ice and plains of sea-ice characteristic of the Southern Ocean appear totally inhospitable, yet they conceal a wealth of life on the sea floor beneath — the benthos, rich in plants and animals, colourful, diverse, unspoilt and hardly known outside the specialist world. Benthic communities contain the true indigenous fauna of the Antarctic, animals adapted to the glacial conditions which have transformed the continent and its ocean during the last 100 million years.

The variety of the marine invertebrate fauna highlights the restricted nature of the terrestrial fauna of the continent and, for this reason, benthic studies have assumed great significance in the biology of Antarctica. The emphasis of these investigations has naturally changed since 1773, when Johann Reinhold Forster became the first naturalist to cross the Antarctic Circle on Cook's second southern voyage. A large number of new species has been described and named, and energies are now being directed to qualitative and quantitative descriptions of 'communities' or 'assemblages' at various locations. Autecological studies over the last twenty years have used important, interesting or readily available species to examine aspects of physiology, metabolism, growth and reproduction in an Antarctic environment. Emphasis is currently being placed on integrated physiological and biochemical work, to

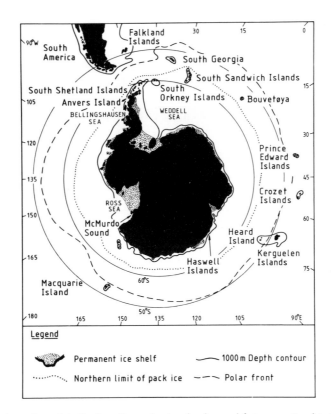

Fig. 4b.1. Antarctica and the Southern Ocean, showing the places and features mentioned in the chapter.

determine the underlying biochemical processes of Antarctic invertebrates, and answer the question of how animals can live in such a cold environment. To understand the nature of the Antarctic benthos in relation to that of other areas, it is first of all necessary to define the extent of the region within which the fauna is considered 'Antarctic'.

## 4b.2. BOUNDARIES OF THE BENTHIC AREA

The limits of the Antarctic benthic region are difficult to define because the Southern Ocean merges with the Atlantic, Pacific and Indian Oceans to the north. Although the Antarctic Convergence is a clear biological boundary for pelagic organisms it is less significant for the benthos. The Antarctic continental shelf is generally narrow, varying in width from 64 to 240 km, and areas of the ocean floor in the Atlantic, Pacific and Indian sectors obviously come under an Antarctic influence. Sea temperatures below 30 m in the Antarctic Ocean and in the abyssal zones of the world's oceans are very similar, so temperature cannot be used as a limiting feature. As will become clear, benthic sediments are important factors in determining the composition and distribution of Antarctic benthic organisms. Since most of these are terrigenous in origin, deposited by ice as it floats north and melts, the northern limit of pack ice may therefore be the most valid limit of the benthos (Fig. 4b.1).

In considering littoral and benthic ecology, it is necessary to include the total area of the sea bed from the continent to the northern limit of the pack-ice. This encompasses the Antarctic continental coastline, and the islands off the Antarctic Peninsula and in the Scotia Arc, but excludes the sub-Antarctic islands Bouvetøya, Campbell Island, Iles Crozet, Heard Island, Iles Kerguelen, Macquarie Island, Marion and Prince Edward Islands. South Georgia does not fit neatly into the overall appraisal. Elements of both Antarctic and sub-Antarctic faunas are present there, and it is recognized as an important link between the two regions. South Georgia is here considered to be sub-Antarctic as far as littoral and sub-littoral communities are concerned, but many animals there display Antarctic adaptations. Within this area there are several unique physical, oceanographic and ecological parameters.

## 4b.3. DESCRIPTION OF THE ENVIRONMENT

Ice moving off the continent carries a large burden of rock and débris, gouged from the underlying land surface, and this is deposited on the sea bed when icebergs break up and melt. An estimated 500 million tons of material are transported out to sea every year, the equivalent of depositing 135 tons on each square kilometre of the sea floor. Benthic sediments in a zone extending from the continent to about the northern limit of the pack-ice are nearly all terrigenous in origin, ranging from greenish-grey silts through fine and coarse sands to boulders more than 50 cm in diameter. These deposits are poorly sorted and, of course, there are no river-borne sediments. Small areas of diatoms, sponge spicules or recent volcanic deposits are present within this terrigenous zone. Glacial deposits cease at about 60°S and are replaced further north by a 1000 – 2000 km wide zone of creamy yellow diatom ooze.

Ice is a major controlling feature of Antarctic benthic biology and its direct influence extends into the benthos to a depth of at least 33 m. In its various forms ice affects water temperature, levels of transmitted light, ocean currents, salinity, the composition of benthic sediments and their stability, and the composition and stability of littoral and sub-littoral communities. The whole continent is isostatically depressed by the ice cap. The continental shelf break occurs at a depth of about 200 m off the European coast, but, in the Antarctic, it is usually 400 – 600 m deep and may be up to 800 m deep in the Ross Sea. At the base of the continental slope lies an abyssal plain with depths ranging from 3700 m to 5000 m, and there is a single deep trench, the South Sandwich Trench, descending to 8620 m. One interesting feature of the submarine topography, especially off Greater Antarctica, is the occurrence of innershelf depressions up to 1700 m deep, separated from the outer continental shelf by sills about 500 m deep.

Visible areas of shoreline may be abraded by fast-ice, pack-ice, brash ice and bergy bits, which may scour littoral and sub-littoral regions to a depth of 15 m (Fig. 4b.2). Icebergs of all sizes can disturb the benthos. Antarctic tabular icebergs float with about 80% of their bulk submerged. Those with recorded heights of 30 m will therefore ground in water less than 150 m deep. When driven by currents icebergs plough furrows in the soft sediment and this causes greater damage to the benthos than simple grounding. Areas of the benthos disturbed in this way take years to re-establish themselves. In truly Antarctic environments ice platelets 2 – 5 mm thick, and up to 15 cm in diameter, may form on the sea bed itself. Such growths of 'anchor ice' can encase plants and animals, eventually tearing them off the substrate; formations capable of lifting 25 kg have been observed (Dayton *et al.*, 1969).

The location and aspect of any shore will determine the duration of ice cover. Littoral and sub-littoral regions have been described in detail at only a few sites and reference will be made to three which illustrate common patterns established by the prevailing ice régime. Fast-ice and pack-ice affect the majority of continental shores (e.g. Anvers Island, Antarctic Peninsula; 64°S 64°W) for more than six months each year, and in the Ross Sea (e.g. McMurdo Sound, 77°S 165°E) fast-ice may persist for 11 months of the year. Signy Island in the South Orkney Islands (60°S 45°W) is a maritime Antarctic island and consequently is less severely affected by ice than sites nearer the continent. The duration of ice impact may be very variable at maritime islands and along the coast of the Antarctic Peninsula. In the 15 years from 1957 to 1973, the period of ice-cover at Signy Island has varied from 70 to 241 days, with an average of 149 days. Despite possible prolonged ice-cover, some shallow rocky sub-littoral sites at Signy Island escape ice damage, because of the topography of the coastline. The mature algal and epifaunal communities established here are examples of the diversity and complexity that may be attained under favourable conditions.

The benthic environment is characterized by low, stable temperatures. Greatest seasonal variation is found in surface waters and shallow sub-littoral zones, particularly in more northerly locations, but even here the extreme range is only from $+4°C$ to $-1.9°C$. Near the sea bed at 10 m depth off Signy Island the annual temperature range is from $-1.8°C$ to $+0.4°C$, and deep areas on the continental shelf at higher latitudes show even smaller variations; at 585 m in McMurdo Sound the range is $0.07°C$ around an average of $-1.89°C$. Although oxygen saturation and salinity are generally high and stable, salinity may be considerably reduced near melting icebergs or in the upper metre of nearshore waters during the spring melt, and oxygen concentrations may be reduced under ice-shelves. Superimposed on this very equable environment are the seasonally fluctuating levels of food availability and light intensity. Incident light levels vary with latitude and season, and light penetration is affected by thickness of ice and snow cover in winter, and by turbidity of the water column in summer. Greatest water clarity is found during winter beneath the ice-cover, but the highest levels of light penetration occur in spring, between the break-out of winter fast-ice and the onset of the phytoplankton bloom.

Primary production is highly seasonal, with a major input in the spring and a smaller one in the autumn. The overall production at Signy Island is 130 g carbon $m^{-2}$, nearly all of which takes place during a 50 day period in spring and early summer (Horne et al., 1969). Primary production at most coastal sites is generally limited to the period of open water. During the rest of the year only redistribution of accumulated energy from one trophic level to another takes place.

Conditions in the Antarctic benthos are therefore unique. A deep shelf and abyssal plain are littered with hard substrates. Water temperature is constantly low, and the input of primary production is large, short-lived and generally regular each year. Ice constantly invades the littoral and shallow sub-littoral regions. Despite the apparent harshness of the environment many invertebrate taxa thrive here, although their communities are very different from those found in other continental shelf assemblages.

Fig. 4b.2. A summer storm at Signy Island drives brash ice and bergy bits ashore, where they cause extensive damage in the shallow sub-littoral zone.

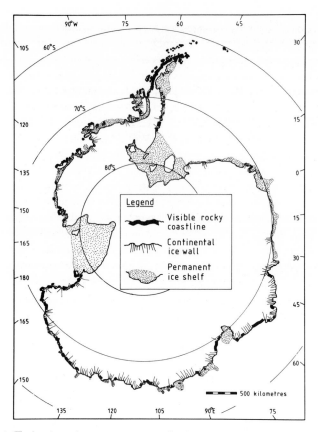

Fig. 4b.3. The location and approximate extents of rocky coast around the Antarctic continent.

## 4b.4. BENTHIC FLORA AND FAUNA

### 4b.4.1. Communities in the littoral region

Most of the Antarctic coastline is hidden beneath the ice flowing over the continental margins. Littoral regions are common on the west coast of Lesser Antarctica and the islands of the Scotia Arc, but rare in Greater Antarctica (Fig. 4b.3). Ice abrasion severely limits the extent and composition of intertidal communities, characteristically producing a 'bare zone' in the littoral region. The upper limit of the zone is approximately high water level, and the depth of the lower limit varies considerably with location (Fig. 4b.4).

Perennial seaweeds are lacking in the intertidal zone of most shores (Fig. 4b.4), and the flora consists of the ephemeral species *Porphyra umbilicalis*, *Adenocystis utricularis*, *Monostroma* sp. and *Chaetomorpha* sp., or the ubiquitous, encrusting calcareous *Lithothamnion* sp. Kelps attain their maximum southerly extent on sub-Antarctic islands. There are extensive littoral and shallow sub-littoral beds of *Macrocystis pyrifera* at 55°S on South Georgia, 53°S on Heard, and 54°S on Macquarie. Mature *Macrocystis* plants may exceed 15 m in length, and, because of their erect growth, the fronds often reach the sea surface. True kelps are absent at higher latitudes (Moe and Silva, 1977), where the largest seaweed is the brown

Fig. 4b.4. The intertidal zone on most Antarctic shores is bare and perennial seaweeds are absent. Low tide on a sheltered rocky shore at Signy Island, mid-summer.

*Himantothallus grandifolius.* Although it grows to 10 m length, *Himantothallus* characteristically lies flat along the sea bed in the depth range 3 − 15 m, thus avoiding the severest effects of the ice.

Sessile animals are rarely found on Antarctic shores and intertidal fauna consists largely of motile forms which can avoid damage by ice impact. Crevices may harbour the chiton, or coat-of-mail shell, *Hemiarthrum setulosum*, and small winkle-like snails, but on the majority of shores the most obvious animal is the 3 − 4 cm long limpet *Nacella concinna*, which grazes freely when submerged at high tide and clings to boulders and vertical rock surfaces when exposed at low tide. Limpets are also abundant on the sub-littoral bare zone, with densities of up to 350 m$^{-2}$. With the approach of winter large numbers of littoral limpets migrate into the sub-littoral to avoid ice damage (Walker, 1972).

Some particularly sheltered locations may occasionally bear a less transient intertidal sessile fauna. On a stable boulder and cobble rubble substrate on Anvers Island the fauna is sometimes dominated by the byssate bivalve *Kidderia subquadratum* (Stockton, 1973), but the duration of this colonization is not known. The mussels and barnacles characteristic of the intertidal region on temperate shores are absent in the Antarctic. A single specimen of the mussel *Mytilus edulis* has been found at South Georgia, but this is considered to be an extreme case of vagrancy. The animal was probably transported by ship from the Falkland Islands, 1330 km away. There are no benthic balanid barnacles (the only members of this group in the Antarctic are found attached to whales) and there is just one verrucomorph barnacle, *Verruca gibbosa*. The majority of the thirty-one species of Antarctic cirripedes are stalked barnacles.

Intertidal rock pools are common at some sites, but they have a limited flora and fauna. Much of the rock surface is bare, and only crevices and vertical faces are covered by *Lithothamnion*, or greyish-pink growths of encrusting red alga *Hildenbrandia* sp. The pools may contain small gastropod snails, amphipods, nemertean worms and small pycnogonids (sea-spiders). Nemerteans are common intertidally and in the shallow sub-littoral, where they are often found in large numbers in the holdfasts of algae or clustered about decaying organisms. The large *Parborlasia corrugatus*, up to 2 m long, is especially abundant and has a circumpolar distribution.

In contrast to the impoverished littoral flora and fauna, sub-littoral communities are diverse and complex. Standing crop in the shallow sub-littoral increases as the effect of ice scour decreases, before declining again in the deeper parts of the shelf.

## 4b.4.2. Communities in the sub-littoral region

In the sub-littoral bare zone, sheltered crevices support small clumps of seaweeds *Adenocystis*, *Enteromorpha*, *Ulothrix* or *Cladophora* in summer, and many other surfaces will have some cover of *Lithothamnion*. Large perennial seaweeds are found only beneath the bare zone, where they are not disturbed by ice. Algae are confined to areas of exposed shoreline and regions of the shallow sub-littoral with a rocky substrate. There are therefore no algal records from continental coasts between 72°W and 173°E, or from 50°W to 40°E. Where they do occur, algae are usually concentrated in a belt from the lower level of the bare zone to a depth of about 37 m, although they have been found in deeper water. The limit of the photic zone has not been defined in the Antarctic and such definition would be difficult because of the effects of variable ice cover. Apparently healthy plants have been collected at depths of up to 668 m in the Ross Sea (Zaneveld, 1968), but light levels at these depths are insufficient for autotrophic production (Drew, 1977). Such specimens were probably carried into deep water by currents or ice.

There is considerable variation in the composition and biomass of the flora at different sites. The location and topography of the site determines incident light levels, water clarity, wave action and the duration and severity of ice abrasion, all of which affect algal growth. Parts of the upper sub-littoral (2 − 20 m) at Signy Island have luxuriant algal growth (Fig. 4b.5). Red seaweeds *Leptosarca simplex* and *Gigartina apoda*, and the green *Adenocystis* are common in the range 2 − 5 m. Below 5 m *Ascoseira mirabilis*, dense thickets of *Desmarstia menziesii* and *D. anceps* up to 1 m high, and *Himantothallus* form a major part of the flora. Of these perennial brown seaweeds *Himantothallus* becomes the most prominent below 9 m. Beneath the canopy formed by these four macrophytes there is often a rich 'understorey' composed of *Iridae obovata*, *Myriogramme mangini*, *Plocamium secundatum*, *Porphyra umbilicalis* and *Gigartina*. Macrophyte biomass at Antarctic sites is significantly lower than that of comparable temperate sites because of the effect of ice abrasion, unstable hard substrates in the shallow sub-littoral zone, and short periods of primary production (Richardson, 1979a).

In association with this algal cover live a large number and variety of invertebrates but, since they tend to be small, cryptic and darkly coloured, few will be noticed by the casual diver on his first visit. Algae provide shelter and food for many animals. The most numerous are weed-dwelling amphipod crustaceans; thousands of individuals of the common species *Bovallia gigantea* and *Orchomonella plebs* cling to algal fronds and rock surfaces. During winter the amphipods *Paramoera walkeri* and *Pontogeneia antarctica* form a unique sympagic community just beneath the ice foot, feeding on diatoms growing within the ice matrix (Rakusa-Susczewski, 1972; Whitaker, 1977). The bright orange isopod *Antarcturus signiensis* hangs on fronds of *Desmarestia* and *Ascoseira*, and closer examination of these algae will reveal numerous, small, red — brown bivalves, *Lissarca miliaris*, attached to the fronds by byssus threads. Many Antarctic lamellibranchs anchor themselves in this way but there are no species which cement their shells to hard substrates (Nicol, 1967).

Numerous small pycnogonids pick their way delicately over fronds and small stones, feeding on hydroids and anemones. Algal fronds also provide a foothold for more than 30 species of prosobranch gastropod, a surprisingly diverse fauna for the depth range 2 — 15 m. With the exception of the limpet *Nacella*, which alone provides 70% of the gastropod biomass, the largest species is a carnivorous *Trophon* with a maximum length of 33 mm. In contrast to the inconspicuous prosobranchs, the common nudibranchs (sea-slugs) in this depth range, *Notaeolidia* sp., are readily seen because of their light colouring which shows up well against the dark foliage. Decapod crustacea are poorly represented. Only three species are found, of which the circum-Antarctic prawns *Chorismus antarcticus* and *Notocrangon antarcticus* are the best known. There are no brachyuran crabs, but a search beneath the algal understorey will reveal the crustacean *Glyptonotus antarcticus*, a large, omnivorous isopod with a circum-Antarctic distribution, which partially fills the niche normally occupied by crabs in the shallow sublittoral (Fig. 4b.6). Starfish are represented by the circum-Antarctic, deep red or purple *Odontaster validus*, and sea urchins by the equally ubiquitous *Sterechinus neumayeri*.

In the depth range 2 — 20 m at Signy Island, vertical faces not dominated by macro-alagae bear an abundant cover of sessile fauna (Fig. 4b.7). Large holothurians of the genus *Cucumaria*, with brightly coloured 'tentacles', mould themselves into angles or ledges on the rock face, and clusters of brachiopods and tunicates anchor themselves to the hard substrate. Concave surfaces in water as shallow

Fig. 4b.5. Areas of the shallow sub-littoral not regularly damaged by ice may bear extensive beds of macro-algae. Here, a diver inspects beds of *Desmarestia* growing at 12 m off Signy Island.

Fig. 4b.6. Two circum-Antarctic eurybathyic species.
a. The giant isopod *Glyptonotus antarcticus*.

as 12 m often bear a profuse growth of the gorgonian coral *Primnoella* sp., a delicate, pale yellow species up to 30 cm long, and numerous individuals of the circum-Antarctic anemone *Utricinopsis antarctica*. Undoubtedly the most extensive cover on such protected vertical and concave surfaces is provided by sponges. In water depths of 12 – 20 m flat, bright yellow growths of *Haliclona* sp. stand out against the dark rock face and purple *Lithothamnion* encrustations. The importance of sponges in this shallow community gives the first hint of their significance in deeper communities all round the Antarctic continent.

Many sites along the west coast of the Antarctic Peninsula have an equally diverse sub-littoral community in the range 2 – 20 m. Arthur Harbour on Anvers Island is typical. The bare or 'upper surf' zone extends to a depth of about 8 m and, below it, there is a diverse algal belt with associated fauna which gives way to sponges and tunicates at 33 – 40 m. On Continental coasts at higher latitudes such shallow water communities become rare. At Haswell Islands, where sub-littoral regions experience open water for only two months of the year, the bare zone extends to a depth of 10 m, and the only recorded alga is *Phyllophora antarctica*. The biota make full use of the ice-free period, however; within four weeks of ice break-out, flora and fauna recolonize the depth range 7 – 10 m, establishing a community with a biomass of about 300 g m$^{-2}$ wet weight (Gruzov, 1977).

Perhaps the most extreme examples of severely ice-affected rocky sub-littoral areas can be found at McMurdo Sound. Three zones have been identified here in the depth range 0 to 180 m, delineated by the effects of ice scour and the depth of anchor ice disturbance (Dayton *et al.*, 1970). The bare zone extends to a depth of 15 m and is briefly recolonized during ice-free periods by *Sterechinus*, *Odontaster*,

*Parborlasia*, *Glyptonotus* and pycnogonids. Soft corals, anemones, hydroids and ascidians (sea-squirts) become increasingly common in the range 15 — 33 m but they are still vulnerable to anchor ice. Below 33 m the bottom is usually covered by shells from the bivalve *Limatula hodgsoni*, which in turn are covered by a layer of sponge spicules up to 2 m thick. Sponges cover 55% of the surface area and are so dominant in the depth range 69 — 180 m that this community has been labelled the 'McMurdo Glass Sponge Assemblage' (Bullivant and Dearborn, 1967). Some species have arborescent growth forms, thus avoiding competition with more prostrate sponges and reducing predation from starfish. Sponges therefore provide most of the vertical structure in the community, and may themselves bear anemones, hydroids, bryozoans, polychaetes and bivalve molluscs. Associated with this assemblage are asteroids, the large nudibranch *Austrodoris macmurdoensis* and the gastropod *Trophon longstaffi*.

Sponges are ubiquitous in the Antarctic and many of the 300 or so Antarctic species, including the common *Rosella antarctica*, *R. racovitzae* and *Mycale tridens* are circumpolar. The fauna is dominated by glass sponges rather than calcareous sponges, and many show a wide depth distribution. *Scolymastra joubini*, a large glass sponge which grows to a height of more than 100 cm may be found at a depth of 10 m in the Antarctic, yet it is typical of species found in water more than 100 m deep in other parts of the world. The interaction of sponges and their predators has been studied in some detail at McMurdo Sound (Dayton *et al.*, 1974). *Mycale acerata* could probably overgrow many other sessile species but is prevented from dominating the space resource by the predation of the asteroids *Perknaster fuscus antarcticus* and *Acodontaster conspicuous*. Despite predation by *Acodontaster* and the sea-lemon *Austrodoris macmurdoensis*, there are large standing crops of the Rossellid sponges *Rossella racovitzae*, *R. nuda* and *Scolymastra joubini*. It is thought that the level of predation is checked by the predation of larval and young *Acodontaster* and *Austrodoris* by the star-fish *Odontaster validus*.

Fig. 4b.6b. The isopod *Serolis polita*, a fine specimen resting on the silt.

Fig. 4b.7. A sheltered vertical rock face at 10 m depth off Signy Island. The surface is completely covered by encrusting sponges and *Lithothamnia*, with clusters of anemones and brachipods. Long stalks of *Primnoella* hang from the face.

In sheltered Antarctic bays up to 30 m deep, soft substrates of muddy sand are virtually devoid of epifauna, although *Glyptonotus* and *Parborlasia* congregate and feed on decaying organisms. Large accumulations of rotting, unattached seaweed invariably bear pycnogonids, molluscs and the amphipods *Cheirimedon femoratus* and *Djerboa furcipes* with densities in excess of 20,000 individuals m$^{-2}$. Large numbers of the isopods *Serolis polita* and *S. cornuta* rest on the sea bed, camouflaged by the fine surface sediment (Fig. 4b.6B). Free-swimming scallops are found at many sites, and of the four species the best known is the large, endemic *Adamussium colbecki*. In terms of biomass, the infauna of these substrates is dominated by bivalve molluscs including *Laternula elliptica*, which grows to about 15 cm long and attains densities of up to 50 m$^{-2}$, the small *Yoldia eightsi* with densities up to 2700 m$^{-2}$ or the tiny *Mysella charcoti* with densities around 75,000 m$^{-2}$. Numerical dominance of the infauna is usually achieved by polychaete worms, including *Maldane sarsi antarctica* (Gallardo *et al.*, 1977), *Aglophamus virginis* and *Apistobranchus typicus*, all with densities around 6000 m$^{-2}$.

Soft substrates are generally characterized by a high density, diversity and biomass of non-attached, deposit feeding polychaetes, molluscs and crustaceans. Species density in the depth range 5 – 75 m in Arthur Harbour is about four times that of the inner continental shelf of North America (Richardson and Hedgpeth, 1977), and diversity is near its theoretical maximum (Lowry, 1975). The environment is physically stable and the low organic content of the sediments and their lack of hydrogen sulphide show that organic matter is rapidly and efficiently utilized by the benthos. At adjacent areas beneath a glacier foot, sediments are periodically disturbed by calving of the glacier. In this unstable environment the

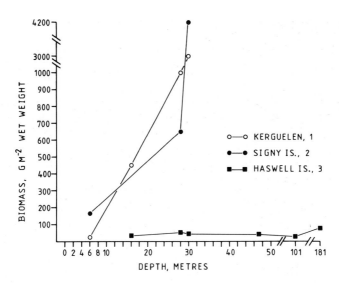

Fig. 4b.8. Biomass estimates of epifauna at several locations. Sources: 1. Desbruyéres and Guille, 1973; 2. White and Robins, 1972; 3. Propp, 1970.

fauna is dominated by a single polychaete species, *Tharyx cincinnatus*, and hydrogen sulphide is present in the sediment. Deeper benthic regions around Antarctica have communities determined largely by the nature of the substrate, and attached fauna resembles that found in deeper parts of the world's oceans. Long-stalked colonial gorgonians, sponges and corals are the principal sessile organisms. Associated with them is an errant fauna of brittle stars, the large starfish *Labidiaster annulatus* and *Perknaster fuscus antarcticus* and feather stars, the most abundant of which is *Promachocrinus kerguelensis*. Ophiuroids may be found in great numbers and local concentrations of 100 million individuals km$^{-2}$ have been reported for *Ophiacantha antarctica* in the depth range 75 – 350 m in the Ross Sea. Amphipods, gastropods and pycnogonids are also found. The Antarctic pycnogonid fauna contains both very small and very large species, a trait which is characteristic of several Antarctic invertebrate taxa. The larger pycnogonid species are normally found below 20 m, and include *Decolopoda australis*, one of the three 10-legged species, and *Dodecolopoda mawsoni*, one of the two 12-legged species unique to the Antarctic.

Three assemblages have been identified in the depth range 200 to 400 m in the Ross Sea (Bullivant and Dearborn, 1967), and these may be similar to those over extensive areas of the Antarctic benthos. On fine sediments with erratic boulders there is a 'Deep Shelf Mixed Assemblage' composed of polychaetes, bryozoans, ophiuroids, crinoids and molluscs. Deeper regions with a substrate of mud, or sandy mud and erratic boulders, have polychaetes, ophiuroids, holothurians, molluscs and asteroids. On a muddy sand substrate with embedded cobbles, bryozoans, gorgonians and ascidians are common. Various other assemblages have been described in the depth range 300 to 1000 m where sponges, the solitary stony coral *Caryophyllia antarctica*, and hydrozoan corals are important constituents of the fauna. One interesting discovery is the recognition that the chthamalid barnacle *Bathylasma corolliforme*, found in water more than 200 m deep, is circumpolar at high latitudes, and, in places, it is abundant enough to constitute a recognizable assemblage.

Estimates of biomass reflect the characteristic features of the Antarctic benthos. Shallow sub-littoral regions with well-developed macro-algae have much higher epifaunal biomasses than severely ice-affected coasts. Deeper benthic regions with hard substrates have a high biomass which is generally greater than that of other comparable areas of the world's oceans (Fig. 4b.8). Probably the principal reasons for this are the abundance of hard substrates on the sea floor, which significantly increase the area suitable for attachment, and also the dominance of large sessile organisms such as sponges and holothurians. Biomass estimates of the infauna based on grab samples are often regarded as too low, because the large, deep-burrowing bivalve *Laternula elliptica* is not sampled adequately by this method. Even with this under-estimate, Antarctic infaunal biomass is similar to that of other sections of the world's oceans, and the addition of estimates for *Laternula* biomass substantially increases the biomass (Fig. 4b.9). Nevertheless, a significant proportion of the Antarctic biomass, perhaps as much as 60%, is in the form of large, sessile organisms which are largely unavailable to the benthic ecosystem or not readily utilized by it.

Clearly, the benthos has a great variety and abundance of animals. The origins of the Antarctic fauna and the continuing process of its evolution are therefore of great interest.

## 4b.5. BIOGEOGRAPHY AND ORIGIN OF BENTHIC FAUNA

Continental Antarctic shores are now 960 km from South America, 4020 km from South Africa and 3380 km from new Zealand, the nearest land masses. The shelf and benthos have connections with the deep ocean benthos of the Indian, Pacific and Atlantic Oceans, and there are three shallow-water 'bridges' to other continents. Most obvious is the Scotia Arc, the island chain from South America through the Falkland Islands, South Georgia, South Sandwich and South Orkney Islands to the Antarctic Penninsula. The Macquarie-Balleny Ridge and the Kerguelen-Gaussberg Ridge rise to within 1800 m and 200 m of the surface respectively, and link Antarctica with the land masses to the north.

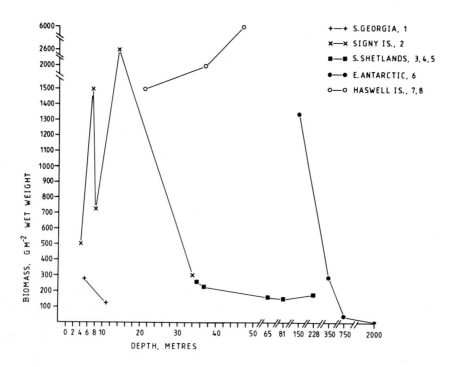

Fig. 4b.9. Biomass estimates of infauna at several locations. Sources: 1. Platt, 1978, 1980; 2. Hardy, 1972; 3. Mills, 1975; 4. Gallardo and Castillo, 1968; 5. Gallardo and Castillo, 1969; 6. Belyaev and Uschakov, 1957; 7. Gruzov et al., 1967; 8. Andriashev, 1968.

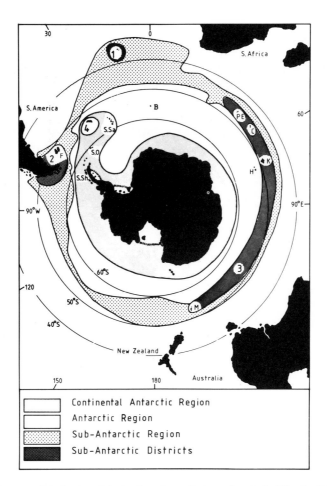

Fig. 4b.10. Biogeographical regions of the benthos of the Southern Ocean. Redrawn from Dell, 1972, after Hedgpeth, 1969. Specific sub-Antarctic districts are: 1. Tristan da Cuhna district; 2. Magellanic sub-region or province; 3. Kerguelenian sub-region or province. The specific Antarctic district is: 4. South Georgian district. Islands identified by letters are B, Bouvetøoya; C, Crozet; F, Falklands; H, Heard; K, Kerguelen; M, Macquarie; PE, Prince Edward and Marion; SO, South Orkney; S.Sa, South Sandwich; S.Sh, South Shetland.

Most taxonomists believe that the benthic fauna is derived from three sources — a relict fauna, the remains of the fauna around the continent before it became glaciated, a fauna derived from the deep basins of the Atlantic, Pacific and Indian Oceans, and thirdly, a shallow-water fauna from South America and Australasia which has crossed to Antarctica. The importance of the Scotia Arc and the two-way process of migration along this island chain is repeatedly emphasized by biogeographers.

Approximate zoogeographic regions have been defined for particular taxa, on the basis of their present distribution and composition, but opinions on precise delineation differ. Distributional data are lacking for large areas of the Antarctic benthos and much of the available data indicate the areas in which major expeditions have worked rather than the distribution of the organisms themselves. Nevertheless, the various schemes usually agree on a number of biogeographical 'provinces' and 'regions' (Fig. 4b.10), which help to characterize the broad composition of the benthos.

The Antarctic fauna apparently consists of many known, non-endemic families and genera which have undergone considerable adaptive radiation. Despite incomplete data, it is clear that a very high proportion of Antarctic species are endemic (Table 4b.1). In determining the degree of endemism, the

TABLE 4B.1. Percentage of Endemic Species and Genera in Selected Antarctic Taxa.

| Taxa | Recorded number of species | % Endemic genera | % Endemic species |
|---|---|---|---|
| Porifera | 300 | — | 50 |
| Scleractinian corals | 10 | — | 60 |
| Actinaria | 31 | — | 74 |
| Polychaeta | 650 | 5 | 57 |
| Brachiopoda | 16 | — | 69 |
| Bryozoa | 321 | — | 58 |
| Pycnogonida | 100 | 14 | 90 |
| Cirripedea | 37 | — | 76 |
| Isopoda + Tanaidacea | 229 | 10 | 66 |
| Echinoidea | 44 | 25 | 77 |
| Holothuroidea | 38 | 5 | 58 |
| Echinodermata | — | 27 | 73 |
| Ascidia | 129 | — | 56 |

Sources: Antarctic Map Folio, 1969; Dell, 1972; Knox and Lowry, 1977.

position of South Georgia is crucial, since it is the northern limit of Antarctic elements and the southern limit of temperate American elements. If the island is considered to be 'Antarctic', then the number of endemic Antarctic species is reduced, because many species found at South Georgia are also found in the South American cool-temperate fauna.

Many Antarctic species living in deep water have a circum-Antarctic distribution but species confined to shallow water may be more restricted. The Scotia Arc may not be such an easy pathway for exchange as was once thought. Speciation has occurred because areas of suitable shallow sub-littoral benthos are widely scattered, geographically separated, and genetically remote since many species lack a pelagic larva. This is just one of the physiological adaptations prevalent in Antarctic invertebrates to combat cold temperatures and low food availability. Invertebrates thrive in the Antarctic and are certainly not just barely clinging to life in an extreme, harsh environment.

## 4b.6. PHYSIOLOGY OF BENTHIC INVERTEBRATES

Low temperature itself is not a limiting condition for Antarctic benthic invertebrates, and few animals are exposed to temperatures below the freezing point of sea water. Because of its intertidal habit the limpet *Nacella* alone has adapted to withstand short periods below $-2°C$. When trapped in anchor ice or intertidal ice scale, it protects itself in a cocoon of highly viscous mucus, where it can survive exposure to $-20°C$ for up to 2 hours (Hargens and Shabica, 1973).

In adapting to polar environments, marine poikilotherms adopt one of two strategies. The mucous cocoon of *Nacella* is a 'resistance adaptation' which increases the animals' tolerance of lower temperatures, whereas 'capacity adaptations', such as the development of low-temperature enzyme systems, allow the animal to function at a relatively high metabolic rate. This is the phenomenon of 'metabolic cold-adaptation', which has received much attention recently. Clearly, Antarctic marine poikilotherms have enzyme systems that are 'adapted' to cold temperature, because they have higher basal metabolic rates at ambient Antarctic temperatures than do species from warmer water. However, they do not have higher rates than those expected by extrapolation of the basal metabolic rates of warmer water species living at their normal temperatures. Antarctic gastropods, for example, do not show unusually high levels of oxygen consumption. It is suggested that this extrapolative technique is the most useful and valid, because in this comparison animals are not exposed to temperatures outside their normal experience (Houlihan and Allan, 1982). If this extrapolative definition of metabolic cold adaptation is adopted, it will be found that very few Antarctic benthic invertebrates are 'cold adapted'.

Growth is generally slow, reflecting the overall low level of food availability throughout the year, and, in many species, is accompanied by longevity. Sponges may be several centuries old and there is good evidence to suggest that the limpet *Nacella* can survive at least 100 years (Shabica, 1971). Other species with less spectacular life-spans live much longer than similar-sized species in temperate and tropical regions. This longevity is invariably accompanied by a correspondingly long period of growth to maturity. Antarctic amphipods, for example, have rates of growth and development that are up to four times slower than those of comparable temperate species. Many species have highly seasonal growth patterns, with marked growth in spring and summer and none in winter. The bivalve *Lissarca miliaris* and the gastropod *Laevilacunaria antarctica* illustrate seasonal growth patterns particularly well (Picken, 1979; Richardson, 1979b). The moulting periods of Antarctic amphipods and isopods are also much longer than those of comparable temperate crustaceans.

Slow-growing invertebrate species generally attain greater size than their fast-growing counterparts. Large species of holothurians, isopods, starfish, ascidians and sponges are found in the Antarctic; of the simple ascidians, the Antarctic *Paramogula gregaria* is the largest known, attaining a length of 33 cm. Conversely, Antarctic taxa also contain some very small species. Bivalves and gastropods in particular are generally small, often with brittle, thin shells, and lacking ornamentation. Of the 51 species of gastropods in the depth range 0 – 100 m at the South Orkney Islands, 75% are less than 10 mm long. Although it has been said that gigantism is a general trait of Antarctic benthic invertebrates, this is too sweeping a statement. Gigantism probably only occurs in those taxa that do not require large amounts of calcium, and this hypothesis is supported by a recent survey of the sizes of the world's amphipod crustaceans (Broyer, 1977). If a 'dwarf' amphipod species is no more than half the genus mean size, and a 'giant' species is at least twice the genus mean size, then the Antarctic has the highest proportion of giant amphipod species and the lowest proportion of dwarf species.

Since food is generally scarce, particularly on the continental shelf, there is a widespread predilection for obligatory and facultative necrophagy among a number of taxa. Species of amphipods, holothurians, gastropods, ophiuroids, echinoids and nemerteans are able to utilize chance sources of food in this way (Arnaud, 1970; 1974). This is obviously of competitive significance, and it also demonstrates the degree of adaptation that has occurred among Antarctic taxa. Necrophagy is not a common trait in the Arctic benthic fauna, which has evolved more recently.

Non-pelagic development, the completion of the life-cycle without the production of a pelagic larva, is predominant in both the Arctic and Antarctic Oceans. Both polar regions are cold, but this is one of the few features they have in common. The Arctic is an ocean basin virtually surrounded by land whereas the Antarctic ocean merges with three other great oceans. Such differences exclude two early hypotheses for the predominance of non-pelagic development at high latitudes, the loss of larvae at sea, and high levels of mortality in low salinity coastal water during spring. It is more likely that this is an adaptation to the predictable but seasonal availability of food. Polar summer phytoplankton blooms are too short to support large numbers of zooplankton, which would require long development times at low temperatures. Of course, the Antarctic krill has a well-documented pelagic developmental history, but this is an open water pelagic crustacean, not a nearshore benthic invertebrate. Few Antarctic invertebrates have a pelagic larval stage; only the larvae of two ophiuroid species, two asteroids, one echinoid and a holothurian have been found in the plankton. The limpet *Nacella* is a broadcast spawner, but it produces relatively few eggs at each annual spawning compared with temperate species that have a pelagic larva. It is quite possible that *Nacella* eggs develop into benthonic larvae, similar in habit to those produced by the Antarctic polychaete *Scoloplos marginatus macleani* or the starfish *Odontaster validus*.

Non-pelagic development in the Arctic is mainly accomplished through the production of large yolky eggs which give rise to fully competent juveniles, but, in the Antarctic, brood protection and viviparity are common (Thorson, 1950). Brooding or viviparity is found in more than 80% of Antarctic sponges. Many species of lamellibranchs, echinoids, ophiuroids, holothurians, echinoids, amphipods, isopods, polychaetes and nemerteans exhibit brood protection (Arnaud, 1974). Some species deposit large yolky eggs in the sub-littoral, however, and this mode is particularly common among shallow water gastropods. Eggs may be laid on the fronds of algae or on the sheltered undersides of stones.

Low levels of food availability prolong the period of oogenesis, and development times are lengthy because of the large size of eggs (Clarke, 1982). In many species the life cycle is geared to make best use of the short season of primary production and take account of the long periods of oogenesis and development. These species show a definite seasonal cycle of reproductive activity, laying eggs in winter to ensure that juveniles hatch during spring and summer to benefit from the phytoplankton bloom. Reproduction in other species is largely aseasonal, and eggs and juveniles may be found throughout the year.

It is now becoming apparent that food is as important a factor as temperature in the physiology and ecology of Antarctic invertebrates. Recent arguments, drawing on data from autecological and physiological studies, propose that the majority of Antarctic invertebrates are able to survive predictable but fluctuating levels of food precisely *because* the water is cold. Low temperature allows the animals to function at low basal metabolic rates, thus reducing their food requirements (Clarke, 1980). Many Antarctic invertebrates have a number of characteristics in common, principally slow growth, deferred maturity, longevity, large size and reduced reproductive effort (Clarke, 1979). These characteristics enable them to compete more successfully for the available limited resources. Periods of sparse food availability are an inevitable hardship of living at high latitudes where the summer periods of phytoplankton production are short. Cold is in many respects advantageous, because it permits lower basal metabolic rates, thus reducing overall energy requirements.

## 4b.7. CONCLUSION

Opportunity to use resources contributing to the development of the Antarctic benthos depends on the location of communities of plants and animals. Areas of visible coast and shallow sub-littoral are generally confined to islands and the west coast of the Antarctic Peninsula. These regions have an impoverished intertidal flora and fauna, vulnerable to repeated destruction by ice. Below the depth of periodic ice-scour, however, mature beds of macro-algae may be found with an associated, diverse epifauna. Sub-littoral assemblages below 30 m are all very similar around the continent. Hard rock debris raining down from melting continental ice furnishes ample substrate for the attachment of epifauna, and well-mixed oceanic water throughout most of the area provides a uniquely equable environment. The fauna here consists predominantly of particle-feeding species which will tolerate only a narrow temperature range, but are found over a wide range of depths. Sponges, corals, anemones and gorgonians are dominant sessile animals, with starfish, crustacea, molluscs and pycnogonids moving over them.

Beyond the range of ice impact the benthic environment is remarkably stable and stress-free. The biomass of both infaunal and epifaunal assemblages is generally high, and equivalent to that of comparable regions in warmer water. Species richness and diversity are also high, and the complex Antarctic benthos is probably less fragile than the more recent, simpler Arctic benthos. Because the sub-littoral environment is stable, ecological niches are small and levels of competition are high. This encourages speciation, which is further enhanced by the predilection for non-pelagic development in many taxa. The benthic fauna is well adapted to its environment, and available energy is utilized very efficiently. It now remains for the significance of the benthos to be established in relation to the pelagic food chains of the Southern Ocean, since it may make a major contribution to the cycle of nutrients fuelling the vast annual primary production.

## ACKNOWLEDGEMENTS

Conversations with my colleagues Dr. D. Houlihan and Dr. P.G. Oliver, aided my thoughts on some topics in this chapter. Many useful comments on the manuscript were made by Mr. W.N. Bonner, Dr. A. Clarke and Mr. M.G. White of the British Antarctic Survey. All the underwater photographs were taken by Mr. D. Allan. Finally, I must thank my wife Anne for translating French texts and unravelling my own tortuous English.

## REFERENCES

Andriashev, A.P. (1968) The problem of life community associated with the lower layers of Antarctic fast ice. In, *Symposium on Antarctic Oceanography*, pp. 147 – 57. Scott Polar Research Institute, Cambridge.

Antarctic Map Folio (1969) *Distribution of selected groups of marine invertebrates in waters south of 35°S latitude. Antarctic Map Folio Series, Folio 11*. Co-edited by J.W. Hedgpeth. American Geophysical Society, New York.

Arnaud, P.M. (1970) Frequency and ecological significance of necrophagy among the benthic species of Antarctic coastal waters. In *Antarctic Ecology*, Ed. M.W. Holdgate, pp. 259 – 67. Academic Press, London and New York.

Arnaud, P.M. (1974) Contribution à la bionomie marine benthique des régions Antarctiques et sub-Antarctiques. *Tethys*, 6(3), 465 – 656.

Belyaev, G.M. and Uschakov, P.V. (1957) Certain regularities in the quantitative distribution of the benthic fauna in Antarctic waters. *Dokl. Akad. Nauk SSSR* 12, 137 – 40.

Broyer, C. de. (1977) Analysis of gigantism and dwarfness of Antarctic and sub-Antarctic Gammaridean amphipoda. In *Adaptations within Antarctic Ecosystems. Proc. 3rd. SCAR Symp. Antarct. Biol.*, Ed. G.A. Llano, pp. 327 – 34. Gulf Publishing Company, Houston, Texas.

Bullivant, J.S. and Dearborn, J.H. (1967) Stanford University invertebrate studies on the Ross Sea, 1958 – 1961: General account and station list. In *The fauna of the Ross Sea. Part 5; General accounts, station lists and benthic ecology. Bull. N.Z. Dep. scient. ind. Res.* 176, 1 – 77.

Clarke, A. (1979) On living in cold water: K-strategies in Antarctic benthos. *Mar. Biol.* 55, 111 – 19.

Clarke, A. (1980) A reappraisal of the concept of metabolic cold adaptation in polar marine invertebrates. *Biol. J. Linn. Soc.* 14, 77 – 92.

Clarke, A. (1982) Temperature and embryonic development in polar marine invertebrates. *Int. J. of Invert. Reprod.* 5, 71 – 82.

Dayton, P.K., Robilliard, G.A. and DeVries, A.L. (1969) Anchor ice formation in McMurdo Sound, Antarctica, and its biological effects. *Science* 163, 273 – 4.

Dayton, P.K., Robilliard, G.A. and Paine, R.J. (1970) Benthic faunal zonation as a result of anchor ice at McMurdo Sound, Antarctica. In *Antarctic Ecology*, Ed. M.W. Holdgate, pp. 244 – 58. Academic Press, London and New York.

Dayton, P.K., Robilliard, G.A., Paine, J.T. and Dayton, L.B. (1974) Biological accommodation in the benthic community at McMurdo Sound, Antarctica. *Ecol. monogr.* 44, 105 – 28.

Dell, R.K. (1972) Antarctic benthos. In *Advances in Marine Biology*, 10, Eds F. Russel and C.M. Younge, pp. 1 – 216. Academic Press, London and New York.

Desbryères, D. and Guile, A. (1973) La faune benthique de l'Archipel de Kerguelen, Premières données quantitatives. *C. r. hebd. Séanc. Acad. Sci. Paris* 276(4), 633 – 6.

Drew, E.A. (1977) The physiology of photosynthesis and respiration in some Antarctic marine algae. *Bull. Br. Antarct. Surv.* 46, 59 – 76.

Gallardo, V.A. and Castillo, J.G. (1968) Mass mortality in the benthic infauna of Port Foster resulting from the eruptions in Deception Island (South Shetland Islands). *Publ. Inst. Antarct. Chile* 16, 1 – 11.

Gallardo, V.A. and Castillo, J.G. (1969) Quantitative benthic survey of the infauna of Chile Bay (Greenwich Is., S. Shetland Is.). *Gayana Zool.* 16, 1 – 18.

Gallardo, V.A., Castillo, J.G., Retamal, M.A., Yáñez, A., Moyano, H.I. and Hermosilla, J.G. (1977) Quantitative studies on the soft-bottom macrobenthic animal communities of shallow Antarctic bays. In *Adaptations within Antarctic ecosystems. Proc. 3rd. SCAR Symp. Antarct. Biol.* Ed. G.A. Llano, pp. 361 – 87. Gulf Publishing Company, Houston, Texas.

Gruzov, Y.N. (1977) Seasonal alterations in coastal communities in the Davis Sea. In *Adaptations within Antarctic ecosystems. Proc. 3rd. SCAR Symp. Antarct. Biol.* Ed. G.A. Llano, pp. 263 – 78. Gulf Publishing Company, Houston, Texas.

Gruzov, Y.N., Propp, M.V. and Pushkin, A.F. (1967) Biological associations of coastal areas of the Davis Sea, based on the observations of divers. *Inf. Byull. sov. antarkt. Eksped.* 65, 125 – 141. (In Russian). English translation (1968); *Sov. Ant. Exp. Inform. Bull.* 6(6), 523 – 33.

Hardy, P. (1972) Biomass estimates for some shallow-water infaunal communities at Signy Island, South Orkney Islands. *Bull. Br. Antarct. Surv.* 31, 93 – 106.

Hargens, A.R. and Shabica, S.V. (1973) Protection against lethal freezing temperatures by mucous in an Antarctic limpet. *Cryobiology* 10, 331 – 7.

Hedgpeth, J.W. (1969) Introduction to Antarctic zoogeography. In *Distribution of selected groups of marine invertebrates in waters south of 35°S latitude. Antarctic Map Folio Series, Folio 11.* Co-edited by J.W. Hedgpeth, pp. 1 – 9. American Geophysical Society, New York.

Horne, A.J., Fogg, G.E. and Eagle, D.J. (1969) Studies *in situ* of the primary production of an area of inshore Antarctic sea. *J. mar. biol. Ass. U.K.* 49, 393 – 405.

Houlihan, D.F. and Allan, D. (1982) Oxygen consumption of some Antarctic and British gastropods: an evaluation of cold adaptation. *Comp. Biochem. Physiol.* 73, A(3), 383 – 7.

Knox, G.A. and Lowry, J.K. (1977) A comparison between the benthos of the Southern Ocean and the north polar ocean with special reference to the Amphipoda and the Polychaeta. In *Polar Oceans. Proceedings of the Polar Oceans Conference, Montreal, May, 1974,* Ed. M.J. Dunbar, pp. 423 – 62. Arctic Institute of North America.

Lowry, J.K. (1975) Soft bottom macrobenthic community of Arthur Harbour, Antarctica. *Ant. Res. Ser.* 23(1), 1 – 19.

Mills, E.L. (1975) Benthic organisms and the structure of marine ecosystems. *J. Fish. Res. Bd. Can.* 32, 1657 – 63.

Moe, R.L. and Silva, P.C. (1977) Antarctic marine flora, uniquely devoid of kelp. *Science* 196, 1206 – 8.

Nicol, D. (1967) Some characteristics of cold-water marine pelecypods. *J. Paleont.* 41, 1330 – 40.

Picken, G.B. (1979) Growth, production and biomass of the Antarctic gastropod *Laevilacunaria antarctica* Martens 1885. *J. exp. mar. Biol. Ecol.* 40(1), 71 – 9.

Platt, H.M. (1978) Assessment of the macrobenthos in an Antarctic environment following recent pollution abatement. *Mar. Pollut. Bull.* 9(6), 149 – 53.

Platt, H.M. (1980) Ecology of King Edward Cove, South Georgia, I. Macrobenthos and the Benthic Environment. *Bull. Br. Antarct. Surv.* 49, 231 – 8.

Propp, M.V. (1970) The study of bottom fauna at Haswell Islands by SCUBA diving. In *Antarctic Ecology,* Ed. M.W. Holdgate, pp. 239 – 41. Academic Press, London and New York.

Rakusa-Suszcsewski, S. (1972) The biology of *Paramoera walkeri* Stebbing (Amphipoda) and the Antarctic sub fast-ice community. *Pol. Arch. Hydrobiol.* 19, 11 – 36.

Richardson, M.D. and Hedgpeth, J.W. (1977) Antarctic soft-bottom, macrobenthic community adaptations to a cold, stable, highly productive glacially affected environment. In *Adaptations within Antarctic Ecosystems. Proc. 3rd SCAR ymp. Antarct. Biol.,* Ed. G.A. Llano, pp. 181 – 96. Gulf Publishing Company, Houston, Texas.

Richardson, M.G. (1979a) The distribution of Antarctic marine macro-algae related to depth and substrate. *Bull. Br. Antarct. Surv.* 49, 1 – 13.

Richardson, M.G. (1979b) The ecology and reproduction of the brooding Antarctic bivalve *Lissarca miliaris. Bull. Br. Antarct. Surv.* 49, 91 – 115.

Shabica, S.V. (1971) The general ecology of the Antarctic limpet *Patinigera polaris. Antarct. J.U.S.* 6(5), 160 – 2.

Stockton, W.L. (1973) An intertidal assemblage at Palmer Station. *Antarct. J.U.S.* 8, 305 – 7.

Thorson, G. (1950) Reproductive and larval ecology of marine bottom invertebrates. *Biol. Rev.* 25(1), 1 – 45.

Walker, A.J.M. (1972) Introduction to the ecology of the Antarctic limpet, *Patinigera polaris* (Hombron & Jacquinot) at Signy Island, South Orkney Islands. *Bull. Br. Antarct. Surv.* 28, 49 – 71.

Whitaker, T.M. (1977) Sea ice habitats of Signy Island (South Orkney Islands) and their primary production. In *Adaptations within Antarctic Ecosystems. Proc. 3rd. SCAR Symp. Antarct. Biol.,* Ed. G.A. Llano, pp. 75 – 82. Gulf Publishing Company, Houston, Texas.

White, M.G. and Robins, M.W. (1972) Biomass estimates from Borge Bay, Signy Island, South Orkney Islands. *Bull. Br. Antarct. Surv.* 31, 45 – 50.

Zaneveld, J.S. (1968) Benthic marine algae, Ross Island to Balleny Islands. *Antarctic Map Folio Series, Folio 10,* pp. 10-12. American Geophysical Society, New York.

# CHAPTER 4c

# Marine Habitats — Antarctic Fish

Bundesforschungsanstalt für Fischerei, Institut für Seefischerei, 2000 Hamburg 50,
Federal Republic of Germany

## CONTENTS

## 4c.1. COMPOSITION AND ORIGIN OF THE ANTARCTIC FISH FAUNA

Since the first fishes from the Antarctic were collected and reported on by Sir James Clark Ross's Expedition (1839 – 1843) (Andriashev, 1976; Richardson, 1844), about 200 deep sea and shallow water bottom dwelling and pelagic fish species have been recorded from south of the Antarctic Convergence

TABLE 4c.1. Number of Genera and Species in Each Family of Antarctic Coastal Fishes

| Family | | Genera | Species |
|---|---|---|---|
| Myxinidae | (hag fish) | 1 | 1 |
| Geotriidae | (lampreys) | 1 | 1 |
| Rajidae | (skates) | 2 | 6 |
| Muraenolepidae | (eel cod) | 1 | 2 |
| Gadidae | (true cod) | 1 | 1 |
| Bovichthyidae | | 1 | 1 |
| Nototheniidae | (Antarctic cod) | 7 | 30* |
| Harpagiferidae | (plunder fish) | 4 | 23 |
| Bathydraconidae | (dragon fish) | 10 | 15 |
| Channichthyidae | (ice or white-blooded fish) | 11 | 16 |
| Congiopodiae | (horse fish) | 1 | 1 |
| Bothidae | (left eye flounders) | 2 | 2 |
| Zoarcidae | (eel pouts) | 7 | 14+ |
| Liparidae | (snail fish) | 4 | 12+ |
| Tripterygiidae | | 1 | 1 |

* The *N. coriiceps-neglecta* group, the *N. larseni* group, *Trematomus bernachii* and *T.b. vicarius* have been considered as representing one single species each.
+ Some of the species (*Notoliparis* already excluded) are deep sea rather than coastal forms.

and isolated islands lying on or near it. This frontal zone has created a marked effect on the evolution and composition of shallow water or coastal living species, whereas it forms no distinct barrier for the distribution of the deep water fish fauna. As a result the relative number of deep sea endemic species is about 25% (probably less) while more than 85% of the coastal species are endemic.

## 4c.1.1 The bottom fish fauna

The bottom, or demersal, fish fauna can be divided into two groups with respect both to their habitats and relationships: the deep sea species inhabiting the continental slope and the deep sea trenches, and the coastal species living on the continental shelf. This simple vertical distributional pattern which is found along almost any coast fronting the open ocean is complicated by the presence of local innershelf depressions especially in Greater Antarctica (usually 400 – 1200 m deep) separated by shallower sills from the continental slope (Andriashev, 1977). Due to the depression of the entire Antarctic Continent and the associated shelf region by accumulated ice during the Pleistocene the shelf is unusually deep, averaging 500 – 750 m compared with the world average of about 132 m. Because of this many typical shelf species have the aspect of deep water fishes. The two faunas seem to overlap about the continent proper, the edge of the continental shelf and the upper part of the continental slope.

The deep sea fauna consists of members of the families Macrouridae (rattails), Eretmophoridae (deep sea cod), Synaphobranchidae (deep sea eels), Halosauridae, Brotulidae, Notacanthidae (spiny-backed eels) and some Liparidae (snail fish), Zoarcidae (eel pouts) and Bathydraconidae (dragon fishes). They have been little studied but most of them are known to be common and widespread in deep water and are not endemic to the Antarctic.

The remaining species may be termed coastal fishes although some are known from depths as great as the deep sea forms. They comprise a diverse group including representatives of fifteen families. Four of these families, the Nototheniidae, Harpagiferidae, Bathydraconidae and Channichthyidae, all belonging to the suborder Notothenioidei, include more than 60% of the species (see Table 4c.1) and over 90% of the individuals. They are by far the best studied group of Antarctic fishes. Most of the physiological and biological characteristics given in the following chapters refer to notothenioids.

The Notothenioidei, which are taxonomically placed within the perciform ('spiny-rayed') fishes, form a group of about 100 species with a large variety of ecomorphological types (Fig. 4c.1). They are adapted to nearly all habitats from shallowest tidal pools to the upper continental slope down to 2000 m depth. Eighty-four species are known from the Antarctic. Their diversity in habit and structure together

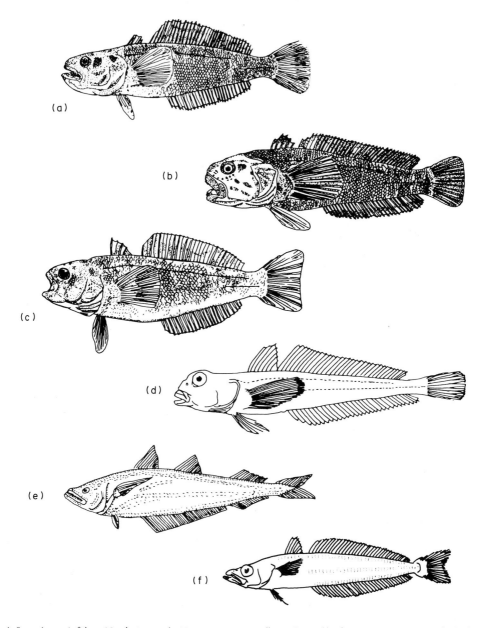

Fig. 4c.1. Some Antarctic fish. a. *Notothenia rossii*; b. *N. coriiceps*; c. *N. magellanica*; d. *N. gibberifrons*; e. *Micromesistius australis*; f. *Pleurogramma antarcticum*;

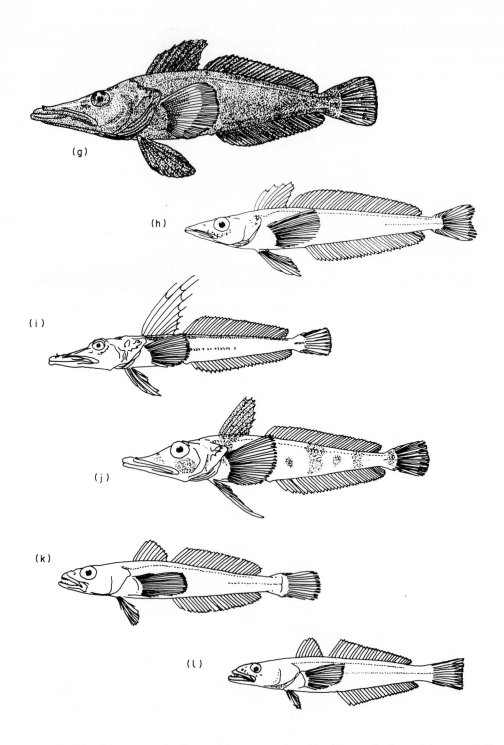

Fig. 4c.1. *cont.* g. *Pseudochaenichthys georgianus*; h. *Champsocephalus gunnari*; i. *Channichthys rhinoceratus*; j. *Chaenocephalus aceratus*; k. *Dissostichus mawsoni*; l. *D. eleginoides*.

with their southern distribution argue strongly for a prolonged isolated evolution in the cool waters of the southern hemisphere. The presence of a cool Southern Ocean as early as the Eocene (about 40 million years ago) gave ample time for the development of such a peculiar cold-adapted fish fauna. During the early Miocene (about 25 million years ago) the Antarctic Convergence was confined to Antarctic high latitudes or perhaps did not exist. The cool water fauna was therefore restricted to the immediate coasts of the Antarctic Continent. The cooling during the Pliocene (about 15 million years ago) resulted in a northward moving of the Convergence to about 40°S, which allowed the northward migration of the cold water fauna and may explain the presence of nototheniods on the Patagonian and the New Zealand shelf area. With post-Pleistocene warming the Convergence moved south again to its present position.

All other families (see Table 4c.1) are not specifically Antarctic and are much less important both in terms of numbers of species and of individuals. Except for the eel cod and the horse fish all belong to families widely distributed over the oceans. They have either penetrated into the region a considerable time ago mainly from the Patagonian — Falkland area and formed endemic species, such as all skates (except *Bathyraja griseocauda*), the eel pouts and the snailfishes, or are probably recent immigrants to the peripheral parts of the Antarctic, for instance the left-eye flounder *Mancopsetta maculata*, and have not yet established endemic forms. The southern blue whiting *Micromesistius australis* (Gadidae) only undertakes feeding migrations from the Patagonian shelf to the Scotia Sea and adjacent islands (Andriashev, 1965; De Witt, 1971).

## 4c.1.2. The pelagic fish fauna

Except for the opah *Lampris guttatus* and the porbeagle *Lamma nasus*, which are obviously occasional invaders from the north, (Kock *et al.*, 1980; Prutko, 1979; Svetlov, 1978) there are no true epipelagic families (i.e. families which are confined more or less to the Antarctic Surface Water during their life cycle) present in the Antarctic. Antarctic pelagic fishes are descendants from several faunal groups of different origins: bathypelagic species, mesopelagic species and species originating from demersal Antarctic families which are secondarily adapted to permanent or temporary midwater life.

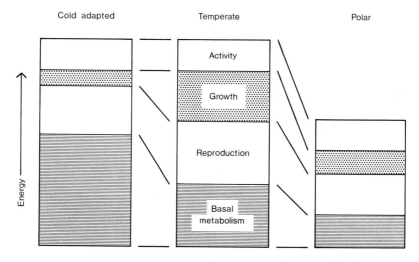

Fig. 4c.2. Diagram of the old and new concepts of 'cold adaptation' in relation to a temperate waters species (after Clarke, 1980).

The bathypelagic fauna, living below the limit of light, consists of a diverse group of families: Argentinidae (argentines), Astronesthidae (snaggletooths), Melamphaidae (bigheads), Macrouridae (rattails), Myctophidae (lantern fishes), Bathylagidae (deep sea smelts), Gonostomatidae (bristlemouths), Scopelarchidae (pearleyed fish), Cetomimidae (whale fish) and Ceratiidae (deep sea angler fish). Most of them are represented by only one or two species, but none of them is endemic to the Antarctic.

The mesopelagic fauna belongs to two groups: one has apparently adapted long ago to the cold waters of the Antarctic and formed a number of Sub-Antarctic − Antarctic endemic species. By far the most numerous are myctophids (*Electrona*, *Gymnoscopelus*, *Protomyctophum* and *Krefftichthys*) and the paralepidids or ionah fishes *Notolepis coatsi* and *N. annulata*. The second group is more northerly distributed but undertakes regular feeding migrations to the peripheral waters of the Antarctic: Notosudidae, Trichiuridae (hairtails), Oreosomatidae (dories), Anotopteridae (daggertooths) and some of the Myctophidae and Paralepididae (Krefft, 1976).

The secondarily pelagic forms comprise a large number of members of the families Nototheniidae, Channichthyidae, Bathydraconidae and even Harpagiferidae which either spend their first year or years in midwater often associated with drifting or fast ice or krill aggregations (see Chapter 4c.6.1) or are adapted to temporary or constant life in the pelagic zone during most of their life (e.g. the Antarctic herring *Pleuragramma antarcticum*).

## 4c.2. PHYSIOLOGICAL ADAPTATIONS

### 4c.2.1. Colourless blood

The oxygen-carrying protein haemoglobin (red blood pigment) in the circulatory system of vertebrates is in general regarded as a necessity to overcome severe limitations in size and activity. Polar fishes in comparison with temperate species show a tendency to decrease the haemoglobin content and the number of red blood cells (erythrocytes). The lack of haemoglobin in the blood of all members of the family Channichthyidae may therefore be considered as the extreme point of this evolutionary development. The Norwegian scientist J.T. Ruud who described this phenomenon in 1954 thought that erythrocytes were absent also, but more recent investigations have recorded relatively large numbers of erythrocyte-like cells which may be considered as evolutionary relicts.

In contrast to red-blooded species such as cod and herring the blood of channichthyids is nearly translucent with a yellowish tint and the gill filaments appear to be creamy white. This has led to their common names 'ice-fish' or 'white-blooded fish'.

Oxygen is transported by simple physical solution in the blood plasma. The oxygen carrying capacity is therefore only about 10% of that of nototheniids in the same environment. This 'disadvantage' misled several authors to think of white-blooded fish as sluggish, sedentary species mainly waiting for prey. In fact channichthyids appear to be a normal fish family reaching sizes of 60 − 75 cm and occupying a range of ecological niches from active swimmers hunting for prey (especially krill) in the pelagic zone, such as the South Georgian ice-fish *Pseudochaenichthys georgianus*, to more sedentary forms such as the Kerguelen icefish *Channichthys rhinoceratus*. They have developed a number of adaptations which compensate for their physiological 'disadvantage' to a large extent. The blood passages in the secondary gill lamellae are larger than those in other teleosts so the respiratory surface of the gills may act more effectively. Cutaneous respiration (especially in the highly vascularized tail section) may play a considerable role at least in juveniles. This would lose its importance as growth progresses because of the more unfavourable surface to volume ratio of the larger fish. The blood volume (8 − 9% of the total volume) is much higher than in comparable nototheniids and the heart and the thickness of cardiac muscles is twice as large as in other teleosts. The cardiac output and the stroke volumes are several times higher than in nototheniids and the vascular resistance is reduced by increasing the bore of the blood vessels and the blood viscosity is lower because of the small number of erythrocytes. The partial pressure gradient of oxygen from water to blood is similar to other fish but the gradient from blood to tissues is higher. The resting metabolism is only $30-50\%$ to that of haemoglobin-containing cold water teleosts (Hemmingsen and Douglas, 1977; Holeton, 1970; Hureau et al., 1977).

Most of these adaptations have been demonstrated during studies on *Chaenocephalus aceratus*, *Pseudochaenichthys georgianus* and *Channichthys rhinoceratus*. It is not yet known if these results are transferable to all channichthyids.

## 4c.2.2. Freezing resistance

The ionic concentration of the blood of marine teleosts is only about one third of that of sea water. Similarly to human blood, the blood of marine teleosts freezes at about $-0.5°C$, which is above the temperatures observed in high polar latitudes in winter in the Arctic or throughout the whole year in the Antarctic. Some species are able to survive these temperatures in a supercooled state if they do not come into contact with ice. In order to survive in ice-laden coastal waters polar fish have evolved macromolecules possessing unique antifreeze properties. In Antarctic notothenioids these are glycoproteins present in the coelomic and intercellular fluid throughout the year. In Arctic and northern temperate species however both glycoproteins, slightly different from those in Antarctic fishes, and peptides of varying structure have been found. They are only present in winter.

Glycoproteins account for more than half of the observed freezing point depression of the blood of Antarctic nototheniids. They are composed of the amino acids threonine and alanine and the sugars galactose and galactosamine. Glycoproteins of various sizes occur in the blood ranging from 2600 to 33,000 daltons in molecular weight. In the nototheniid *Pagothenia borchgrevinki* for example, eight glycoproteins of different sizes have been found. These different sizes can be related to the presence of different numbers of the basic repeating unit alanine $-$ alanine $-$ threonine and the disaccharide N-acetyl-galactosamine $-$ galactose.

Notothenioids do not freeze when cooled down to $-2°C$ even when they come into direct contact with ice or when ice crystals are experimentally introduced into their blood. The mechanism by which the glycoproteins lower the freezing point of water as well as preventing ice from forming in the body fluids is not yet fully understood. There are two ways in which it might occur. One is that if the polar hydrophilic groups bind the glycoproteins to the oxygens of the ice lattice, then the non-polar hydrophobic polypeptide backbone could possibly cause the water around it to become oriented in such a way that it would form an unfreezable barrier. In this way uniform ice crystal growth might be impeded. The freezing point would be the temperature at which the glycoprotein-covered ice could penetrate the layer of oriented water. The other way in which the glycoproteins may prevent ice from growing is by increasing the surface free energy. When the glycoproteins bind to the surface of an ice crystal they increase the surface free energy by dividing the surface of the crystal into many small microcrystals. Because of their higher surface free energy these have lower melting points than large crystals. This means that, in order for a microcrystal to increase in size the temperature has to be lowered substantially (about $1°C$). So glycoproteins binding to a larger ice crystal may in fact divide the surface into many smaller units analogous to microcrystals, thus lowering the freezing point (De Vries, 1977; De Vries and Lin, 1977).

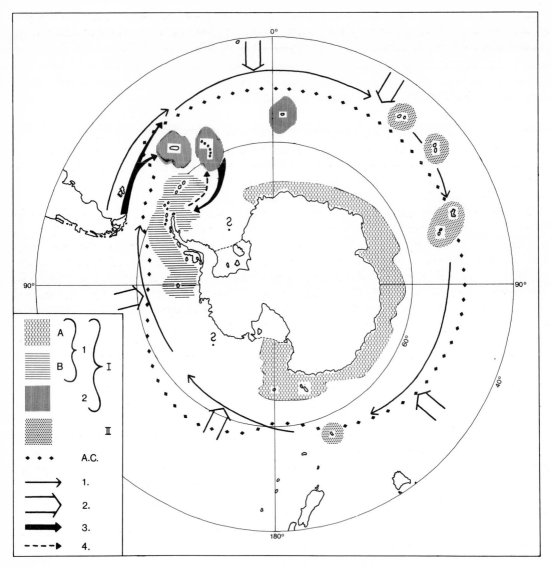

Fig. 4c.3. Zoogeographic subdivision of the Southern Ocean based on the distribution of coastal fishes and origin and dispersal of bottom fishes (modified after Andriashev, 1965). I. Glacial subregion; 1. Continental Province, A. Greater Antarctic District; B. Lesser Antarctic District; 2. South Georgian Province. II. Kerguelen Subregion; A.C. Mean position of the Antarctic Convergence; 1. West Wind drift dispersal; 2. Migration of deep sea species; 3. Migration via Scotia arc; 4. Migration to South America via Scotia arc.

## 4c.2.3. Kidney structure

In most marine teleosts glomerular kidneys are present which are able to filter out molecules up to about 40,000 daltons in size and excrete them into the urine. Antarctic waters, as compared with the remainder of the world ocean, contain a disproportionately high percentage of aglomerular species.

Urine is formed by secretion. This development of non-glomerular kidneys in nototheniids may be considered as an evolutionary adaptation to reduce glycoprotein loss to a minimum and thus save energy otherwise lost for the resorption of glycoproteins (Dobbs and De Vries, 1975).

## 4c.2.4. Cold adaptation

Temperate water teleosts are able to adjust even for large changes in their environmental temperature. This 'acclimation' needs several weeks and is observed as a common phenomenon. Fish of Antarctic high latitudes (e.g. *Trematomus, Dissostichus*), are however, only able to survive within a narrow range of temperatures and cannot tolerate temperatures beyond $+4°$ to $+6°C$. These limits do not change even after long periods of cold and warm acclimation.

Temperature is a major factor governing the rate of metabolism as well as all other chemical reactions. The rate of reaction in biochemical systems normally slows down by a factor of 2.5 per $10°C$, i.e. the difference in metabolic rate of tropical fish ($30°C$) can be expected to be nearly 20-fold that of Antarctic fish. Wohlschlag (1964), however, found the basal oxygen consumption of polar fish (Antarctic as well as Arctic) to be five to tenfold higher than expected from extrapolating temperate rates to polar temperatures using the factor 2.5. Although other species such as sea snails and eel pouts had more or less the expected rates, this elevated metabolic rate, termed 'cold adaptation', was considered as an evolutionary adaptation to overcome the rate-depressing effects of temperature. It also meant that more energy was required for maintenance and less was therefore available for growth and activity (Fig. 4c.2). Thus this concept of 'cold adaptation' offered a good explanation for the slow growth rates in many polar fish. Holeton (1974), however, demonstrated that Wohlschlag's measurements were influenced by physiological stress and thus do not reflect true metabolic rates. Corrected metabolic rates were in accordance with results from polar marine invertebrates and Wohlschlag's concept of cold adaptation may have to be abandoned.

Compensation for decreasing temperature has been recently observed in the biochemical systems of Antarctic fish. Protein synthesis rate in liver was twice, and in white muscle three times, that predicted from the theoretical temperature dependency relationship based on the factor of 2.5. The low protein synthesis rate of 6% liver protein per day compared with values of about 20% measured in tropical fish (Smith *et al.*, 1980) nevertheless indicates a reduced basal metabolic rate. Under the assumption that the energetic costs of feeding and activity are comparable to temperate water species, evidence of reduced basal metabolism, reduced growth and reduced reproductive effort leads to a new concept of cold adaptation (Clarke, 1980, Fig. 2).

## 4c.2.5. Buoyancy adaptations

Notothenioids are primarily bottom dwelling fish lacking swim bladders. Some of the nototheniids and channichthyids, however, are adapted to temporary or even permanent pelagic life thus utilizing the large food resources of the midwater zone, especially krill. Although originally heavier than sea water they evolved a number of specializations of the digestive, muscular, integumentary and skeletal system which made them nearly neutrally buoyant. In *Pleuragramma antarcticum* and *Dissostichus mawsoni*, for example, the skeleton has been found to be weakly calcified, containing a considerable amount of cartilage. Scales were also incompletely mineralized. A subcutaneous lipid layer in *Dissostichus* and subcutaneous lipid sacs in *Pleuragramma* and high lipid content in white muscle provide buoyancy (Eastman and De Vries, 1981). Lipid deposition in the white muscle and intestines as well as a high

TABLE 4c.2. Spawning Time of Antarctic Bottom Fishes

| Spring<br>November — December | Summer<br>January — February | Autumn<br>March — June | Winter<br>July — October |
|---|---|---|---|
| *Notothenia kempi*<br>(South Georgia,<br>Elephant Is.) | *Notothenia neglecta*<br>(Terre Adélie) | *Notothenia neglecta*<br>(South Georgia,<br>S. Orkney Is.,<br>Elephant Is.,<br>S. Sandwich Is.) | *Notothenia gibberifrons*<br>(South Georgia) |
| | *Notothenia squamifrons*<br>(South Georgia) | | *Dissostichus eleginoides*<br>(Patagonia) |
| *Notothenia squamifrons*<br>(Kerguelen, Crozet Is.) | | *Notothenia rossii*<br>(S. Shetland Is.,<br>South Georgia,<br>Kerguelen) | *Dissostichus mawsoni*<br>(Amundsen Sea) |
| *Trematomus bernacchii*<br>(Ant. Peninsula, Terre Adélie,<br>McMurdo Sound, Enderby-Land<br>Queen Mary Land) | | *Notothenia larseni*<br>(South Georgia, S. Orkney Is.<br>S. Sheltand Is., Crozet Is.) | |
| *Trematomus hansoni*<br>(Terre Adélie) | | *Notothenia nudifrons*<br>(South Georgia) | *Micromesistius australis*<br>(Patagonia) |
| | *Channichthys rhinoceratus*<br>(Kerguelen) | *Notothenia angustifrons*<br>(S. Sandwich Is.) | |
| | | *Notothenia magellanica*<br>(Kerguelen) | |
| | | *Notothenia cyanobrancha*<br>(Kerguelen) | |
| | | *Dissostichus eleginoides*<br>(Kerguelen, South Georgia) | |
| | | *Cryothenia peninsulae*<br>(Ant. Peninsula) | |
| | | *Pagothenia borchgrevinki*<br>(Queen Mary Land) | |
| | | *Parachaenichthys georgianus*<br>(South Georgia) | |
| | | *Psilodraco breviceps*<br>(South Georgia) | |
| | | *Harpagifer bispinis*<br>(Ant. Peninsula) | |
| | | *Champsocephalus gunnari*<br>(South Georgia, S. Orkney Is.,<br>S. Shetland Is., S. Sandwich<br>Is., Kerguelen) | |
| | | *Chaenocephalus aceratus*<br>(South Georgia, S. Orkney Is.,<br>S. Shetland Is.) | |
| | | *Pseudochaenichthys georgianus*<br>(South Georgia) | |
| | | *Chionodraco hamatus*<br>(S. Orkney Is., S. Shetland Is.) | |
| | | *Paraliparis antarcticus*<br>(S. Shetland Is.) | |
| | | *Paraliparis gracilis*<br>(South Georgia) | |

cartilage content in the ice fish *Pseudochaenichthys georgianus* and *Champsocephalus gunnari* may be interpreted in the same way. The lack of a swim bladder and the development of large pectoral fins may allow vertical migration over several hundred metres even in a short time.

## 4c.3. GEOGRAPHICAL DISTRIBUTION

### 4c.3.1. Bottom fish

About 65% of coastal fish species are confined to Antarctic high latitudes with either a circum-Antarctic or a Lesser Antarctica or Greater Antarctica type of range (Fig. 4c.3). Most of the channichthyids, harpagiferids, bathydraconids and nototheniids except the genus *Notothenia* belong to this group. Some of the species known so far only from Greater Antarctica have recently been found in the inner Weddell Sea. An increase in sampling effort in the unexplored areas of the Weddell Sea and Amundsen Sea may therefore show that more species than so far known have a circum-Antarctic range of distribution (e.g. all members of the genus *Trematomus*).

South Georgia is obviously an important transition zone between the Patagonian and Antarctic ichthyofaunas. It is characterized by immigrants from Patagonia (e.g. the left-eye flounder *Mancopsetta maculata*) as well as from the West Antarctic (e.g. the ice-fish *Chaenocephalus aceratus*) and endemic species evolved during the long geographical isolation (e.g. the dragonfish *Psilodraco breviceps*). Although separated by vast deep sea areas from the shelf zones of South Georgia and Greater Antarctica, the fish from Kerguelen show a much closer relationship to those from South Georgia than those from the Continent. About 14 species (e.g. *Dissostichus eleginoides*, *Notothenia squamifrons*) have so far been discovered off South Georgia as well as off Kerguelen. As both islands are located in the Antarctic Circumpolar Current (West Wind Drift) the prevailing easterly flow may provide an exchange between the two areas, principally from South Georgia to Kerguelen. About 16 species have been exclusively found around Kerguelen.

The vast deep sea basins and trenches between the narrow shelf zones have been poorly sampled so far. No distributional pattern of deep sea species can therefore be suggested.

### 4c.3.2. Pelagic fish

Most Antarctic mesopelagic and bathypelagic fishes presumably exhibit a circum-Antarctic distribution. Though the distribution pattern of bathypelagic species often seems to be governed by parameters other than those responsible for the distribution of mesopelagic fishes, the patterns themselves appear to correspond (Krefft, 1976). Most South Atlantic lantern fish (e.g. *Krefftichthys anderssoni*, *Gymnoscopelus nicholsi*) occur from south of the Antarctic Convergence to as far north as the Subtropical Convergence and are sexually mature throughout their geographical range. Other species spawn only south of the Antarctic Convergence; two species have been described so far belonging to this group: *Electrona antarctica* (mesopelagic) and *Gymnoscopelus opisthopterus* (bathypelagic).

## 4c.4. VERTICAL DISTRIBUTION

### 4c.4.1. Bottom fish

The vertical distribution pattern of bottom fish can be described using three groupings, although given depth limits may be somewhat artificial and a large degree of overlap occurs. The first group of species is restricted to nearshore waters from depths of 0 to about 100 m (stenobathic species; e.g.

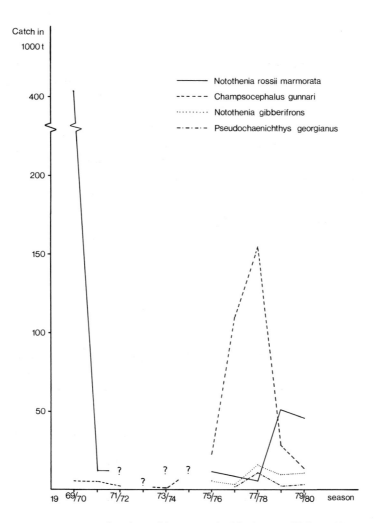

Fig. 4c.4. Catches of main fish species in the Atlantic sector (FAO area 48).

*Harpagifer* spp.). The second group, comprising the bulk of coastal species form a eurybathic group inhabiting the entire continental shelf, its inner-shelf depressions and the edge of the continental shelf. The depth range of these species is obviously affected by the depth of the shelf as a number of circum-Antarctic species are found considerably deeper in Greater Antarctica than in Lesser Antarctica. The dragonfish *Bathydraco antarcticus* and the eel pout *Lycenchelys antarcticus* have been discovered down to 2300 m and 3200 m respectively. A third group includes the fish of the continental slope and the deep sea basins and trenches. The deepest living species so far recorded are the brotulid *Holcomycteronus brucei* from 4572 m in the Weddell Sea and the liparid *Notoliparis kurchatovi* from 5474 m in the South Orkney Trench (Andriashev, 1975; De Witt, 1971).

Number of species and abundance (biomass) is greatest from 150 to 350 m off South Georgia and in parts of Lesser Antarctica, while the largest number of species in Greater Antarctica is found from 350 to 600 m, obviously due to the depression of the whole shelf area.

## 4c.4.2. Pelagic fish

The vertical distribution of mesopelagic fish is related to the Antarctic Surface Water, the derived Antarctic Intermediate Water and the Warm Deep Water, which is centred near 500 m. South of the Antarctic Convergence lantern fishes, for example, may be taken from near the surface to several hundred metres depth but are found significantly deeper northwards. Near the Subtropical Convergence they are caught only at depths of 2000 m or more (Hulley, 1981).

Information on the vertical distribution range and pattern of bathypelagic fish, especially their lower depth limits, is scarce. Vertical non-migrators are probably confined to the Warm Deep Water and its transition zone to the Antarctic Bottom Water. Some species, however, carry out extensive vertical migrations; the deep sea angler fish *Oneirodes notius* has been recently encountered in krill swarms (Abe, 1979).

Pelagic juveniles of nototheniods are obviously restricted to the Antarctic Surface Water. The frequent occurrence of adult *Dissostichus mawsoni* in sperm whale stomach contents may indicate the abundance of the species even as deep as the Warm Deep Water.

## 4c.5. REPRODUCTION

Sexual maturity in most coastal species so far studied (primarily nototheniids and channichthyids) is not reached until the fish are 5 to 7 years old. Exceptions are the ice-fish *Champsocephalus gunnari* and the Antarctic cod *Notothenia larseni*, which are already mature in their third year.

The development of mature ova in several channichthyids and nototheniids as well as in the sea snail *Paraliparis antarcticus* is a biennial process. The steady build up of eggs of about the same size indicates a spawning season of limited duration and a single spawning within a season. No intermittent spawning has been observed so far.

As in most other fish, length and weight are the major variables governing fecundity. Overall fecundity ranges from about 100 eggs in *Paraliparis antarcticus* (Lisovenko and Svetlov, 1980) to more than 500,000 in *Dissostichus eleginoides*. Relative fecundity (i.e. the number of eggs per gram body weight) is in general 4 − 30 for all channichthyids, most nototheniids, *Parachaenichthys georgianus* and *Harpagifer* spp. A higher relative fecundity of about 50 − 150 eggs is found only in several late winter and spring spawning species of *Notothenia*, in the dragon fish *Psilodraco breviceps* and in the eel cod *Muraenolepis microps*. Fecundity in different populations of one species as well as in different genera of one family normally decreases with higher latitudes (e.g. *Champsocephalus gunnari*; *Notothenia* and *Trematomus*) (Kock, 1981; Permitin and Silyanova, 1971).

Spawning migrations have been described for several species. The marbled notothenia, *Notothenia rossii rossii*, undertakes regular migrations from the feeding grounds in the northeast to the spawning area in the southeast of the Kerguelen Plateau, obviously supported by a prevailing southward flowing current (Meissner *et al.*, 1974). South Georgian ice-fish in spawning condition were caught only in nearshore waters whereas in offshore trawled specimens maturation was far less advanced, indicating that the species migrates towards the coast to spawn. Males start their spawning migration earlier than females (Kock, 1981). Several species are known to form large congregations prior to spawning.

Spawning of most species takes place in austral autumn and winter. Spring or summer spawning is only known from two species of *Trematomus*, three species of *Notothenia* and the Kerguelen ice-fish *Channichthys rhinoceratus* (Table 4c.2). The weight of the ovaries has then reached from 15% to more than 20% of the body weight. The weight of the testes is in general much less (6 − 8%).

Two reproductive strategies occur in Antarctic coastal species which may overcome postlarval mortality and reduce inter-specific competition for food. Some fish lay a few but large yolky eggs. The larvae are then larger at hatching and in an advanced developmental stage with well developed jaws and large pectoral fins. This gains certain 'advantages' in avoiding density dependent effects, such as a smaller relative food requirement, an advanced capability for swimming in search of food and for avoiding predators and a utilization of less frequent but larger plankton organisms as food. This is the case in all autumn spawning species which produce large yolky eggs of 3 – 6mm diameter and have a prolonged incubation period of up to five months over the austral winter e.g. *Harpagifer bispinis*, *Notothenia neglecta* (Daniels, 1979; White *et al.*, 1982).

Other species lay many but small eggs. The larvae are smaller, more numerous and less developed at hatching and are more vulnerable to density dependent effects. This may result in a higher postlarval mortality which may be compensated by a higher fecundity of the species. This has been observed in late winter and spring spawning species, where incubation periods of 4 – 8 weeks are suspected.

Based on the nature of the egg's chorion and the lack of oil droplets, the spawn of most species is thought to be deposited on the bottom (demersal) or to float loosely close to the sea floor (benthopelagic) where they are protected against lower surface water salinity (e.g. in fjords) and destruction by drifting or freezing ice. Eggs of *Notothenia neglecta*, for example, are free-floating but become negatively buoyant after 14 days, and then settle on the bottom (White *et al.*, 1982). Eggs of *Paraliparis gracilis* have been found to be attached to a hydroid. Direct protection by nest guarding has been described in *Harpagifer bispinis* and *Trematomus bernacchii* (Daniels, 1979; Moreno, 1980).

Mortality of the eggs due to predation by benthic feeding nototheniids and Weddell seals has been reported. First larvae of autumn spawning species off South Georgia appeared in late August, those of late winter and spring spawners in November and January respectively (Efremenko, 1979a,b).

## 4c.6. FEEDING ECOLOGY AND SIGNIFICANCE IN THE ECOSYSTEM

### 4c.6.1. Feeding ecology

In the last 20 years feeding has been investigated in a variety of Antarctic fish species (Kock, 1981; Targett, 1981). Most of these studies, however, were confined to single species and/or qualitative analysis of the diet in the austral summer. Except for krill, prey identification was often limited to higher taxonomic categories (e.g. family), so some of the studies are of little use in an interpretation of habitat and food resource partitioning. The approach given below is therefore somewhat speculative.

Being primarily bottom dwelling species, notothenioids have successfully occupied various ecological niches. Although an ecological niche is in general determined by a complex set of dimensions (e.g. habitat, prey size, prey species), a gross partitioning of the available food resources along a vertical distribution axis is already observed on a high taxonomic level: ice-fish first of all feed on pelagic organisms, especially krill, and on fish. Plunder fish and probably the little-studied dragon fish prey on motile benthic invertebrates while Antarctic cod exploit pelagic as well as benthic habitats. This simple scheme is complicated by the opportunistic feeding habits of many species, especially nototheniids, which leave their usual trophic niche to feed on krill when it is (temporarily) abundant.

Most species change their niche at least once during their life cycle. After hatching *Notothenia rossii marmorata*, for example, spend part of their first year in subsurface waters far from the coasts feeding on krill and copepods. They then migrate towards the coasts and live for the next 5 – 6 years close to the bottom in nearshore kelp-beds preying primarily on weed-dwelling amphipods. After reaching maturity they leave the inshore waters for the offshore feeding grounds at 150 – 350 m depth from where they ascend to the midwater zone to prey on krill, fish and salps.

Different pelagic and benthic feeding communities have been distinguished:

*A cryopelagic community* closely associated with drifting or fast ice in Antarctic high latitudes is formed mainly by juvenile and adult *Trematomus* which feed on ice-dwelling amphipods and copepods. The fish cling to the surface of the ice when resting (see Section 4c.2.2) and hide in ice crevices and holes when endangered. A similar ecological niche in Arctic waters is occupied by polar and glacial cod (Andriashev, 1970).

A specific *epipelagic krill-fish community* has been suspected by several authors (e.g. Rembiszewski *et al.*, 1978), as about 40 notothenioid and mesopelagic species have been encountered in association with krill. This community is built up mainly by juvenile notothenioids living (probably) permanently at the border of krill swarms, and by various ice-fish, Antarctic cod and mesopelagic species undertaking temporary but regular migrations from the bottom or deeper layers to subsurface waters to feed on krill. This vertical migration has been confirmed by echo sounder and visual observations (Basalaev and Petuchov, 1969; Olsen, 1955).

Various *nearshore* and *offshore communities* have been distinguished (Richardson, 1975; Targett, 1981). The dominant species in each community obviously partition the prey resources by vertical habitat separation. Feeding modes are consistent from community to community although specific prey groups may change. It seems likely that the species consume whatever prey is available in their particular feeding habitat. As a result, prey overlap is in general low and thus competition among the species minimized. High prey overlap is observed mainly in offshore communities where krill is at least temporarily abundant and forms the staple food of most species. Krill should therefore be considered then as a non-limited resource for fish and so interspecific competition is reduced to a minimum. Off South Georgia, for example, where krill is only temporarily

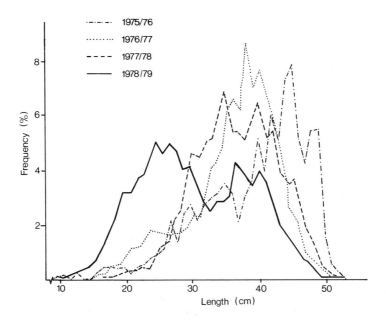

Fig. 4c.5. Length compositions of catches of the green notothenia *Notothenia gibberifrons* off South Georgia in four successive fishing seasons (compiled from Kock, unpubl. and Skora, 1980).

abundant, the ice-fish *Champsocephalus gunnari* and *Pseudochaenichthys georgianus* feed nearly exclusively on krill when it is available. When a shortage of krill occurred, as in 1977/78, *Champsocephalus gunnari* substituted part of their diet by hyperiids and mysids while *Pseudochaenichthys georgianus* increased the proportion of fish in its diet (Kock, 1981).

Benthic feeding species, such as the plunder fishes and some of the Antarctic cod generally take the same prey groups, so food analysis based on high taxonomic groups indicates extensive overlap. Fine division of the within-habitat resources, however, show that interspecific competition is lowered by taking different proportions of the same prey groups. In the morphologically-similar *Notothenia gibberifrons* and *N. nudifrons*, for example *N. gibberifrons* primarily exploits sedentary polychaetes, while *N. nudifrons* feeds on errant polychaetes (Targett, 1981).

In addition to many endemic fish preying on krill a considerable number of species migrate from north of the Convergence into Antarctic waters in summer to feed on krill. This has been observed for various mesopelagic species, e.g. the daggertooth *Anotopterus pharao*, the paralepidid *Notolepis rissoi*, the nototheniid *Notothenia magellanica* and part of the Patagonian population of the southern blue whiting *Micromesistius australis*. It gives further evidence of the non-limiting nature of krill for fish in the Antarctic.

### 4c.6.2. The significance of fish in the ecosystem

As has been demonstrated qualitatively in various studies euphausiids, especially krill, mysids, amphipods and polychaetes are important food resources for Antarctic fish. Reliable estimates of diet and annual feeding rhythms, gastric evacuation rates and biomass and productivity of fish stocks are nevertheless scarce. These data, however, are essential in order to quantify prey consumption in terms of fish populations and geographic areas and thus to understand the impact of fish on the different prey resources in the ecosystem. The crude estimates of krill consumption given below should therefore be considered as a first approach.

Based on fish stock assessments derived from 'swept area' abundance estimates, a daily food intake of about 2% of the body weight and gastric evacuation rates of about 48 hours, the krill consumption of the ice-fish *Champsocephalus gunnari* of South Georgia and in Lesser Antarctica waters was estimated to be about 4 million tonnes in 1975/76. Krill consumption of all demersal fish stocks off South Georgia was probably of the same magnitude for 1975/76 (Kock, 1981). A total of about 15 − 20 million tonnes of krill consumed by Antarctic demersal fish before the onset of commercial fishing may therefore be a realistic estimate. Their importance as krill predators seems to be less than that of seals (about 65 million tonnes), whales (about 40 million tonnes) and birds (about 35 million tonnes) (Laws, 1977; Mougin and Prévost, 1980). The significance is decreasing in recent years due to extensive commercial exploitation directed mainly at the krill feeding ice-fish *Champsocephalus gunnari* and the marbled notothenia *Notothenia rossii marmorata*.

Besides their importance as krill predators, fish themselves play at least locally a significant role in the diet of Antarctic seals and birds and also other fish. Predation of seals (Weddell, leopard, elephant) and blue-eyed shags, for example, contributes to most of the mortality observed in a population of *Notothenia neglecta* off Signy Island (Everson, 1970). The southern lamprey *Geotria australis* forms an important part of the diet of grey-headed albatros off Bird Island (Prince, 1980). The annual fish consumption in the Southern Ocean by seals and birds is estimated to be about 15 million tonnes.

## 4c.7. AGE, GROWTH AND SIZE

Age determination in Antarctic fish has been tried using various bony structures. Scales and otoliths proved to be more or less successful in a number of nototheniids, revealing maximum ages of 9 – 10 years in smaller species such as *Notothenia larseni* and *Trematomus hansoni* and of 20 – 22 years in larger ones such as the Patagonian toothfish (*Dissostichus eleginoides*) and the marbled notothenia (Frolkina, 1977; Scherbick, 1976). Cross sections of the first pelvic fin ray have been recently shown to give reliable estimates of age at least in the ice-fish *Chaenocephalus aceratus* from South Georgia (Gubsch, 1980). Maximum ages in the ice-fish *Champsocephalus gunnari*, *Chaenocephalus aceratus* and *Pseudochaenichthys georgianus* derived from otolith and fin ray cross section readings were 13 – 15 years.

Growth rates tend to be slower than in temperate water species. They are comparable to those of Newfoundland cod living at temperatures of 1.5 – 2°C. Off South Georgia, for example, various ice-fish and Antarctic cod exhibit growth rates of 12 – 15% of their maximum attainable length per year before the onset of sexual maturity. In the nototheniids *Trematomus bernacchii* and *T. hansoni* from McMurdo Sound three-quarters of the protein synthesis of white muscle under normal summer feeding conditions (see Section 4c.2.4) is directed towards growth. If the effective growing season is restricted to four months the observed daily growth rates of 0.15% of body weight correspond well with average protein synthesis rates of 0.17% per day (Smith and Haschemeyer, 1980).

Although experiments indicate that there is no absolute temperature dependence of protein synthesis in polar fish, observed rates are much slower than in tropical or temperate water species (Haschemeyer, 1980). It seems therefore possible that a reduced basal metabolism, which may be evolutionarily advantageous because of the reduced energy requirement, brings with it a reduced growth rate, as presumably the same set of ribosomal systems is used for both basal and growth protein synthesis (see Section 4c.2.4).

Small species up to 40 cm predominate in the Antarctic fish fauna. Around Lesser Antarctica, South Georgia and Iles Kerguelen more than 15 species attain sizes of more than 40 cm whereas from Greater Antarctica no more than 6 species are known to exceed 40 cm. It is interesting to note that about 50% of these species are ice-fish, a family possessing neither functional erythrocytes nor any respiratory pigment. By far the biggest species are the Patagonian and Antarctic toothfish (*Dissostichus eleginoides*, *D. mawsoni*). Maximum sizes of 206 cm (76 kg) and 174 cm (about 70 kg) respectively have been recorded.

## 4c.8. EXPLOITATION

Attempts to exploit Antarctic fish stocks commercially date back to the beginning of this century. Vanhoeffen, the biologist of the German 'Gauss' Expedition (1901 – 1903) hoped to find commercially utilizable fishes in the Southern Ocean. He proposed to increase the fish harvest from Kerguelen waters by introducing new species from the islands of St. Paul and Amsterdam. His idea has recently been taken up again by introducing trout to rivers on Iles Kerguelen and by the suggestion of seeding South Chilean rivers with Pacific salmon from whence they could undertake feeding migrations to waters south of the Antarctic Convergence. At the same time as Vanhoeffen's ideas, whalers discovered large concentrations of the marbled notothenia off South Georgia. A considerable number of barrels of salted fish were sold to Buenos Aires and the Norwegian biologist Sørling mentioned the possibility of establishing an important fishery off the island. Due to the abundance of whales and the distance to a market his words were soon forgotten. Repeated fishing trials with purse seines and small bottom trawls by Argentine, Norwegian and Japanese companies in the thirties, fifties and sixties failed.

Exploratory fishing by the Soviet Union in the Scotia Sea during the mid-sixties developed into commercial fishing off South Georgia in the late sixties. The main target species was the marbled notothenia. Catches soon reached a maximum of 403,000 tonnes in the season 1969/70, which was followed by a rapid decline to a few thousand tonnes in the following years (Fig. 4c.4). This can be attributed to a lower fishing effort (e.g. fewer vessels) and a sharp decline in catch per unit effort (Everson, 1977). A repeated increase in fishing effort from 1975/76 onwards, the entry of Poland, the German Democratic Republic and Bulgaria into the fishery since 1976/77 and the extension of the fishery to other shelf areas of the Scotia Arc led to a rapid build up of catches up to more than 150,000 tonnes in the following two seasons. Catches consisted then mainly of ice-fish (*Champsocepalus gunnari*) and also marbled notothenia in 1978/79 (Fig. 4c.4).

As in the Scotia Sea, the commercial exploitation of fish stocks on the Kerguelen Plateau by the USSR, which started in 1970, was preceded by three years of semi-commercial fishing. The progress of the fishery was similar to that in the Atlantic sector. A rapid build up of the catches in 1971 with 229,500 tonnes (about 65% marbled notothenia) was followed by low catches of 12 − 13,000 tonnes in the next two years. A second and third maximum with 101,000 and 90,000 + M = tonnes respectively in 1974 and 1977 were followed by years of low catches. Main target species were now the ice-fish *Champsocephalus gunnari* and the grey notothenia *Notothenia squamifrons*, which made up 75 − 85% of the catches. In 1978 the French Government established a 200 nautical mile exclusive economic zone around Iles Kerguelen. The fishing fleet of the Soviet Union was limited to a maximum of 7 vessels with a catch quota of 35,000 tonnes within 18 months. In 1981 three French deep sea trawlers were also fishing in the zone (Duhamel and Hureau, 1981).

The assessment of the impact of fishing and the actual status of the fished stocks is hampered by the almost complete lack of fishery related information (catch and effort data grouped by fishing grounds and months, length and age compositions of the species in the catches and the amount of discards) at least in the Atlantic part of the Southern Ocean.

Standing stock biomass of demersal fish has been estimated to be 130,000 tonnes for Kerguelen (Hureau, 1979) and about 500,000 tonnes for South Georgia (Everson, 1977). Corresponding values of maximum sustainable yields were 20,000 and 50,000 tonnes respectively. Although both predictions are probably underestimates, they may indicate the order of magnitude. Reported catches of even single species (Fig. 4c.4) greatly exceed those values.

Due to a relatively high age at first maturity and low growth rates at least of mature specimens, Antarctic fish are vulnerable to overfishing even at low levels of fishing activity. The green notothenia (*Notothenia gibberifrons*) off South Georgia, for example, is a typical by-catch species. Catches rarely exceeded ten thousand tonnes (Fig. 4c.4). The fish reach a maximum size of 52 cm when they are about 20 years old. Most specimens become sexually mature at about 32 − 35 cm (6 − 8 years old, Boronin and Frolkina, 1976). Length compositions of the catches in four successive fishing seasons (Fig. 4c.5) show a rapid decline of large (i.e. old) specimens from 1975/76 onwards when the stock was obviously in a quasi-virgin state. Mean lengths in the catches decreased from 41 cm in 1975/76 to about 30 cm in 1978/79 which is about the length at first maturity. Similar results have been obtained for the ice-fish *Champsocephalus gunnari* off South Georgia by comparing research vessel data from 1975/76 and 1977/78 (Kock, 1981).

The lack of any mesh size regulations and the availability of both juveniles and adults of several species on the fishing grounds increases their vulnerability to recruitment overfishing, i.e. part of the stock is already exploited prior to the onset of sexual maturity. This may be indicated by the growing proportion of juveniles in the catches of the green notothenia (especially in 1978/79, Fig. 4c.5). Catches of the Patagonian toothfish *Dissostichus eleginoides*, for example, consisted of more than 90% juveniles.

Although published data allow only fragmentary analyses it seems quite obvious that the large catches represented the removal of accumulated stocks and that only a relatively small annual catch can be sustained. An international agreement for an effective management of the fish stocks in the Atlantic sector is therefore urgently needed. A limit on the number of vessels fishing, as in the Kerguelen area, may be better than to set a Total Allowable Catch. The situation of the fish stocks on the Kerguelen Plateau may have improved since the establishment of the exclusive economic zone and its concomitant regulations.

# REFERENCES

Abe, T. (1979) A record of the ceratioid anglerfish *Oneirodes notius* Pietsch caught along with the Antarctic Krill, *Bull. Biogeogr. Soc. Japan* 34, 1−3.

Andriashev, A.P. (1965) A general review of the Antarctic fish fauna. In *Biogeography and Ecology in Antarctica*, Monographiae Biologicae 15, Eds P. van Oye and J. van Mieghem, pp.491−550. Junk, The Hague.

Andriashev, A.P. (1970) Cryopelagic fishes of the Arctic and Antarctic and their significance in polar ecosystems. In *Antarctic Ecology* Vol.1, Ed. M.W. Holdgate, pp.297−304. Academic Press, London.

Andriashev, A.P. (1975) New radal fish *Notoliparis kurchatooi* gen. et sp.n. (Liparidae) from the South Orkney Trench (Antarctica) (in Russian). *Trudy Inst. Okeanol.* 103, 313−19.

Andriashev, A.P. (1976) On the first fishes from the Antarctic collected by J.C. Ross' expedition and some problems of marine cryobiology: 1. Identification of *Sphyraena* and *Notothenia phocae* (in Russian). *Zool. Zhurn.* 55(6), 866−78.

Andriashev, A.P. (1977) Some additions to schemes of the vertical zonation of marine bottom fauna. In *Adaptations within Antarctic Ecosystems*, Ed. G.A. Llano, pp.351−60. Smithsonian Institution, Washington D.C.

Basalaev, V.N. and Petuchov, A.G. (1969) Experimental poutassou fishing in the Scotia Sea from the research factory ship 'Academik Knipovich' (in Russian). *Trudy VNIRO* 66, 307−10.

Boronin, A.V. and Frolkina, Zh.A. (1976) Age determination in the green notothenia (*Notothenia gibberifrons* Lönnb.) from the Southwest Atlantic (in Russian). *Trudy Atlant. NIRO* 70, 29−37.

Clarke, A. (1980) A reappraisal of the concept of metabolic cold adaptation in polar marine invertebrates. *Biol. J. Linn. Soc. Lond.* 14, 77−92

Daniels, R.A. (1979) Nesting behaviour of *Harpagifer bispinis* in Arthur Harbour, Antarctic Peninsula. *J. Fish Biol.* 12(5), 465−74.

DeVries, A.L. (1977) The physiology of cold adaptation in polar marine poikilotherms. In *Polar Oceans*, Ed. E.J. Dunbar, pp.409−22. Arctic Institute of North America, Calgary, Alberta, Canada.

DeVries, A.L. and Lin, Y. (1977) The role of glycoprotein antifreezes in the survival of Antarctic fishes. In *Adaptations within Antarctic Ecosystems*, Ed. G.A. Llano, pp.439−58. Smithsonian Institution, Washington D.C.

De Witt, H.H. (1971) Coastal and deep-water benthic fishes of the Antarctic. *Antarctic Map Folio Series* 15, 1−10. American Geophysical Union, Washington.

Dobbs, G.H. and De Vries, A.L. (1975) The aglomerular nephron of Antarctic teleosts: a light and electron microscopic study. *Tissue Cell.* 7(1), 159−70.

Duhamel, G. and Hureau, J.C. (1981) La situation de la pêche aux îles Kerguelen en 1981. *La Pêche maritime* 5, 1−8.

Eastman, J.T. and De Vries, A.L. (1981) Buoyancy adaptations in a swimbladderless Antarctic fish. *J. Morph.* 167, 91−102.

Efremenko, V.N. (1979a) The larvae of six species of the family Nototheniidae from the Scotia Sea. *J. Ichthyol.* 19(6), 95−104.

Efremenko, V.N. (1979b) Descriptions of larvae of six species of the family Chaenichthyidae from the Scotia Sea. *J. Ichthyol.* 19(3), 65−75.

Everson, I. (1970) The population dynamics and energy budget of *Notothenia neglecta* Nybelin at Signy Island, South Orkney Islands. *Bull. Br. Antarct. Surv.* 23, 25−50.

Everson, I. (1977) The living resources of the Southern Ocean. *FAO GLO/SO/77/1*, pp.1−156, Rome.

Frolkina, Zh.A. (1977) A method for age determination in Patagonian toothfish (in Russian). *Trudy Atlant. NIRO* 73, 86−93.

Gubsch, G. (1980) Untersuchungen zur Altersbestimmung und zum Wachstum beim Eisfisch *Chaenocephalus aceratus* (Lönnberg). *Fischereiforschung* 18(1), 7−10.

Haschemeyer, A.E.V. (1980) Temperature effects on protein metabolism in cold-adapted fishes. *Antarct. J. U.S.* 15, 147−9.

Hemmingsen, E.A. and Douglas, E.L. (1977) Respiratory and circulatory adaptations to the absence of hemoglobin in chaenichthyid fishes. In *Adaptations within Antarctic Ecosystems*. Ed. G.A. Llano, pp.479−87. Smithsonian Institution, Washington D.C.

Holeton, G.F. (1970) Oxygen uptake and circulation by a hemoglobinless Antarctic fish *Chaenocephalus aceratus* Lönnberg) compared with 3 red-blooded Antarctic fish. *Comp. Biochem. Physiol.* 34(2), 454−71.

Holeton, G.F. (1974) Metabolic cold adaptation of polar fish: fact or artefact? *Physiol. Zool.* 47(3), 137−52.

Hulley, P.A. (1981) Results of the research cruises of FRV "Walthe Herwig" to South America. LVIII. Family Myctophidae (Osteichthyes, Myctophiformes). *Arch. FischWiss.* 31 (Beih.1), 1−300.

Hureau, J.C. (1979) La faune ichthyologique du secteur indien de l'ocean Antarctique et estimation du stock de poissons autour des îles Kerguelen. *Mem. Mus. natn. Hist. nat.* 43, 235−47.

Hureau, J.C., Petit, D., Fine, J.M. and Marneux, M. (1977) New cytological, biochemical and physiological data on the colorless blood of the Channichthyidae (Pisces, Teleosteans, Perciformes). In *Adaptations within Antarctic Ecosystems*. Ed. G.A. Llano, pp.459 – 77, Smithsonian Institution, Washington D.C.

Kock, K.-H. (1981) Fischereibiologische Untersuchungen an drei antarktischen Fischarten: *Champsocephalus gunnari* Lönnberg 1905, *Chaenocephalus aceratus* (Lönnberg, 1906) und *Pseudochaenichthys georgianus* Norman, 1937 (Notothenioidei, Channichthyidae). *Mitt. Inst. Seefisch. Hamburg* 32, 1 – 226.

Kock, K.-H., Nast, F. and Stein, M. (1980) Some interactions between hydrography, krill and fish near Elephant Island in 1977/78. *ICES C.M. 1980* L:3 (mimeogr.).

Krefft, G. (1976) Distribution patterns of oceanic fishes in the Atlantic Ocean. *Rev. Trav. Inst. Pêches marit.* 40(3 – 4), 439 – 60.

Laws, R.M. (1977) Seals and whales of the Southern Ocean. *Phil. Trans. R. Soc. Lond. B* 279, 81 – 96.

Lisovenko, L.A. and Svetlov, M.F. (1980) Some data on the biology of *Paraliparis* (Family Liparidae) off the South Shetlands. *J. Ichthyol.* 20(1), 131 – 5.

Meissner, E.E., Tom, V.S. and Silon, V.N. (1974) Characteristics of the behavior and biological cycles of the marbled Notothenia, *Notothenia rossi*, in relation to bottom topography, bottom materials and currents. *J. Ichthyol.* 14(4), 706 – 8.

Moreno, C.A. (1980) Observations on food and reproduction in *Trematomus bernacchii* (Pisces, Nototheniidae) from the Palmer Archipelago. *Copeia 1980* (1), 171 – 3.

Mougin, J.L. and Prévost, J. (1980) Evolution annuelle des effectifs et des biomasses des oiseaux antarctiques. *Rev. Ecol. (Terre Vie)* 34, 101 – 33.

Olsen, S. (1955) A contribution to the systematics and biology of Chaenichthyid fishes from South Georgia. *Nytt. Mag. Zool. Oslo* 3, 79 – 93.

Permitin, Yu. Ye. (1977) Species composition and zoogeographical analysis of the bottom fish fauna of the Scotia Sea. *J. Ichthyol.* 17(5), 710 – 26.

Permitin, Yu. Ye. and Silyanova, Z.S. (1971) New data on the reproductive biology and fecundity of fishes of the genus Notothenia (Richardson) in the Scotia Sea (Antarctica). *J. Ichthyol.* 11, 693 – 705.

Prince, P.A. (1980) The food and feeding ecology of grey-headed albatros *Diomedea chrysostoma* and black-browed albatros *D. melanophris*. *Ibis* 122, 476 – 88.

Prutko, V.G. (1979) On the occurrence of the opah *Lampris guttatus* (Osteichthyes, Lampridae) in the southern part of the Indian Ocean. *J. Ichthyol.* 19(4), 140 – 1.

Rembiszewski, J.M., Krzeptowski, M. and Linkowski, T.B. (1978) Fishes (Pisces) as by-catch in fisheries of krill *Euphausia superba* Dana (Euphausiacea, Crustacea). *Pol. Arch. Hydrobiol.* 25, 677 – 95.

Richardson, J. (1844) The zoology of the voyage of HMS 'Erebus' and 'Terror' under the command of Capt. Sir James Clark Ross during the years 1839 to 1843. *II. Ichthyology*, 139pp., London.

Richardson, M.G. (1975) The dietary composition of some Antarctic fish. *Bull. Br. Antarct. Surv.* 41/42, 113 – 20.

Scherbig, L.V. (1976) Length age composition and growth of *Notothenia rossii marmorata* (in Russian). *Trudy Atlant. NIRO* 65, 151 – 9.

Skora, K.E. (1980) Changes in the composition of *Notothenia gibberifrons* Løonnberg population of the shelf of South Georgia in the years 1977 – 1979. *Pol. Polar Res.* 1(4), 155 – 62.

Smith, M.A.K. and Haschemeyer, A.E.V. (1980) Protein metabolism and cold adaptation in Antarctic fishes. *Physiol. Zool.* 53(4), 373 – 82.

Smith, M.A.K., Mathews, R.W., Hudson, A.P. and Haschemeyer, A.E.V. (1980) Protein metabolism of tropical reef and pelagic fish. *Comp. Biochem. Physiol.* 65B, 415 – 18.

Svetlov, M.F. (1978) The porbeagle *Lamna nasus* in Antarctic waters. *J. Ichthyol.* 18(5), 850 – 1.

Targett, T.E. (1981) Trophic ecology and structure of coastal Antarctic fish communities. *Mar. Ecol. Prog. Ser.* 4, 243 – 63.

White, M.G., North, A.W., Twelves, E.L. and Jones, S. (1982) Early development of *Notothenia neglecta* from the Scotia Sea, Antarctica. *Cybium 3e ser.* 6(1), 43 – 51.

Woheschlag, D.E. (1964) Respiratory metabolism and ecological characteristics of some fishes in McMurdo Sound, Antarctica. *Antarctic Research Series* 1, 33 – 62.

# CHAPTER 4d

# Marine Habitats — Antarctic Cephalopods

## MALCOLM R. CLARKE

Marine Biological Association of the U.K., Citadel Hill, Plymouth, U.K.

## CONTENTS

## 4d.1. INTRODUCTION

Squids, cuttlefish, octopuses and the pearly *Nautilus* all belong to the molluscan class Cephalopoda. Many millions of years ago they took off from the sea bottom by the development of buoyant chambers in their shells and a new form of locomotion, jet propulsion. Ever since this early leap from the two-dimensional into the three-dimensional world, buoyancy and jet propulsion have been all important in the subsequent evolution of cephalopods. Two major groups, the ammonites and the nautiloids, evolved vigorously in many directions to produce numerous diversely-shaped shells which, except for the single genus *Nautilus*, only survive as fossils. The shells became buoyancy chambers containing a partial vacuum to compensate for the dense body of the animals (Denton *et al.*, 1973). While the shells gave buoyancy as well as providing protection to the animals, they were cumbersome for swimming and limited the depth to which the animals could go since high pressures at great depth would crush the shells. Streamlining of the shells gave rise to the belemnites, the bullet rocks of Carboniferous and later fossil strata, and this allowed the jet propulsive apparatus to become much more efficient.

Further evolution to the present day resulted in a reduction and lightening of the shells to give the cuttlefishes with their internal buoyant shells, the squids with their shells reduced to thin, non-buoyant chitinous stiffening 'pens', resembling feathers more than shells, and the octopuses often with no vestige of shells at all. Cuttlefish are limited to the shallow waters of the continental shelves but loss of the buoyancy function of the shell has permitted the colonization of the deepest oceans by squid and

193

194                                                              M. R. CLARKE

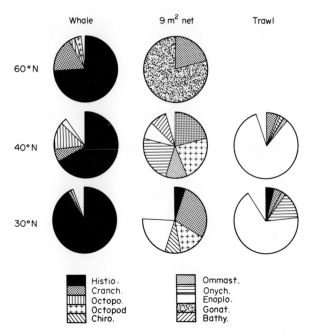

Fig. 4d.1. The families of oceanic cephalopods (mainly squid) caught by sperm whales, research nets (9 m² mouth) and commercial trawls in three areas of the North Atlantic. The whales eat mainly histioteuthids, cranchiids and octopoteuthids. Nets catch mainly squids from other families.

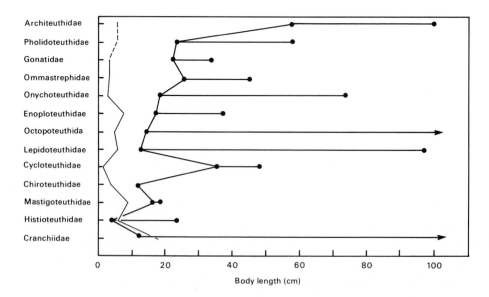

Fig. 4d.2. The body lengths of squids of various families caught in nets and by sperm whales. The range of specimens found in sperm whale stomachs is shown by a horizontal line for each family; the lower ends of the ranges are connected by a heavy line. A second line joins the maximum body length for squids of each family caught in almost 600 research net hauls (9 m² mouth).

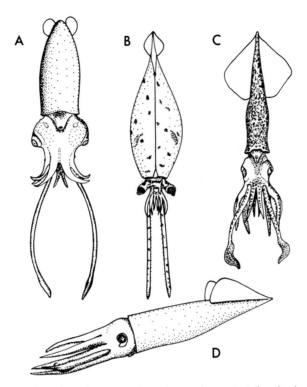

Fig. 4d.3. The four commonest squids caught in the Antarctic Ocean in research nets. A, *Bathyteuthis abyssicola* (mantle length <6 cm); B, *Galiteuthis glacialis* (ML <30 cm); C, *Brachioteuthis picta* (ML <5 cm); D, *Gonatus antarcticus* (ML >34 cm). A – C after Chun 1910, D after Pfeffer 1912.

octopuses. While some squids are denser than water and have to swim continuously to stay at any level in the sea, others have redeveloped buoyancy systems which need no shell. Many of the one thousand species, instead of urinating, retain low density ammonia in solution in their tissues or in a large internal bag and this buoys up the body in much the same way as do air bladders in fish. Some other squids accumulate a special low density oil in their livers for the same purpose and some deep sea octopuses have extremely gelatinous, watery tissues which reduce their sinking rates.

## 4d.2. FOOD AND FEEDING

All living cephalopods are carnivorous and many are known to have a mixed diet of fish, crustaceans and other species of cephalopod. To catch and immobilize these they have an unusually formidable weaponry including eight (octopuses) or ten (squids) muscular, contractile tentacles each with two or more rows of over a hundred powerful suckers which often have apple-corer like chitinous rings round their margins and can cut a circle through whale or human skin. These rings, sometimes more than 4 cm in diameter may bear needle-like teeth or be modified into a large 'fish hook' which may be as much as 2 cm long. Having grasped the prey, the squid pulls it to the jaws at the centre of its tentacles. These horny jaws, usually called 'beaks' because of their superficial resemblance to parrots' beaks, are buried in

TABLE 4d1. The Percentages by Number of the Species of Antarctic Cephalopods in the Diet of Some Predators in the Antarctic Ocean (Extracted from Clarke, 1980; Clarke and Prince, 1981; Clarke *et al.* 1981, 1982 a & b and (*) Clarke unpublished).

| | Sperm Whales | | Seals | | | | Albatrosses | | |
|---|---|---|---|---|---|---|---|---|---|
| | Antarctic | South Georgia | Elephant | Weddell | Fur* | Leopard* | Wandering | Grey-Headed | Black-browed |
| **Cranchiidae** | | | | | | | | | |
| *Mesonychoteuthis hamiltoni* | 16 | 30 | | | | | | | |
| *Galiteuthis glacialis* | | | 2 | | 44 | | 4 | 8 | 25 |
| *Taonius* | | | | | | | 17 | | |
| **Onychoteuthidae** | | | | | | | | | |
| *Kondakovia longimana* | 18 | 28 | 4 | <1 | 14 | 4 | 40 | <1 | 1 |
| *Moroteuthis knipovitchi* | | | | | | | | | |
| *Moroteuthis robsoni* | 51 | 23 | 14 | 31 | 14 | 92 | 2 | | |
| **Gonatidae** | | | | | | | | | |
| *Gonatus* | 4 | <1 | 42 | 1 | | | 6 | | 1 |
| **Octopoteuthidae** | | | | | | | | | |
| *Taningia danae* | — | 14 | | | | | 1 | | |
| **Ommastrephidae** | | | | | | | | | |
| *Todarodes ? sagittatus* | — | 3 | | | 28 | | | 88 | 68 |
| **Brachioteuthidae** | | | | | | | | | |
| *Brachioteuthis picta* | | | | 1 | | | | | |
| **Histioteuthidae** | | | | | | | | | |
| *Histioteuthis* spp. | 11 | 2 | | | | | 15 | <1 | |
| **Mastigoteuthidae** | | | | | | | | | |
| *Mastigoteuthis* | | | | | | | 3 | <1 | |
| **Psychroteuthidae** | | | | | | | | | |
| *Psychroteuthis glacialis* | | | 6 | 29 | | | | | |
| **Neoteuthidae** | | | | | | | | | |
| *Alluroteuthis* sp. | | | 2 | 2 | | | 4 | | |
| **Octopoda** | | | 10 | 35 | | | 2 | | 1 |

a ball of muscle and, in a two metre squid, have the shearing force to cut through a fisherman's trace wire or the scaly armour of a fish. A toothed salivary papilla is used by some octopods to drill holes in the hard shells of crustaceans and rapidly acting toxins are often secreted to immobilize prey.

While the non-buoyant, most powerful, jet-propelling squid can chase and capture fish, the buoyant kinds have often developed a more passive, short duration jet propulsion and powerful fins for sustained swimming. These squids probably wait for prey to come close to them before short, rapid jetting aids capture. A few species have two enormously elongated tentacles with thousands of small suckers at their tips and narrow contractile stems looking for all the world like fisherman's lines. The suckers are sometimes luminous and there is no doubt that these attract, secure and dispatch any passing prey organisms.

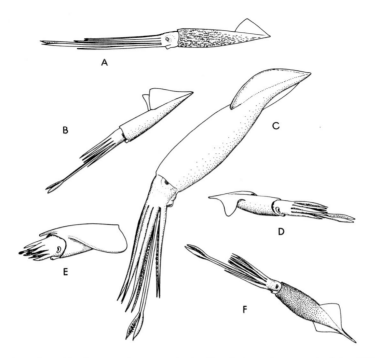

Fig. 4d.4. The three Antarctic squids of greatest importance in the diet of the predators examined (see Table 3). A, *Kondakovia longimana* (ML >80 cm); B, *Moroteuthis knipovitchi* (ML >40 cm); C, *Mesonychoteuthis hamiltoni* (ML >200 cm); and the three squids which extend into the Antarctic from the North, D, *Todarodes* sp. (ML <75 cm); E, *Taningia danae* (ML >140 cm); F, *Moroteuthis robsoni* (ML <9 cm).

## 4d.3. LIGHT ORGANS

The majority of deep sea squid bear luminous organs on their heads, bodies and/or tentacles. Besides commonly using the light as a lure, some squid use it to see by, attract mates, keep shoals together and, surprisingly, to hide from predators. This last function is achieved by light organs arranged on the underside of squids living in or migrating into the upper 800 m of water reached by daylight. The light produced is carefully regulated to match the down-welling daylight and to break up or to hide the squid's silhouette from below (Young *et al.*, 1977).

## 4d.4. GROWTH AND REPRODUCTION

Some species of squid reach less than 2 cm before maturing and dying but others are known to reach over 20 m. Little is known about the life cycle of oceanic cephalopods but it seems that many lay egg masses on the continental slopes deeper than 1000 m. Some are very quick growing, taking one or two years to grow from the egg to sexual maturity.

## 4d.5. ECOLOGICAL RELATIONSHIPS

Whether or not cephalopods have a large effect, by predation, on populations of fish and crustaceans in the ocean depends upon the numbers and sizes of cephalopods. While commercial fisheries for squid are flourishing the world over they are almost entirely carried out on continental shelves and only involve species which either live near land such as cuttlefish and loligos or move inshore seasonally such as some of the muscular ommastrephids. Because the latter are very numerous in some years and often grow quickly to over a metre in length (over 4 m in one species, *Dosidicus gigas* off Peru and Chile) they must have a very considerable local predatory effect on fish populations, including many commercial species. They are also important in the diet of many fish.

In the Antarctic Ocean, however, the inshore species have not been exploited commercially. In fact the cuttlefishes, loligos and, with one exception, the ommastrephids are absent from or unimportant in the region. Here, the cephalopods of by far the greatest ecological importance are the oceanic squids. Oceanic squids are not caught commercially except when they concentrate on continental shelves and this has not been observed in the Antarctic. We only know of the existence of such species from net catches by research ships and from their presence in the stomachs of squid-eating predators such as sperm whales, seals, albatrosses and penguins.

Few collections of squids from net hauls in the Antarctic Ocean have been examined but sufficient has been done to show that squids are only a small fraction, by volume or by number, of the animals caught. This is the same the world over in oceanic net catches. However, there is now good evidence to show that these catches give a very poor insight into what squid are present in the ocean. The 'standard' research nets have a mouth less than 9 m$^2$ in area. In the last 20 years use of large commercial midwater trawls in oceanic water, including the Antarctic, has shown that bigger nets catch larger squid which are rarely or never caught in 'standard' nets. However, even these do not tell the whole story. If the stomachs of sperm whales caught by commercial activities in any ocean are examined, large numbers of squid beaks are found; as many as 18,000 have been collected from a single stomach. From these beaks and other remnants of squids from stomachs it has been shown that sperm whales eat many squid species which have been rarely or never caught in nets. Throughout the world, comparisons have been made between squids caught in research net hauls, commercial trawl hauls and those eaten by predators (Clarke, 1977) and an example is shown in Fig. 4d.1. Whereas squids in the family Histioteuthidae are dominant in the whale samples, enoploteuthids are dominant in trawl hauls and various other families are important in 9 m$^2$ nets.

The size of squids caught in research nets is also very different from those eaten by sperm whales. If the body lengths of squids of various families from research nets are compared with those of squids collected from sperm whales there is little overlap (Fig. 4d.2). In fact, some squid *beaks* are more than 10 cm long, exceeding the maximum *body length* of squids from all but two of the families from nets! Although commercial trawl hauls catch very different proportions of the squid families from the whale they do extend the size range of histioteuthids and cranchiids caught into the lower range of those caught by whales. An exactly similar situation is found in the Antarctic Ocean.

Research midwater nets used in the Antarctic Ocean (Roper, 1969) show a predominance of a small, dark red squid called *Bathyteuthis abyssicola* which grows to about 6 cm and comprises 39% of squids caught (Fig. 4d.3A). Other important squids (Fig. 4d.3) are the bladder-like *Galiteuthis glacialis* (23% and <30 m), *Brachioteuthis picta* (12% and <5 cm) and the hooked and oily *Gonatus antarcticus* (9% and <34 cm). Only juveniles of *Galiteuthis glacialis* and *Gonatus antarcticus* are caught in nets but adults are eaten by albatrosses, sperm whales and seals (Table 4d.1).

Many cetaceans, seals and birds living in the Antarctic are known to include cephalopods in their diet but detailed investigations of the species involved have been carried out only for a few predators. Table 4d.1 summarizes this data and shows that the most important squids eaten by these are the cranchiids

*Mesonychoteuthis hamiltoni* (Fig. 4d.4C) and *Galiteuthis glacialis* and the onychoteuthids *Kondakovia longimana* (A) and *Moroteuthis knipovitchi* (B). Other species may be particularly important in the diet of one or other of the predators. Octopuses are only consumed in large quantities by seals. Three species *Taningia danae* (Fig. 4d.4E), *Todarodes ? sagittatus* (D) and *Moroteuthis robsoni* (F) occur off South Georgia and are broadly distributed in warmer, more northerly waters. Their range possibly extends south in summer months and may comprise an important annual input of nutrient into the Antarctic Ocean. Other species listed are Antarctic endemics or do not extend further North than 40°S. We may well ask 'How important are cephalopods in the Antarctic ocean ecosystem?' To answer this we could perhaps try to estimate the weight or 'biomass' of cephalopods which must be consumed to sustain the populations of their predators. Such a calculation can only be extremely rough and based on minimum estimates. From what we know of the populations and food of marine mammals in the Antarctic, the sperm whale is by far the greatest consumer of squids because its diet includes very little else, it grows to a large size (up to 60 tons) and a large proportion of its male southern population visits the region (Laws, 1977). However, of the 46 species of Antarctic birds little can be said except that penguins are estimated to eat over 80% and albatrosses only 3% of all the cephalopods eaten by birds (Mougin and Prévost, 1980; Prévost, 1981).

A very crude estimate (Clarke, in press) of the consumption of cephalopods by whales, seals and birds suggests that over 30 million tonnes are consumed annually. This neglects predation by fish which must be extremely important throughout the whole water column.

A rough estimate of cephalopod biomass can also be made from research net catches in the Antarctic. The weight of squid caught when a known large volume of water is filtered by many nets used throughout the water column is related to the total volume of the ocean. This works out to be an order of magnitude less than the estimate of predators' consumption and reflects the low catching efficiency of our methods. Of course, in much the same way that fisheries can only take a proportion of the fish stocks if they are to continue for any length of time, the predators can only eat a proportion of the squids which are present and very large stocks must be present to support the predation. We can only guess at the stocks of squid necessary to support predation of over 30 million tonnes a year but it may well be over 100 million tonnes which is a very large biomass by oceanic standards.

There has been much discussion on the effect that depletion of whales caused by whaling must have had on the populations of krill and consequently on the populations of other krill eaters. Reduction of sperm whale numbers must also have had an influence on the availability of cephalopods for other predators. While other toothed whales, seals and sea birds might be expected to benefit and perhaps increase, many of the species eaten by the deep diving sperm whales probably live deeper than some cetaceans, seals or birds can dive and any increase in population is likely to influence deep sea fish more than air breathing predators. Many of these squid probably accumulate to spawn on the sea bed of the continental slopes where bottom fishes abound in many regions of the world.

In summary, we can be certain that cephalopods are a very important factor in the ecology of the Antarctic Ocean but much greater efforts must be made to catch and to study them before we can properly quantify their importance.

# REFERENCES

Clarke, M.R. (1977) Beaks, nets and numbers. *Symp. zool. Soc. Lond.* 38, 89 – 126.

Clarke, M.R. (1980) Cephalopoda in the diet of sperm whales of the southern hemisphere and their bearing on sperm whale biology. *Discovery Rep.* 37, 1 – 324.

Clarke, M.R. (in press) Cephalopod biomass-estimation from predation. *Proc. Vict. Mus.*

Clarke, M.R., Croxall, J.P. and Prince, P.A. (1981) Cephalopod remains in regurgitations of the wandering albatross *Diomedea exulans* L. at South Georgia. *Br. Antarct. Surv. Bull.* 54, 9 – 21.

Clarke, M.R. and MacLeod, N. (1982a) Cephalopods in the diet of elephant seals at Signy Island, Antarctic. *Br. Antarct. Surv. Bull.* 57, 27 – 32.

Clarke, M.R. and MacLeod, N. (1982b) Cephalopod remains in the stomachs of eight Weddell seals. *Br. Antarct. Surv. Bull.* 57, 33 – 40.

Clarke, M.R. and Prince, P.A. (1981) Cephalopod remains in regurgitations of black-browed and grey-headed albatrosses at South Georgia. *Br. Antarct. Surv. Bull.* 54, 1 – 7.

Denton, E.J. and Gilpin-Brown, J.B. (1973) Floatation mechanisms in modern and fossil cephalopods. *Adv. mar. Biol.* 11, 197 – 268.

Laws, R.M. (1977) Seals and whales of the Southern Ocean. *Phil. Trans. R. Soc. Lond.* B 279, 81 – 96.

Mougin, J.-L. and Prévost, J. (1980) Evolution annuelle des effectifs et des biomasses des oiseaux antarctiques. *Rev. Ecol. (Terre Vie)* 34, 101 – 33.

Prévost, J. (1981) Population, biomass and energy requirements of Antarctic birds. In, *Biological Investigations of Marine Antarctic Systems and Stocks (BIOMASS)* Vol. 11: Selected Contributions to the Woods Hole Conference on Living Resources of the Southern Ocean 1976. Ed. Sayed Z. El-Sayed, pp. 125 – 37. Publ. SCAR and SCOR.

Roper, C.F.E. (1969) Systematics and zoogeography of the Worldwide bathypelagic squid *Bathyteuthis* (Cephalopoda: Oegopsida). *Bull. U.S. natn. Mus.* 291, 1 – 208.

Young, R.E. and Roper, C.F.E. (1977) Intensity regulation of bioluminescence during countershading in living midwater animals. *Fishery bull.* 75, 239 – 52.

# CHAPTER 5

# Birds and Mammals — Introduction

## W. N. BONNER and D. W. H. WALTON

British Antarctic Survey, Natural Environment Research Council, High Cross, Madingley Road, Cambridge CB3 0ET, U.K.

To many people, the larger vertebrates constitute the only part of the Antarctic wildlife of which they are aware and no one can deny the impressiveness of a penguin rookery, a seal breeding beach or one of the great whales glimpsed at sea. In this chapter we deal with the higher vertebrate predators of the Antarctic.

Seals, reviewed here by Nigel Bonner, are an integral and characteristic part of the scene in both polar regions. Their aquatic modifications enable them to withstand the coldest climates successfully and the Antarctic, with its complete absence of terrestrial predators, has provided a very favourable environment for them. The crabeater seal is by far the most abundant seal species in the world and represents, in terms of biomass, the largest consumer of krill. Antarctic fur seals, which also feed on krill, were once very numerous but were nearly completely eradicated by nineteenth-century sealers. However, during the course of this century they have shown a remarkable recovery in many parts of their range.

To some extent it seems that the krill-eating seals have benefited from the great reduction in numbers of the baleen whales. Ray Gambell describes the life histories of these impressive mammals. The rorquals have developed a remarkable pattern of life in which they divided their lives between a period spent in warmer, less fertile, waters in the tropics where they gave birth and raised their young, and fasted while they did so, and a period in the Antarctic when they could feast on the dense swarms of krill. The vast blanket of blubber that they accumulated during this feeding period was an irresistible attraction to Man, and the resulting whaling industry devastated the whale stocks and brought about a radical change in the structure of the higher predator components of the Antarctic marine ecosystem.

In the section on oceanic birds Roy Siegfried examines some of the general questions posed by the birds of the Antarctic. There are more species north of 50°N latitude than south of 50°S latitude, though the Antarctic is richer in truly oceanic species. Despite the appearance of great abundance in some localities the Southern Ocean as a whole supports a relatively low abundance of resident breeding sea birds, though these include an unexpectedly high proportion of large species.

No visitor to the Antarctic can fail to be aware of penguins, the archetypal bird of the high Antarctic. Penguins may nest in dense colonies numbering tens of thousands, or even more, and such colonies are features of many of the rock exposures which have in the past been chosen as sites for research stations. In the final section of this chapter Bernard Stonehouse provides a general introduction to these birds and an account of each of the seven species found in the Antarctic, relating them to their environment and the food web in which they form by far the greatest part of the avian biomass.

# CHAPTER 5a

# Birds and Mammals — Antarctic Seals

## W. NIGEL BONNER

British Antarctic Survey, Natural Environmental Research Council, Madingley Road,
Cambridge CB3 0ET, U.K.

## CONTENTS

## 5a.1. INTRODUCTION

The wide productive expanses of the Southern Ocean provide an ideal habitat for seals. When the ancestors of seals moved from the land into the sea sometime during the Miocene about twenty million years ago they were faced with the problem of maintaining their heat balance. Sea-water is appreciably colder than the approximately 40°C at which mammals maintain their core temperature. Moreover, its thermal capacity is nearly five thousand times as great as that of an equal volume of air. It was necessary for the proto-seals to develop insulation in the form of dense fur or a layer of blubber beneath the skin, to prevent too great a loss of heat from the body surface when submerged in the sea.

These insulating layers are, of course, effective in air as well as in water, and consequently seals suffer little heat loss in air even in the coldest conditions, while conversely they can suffer severely from thermal stress in warm climates.

Because of their effective thermal insulation, seals are well able to exploit the polar seas. As far as water temperature is concerned, the extra heat gradient between the animal's core temperature and its surroundings where temperate and polar oceans are compared is not a critical factor, and seals are found in the most frigid seas, both north and south.

## 5a.2. THE ORIGIN OF THE ANTARCTIC SEALS

The order Pinnipedia, or seals in the broad sense, evolved in the Northern Hemisphere probably from two separate stocks. One stock, arising from a dog-like ancestor in the North Pacific, gave rise to the walruses (family Odobenidae) and the related eared seals, the Otariidae. Another stock, derived from an otter-like ancestor in the North Atlantic, produced the true seals, the Phocidae. It is the Phocidae that provide the seals of the Antarctic proper, though one species of otariid has successfully colonized the oceanic islands that lie to the south of the Antarctic Convergence.

Early in the radiation of the phocid seals during the Miocene, a stock split off that was associated with temperate, rather than cold polar water. This stock, the Monachinae, was the origin of the monk seals whose living survivors are found today around the Mediterranean and the Hawaiian Islands. Another group of monachine seals, the Lobodontini, colonized the Antarctic during the late Miocene or early Pliocene, at a time when the region was much warmer than it is today. This group remained associated with the Antarctic and radiated to produce the four very distinct and highly successful Antarctic seals that we have today. These are the crabeater seal, *Lobodon carcinophagus*; the leopard seal, *Hydrurga leptonyx*; the Ross seal, *Ommatophoca rossii*; and the Weddell seal, *Leptonychotes weddellii*.

The elephant seals were another group of monachine seals that invaded the Southern Hemisphere. Their affinities, however, remained with the cool temperate and sub-Antarctic regions and they developed no special adaptations to an ice environment. One of the two extant forms, in fact, is found in quite warm conditions on islands off the Pacific coast of Mexico and California. The more numerous species, the southern elephant seal, *Mirounga leonina*, occurs around the southern part of South America, on the Falkland Islands, Gough Island, the Prince Edward Islands, Macquarie Island, and on the sub-Antarctic islands of New Zealand, as well as on the oceanic islands south of the Antarctic Convergence, with the largest concentrations at South Georgia and Macquarie Island.

Otariid seals originated in the north-east Pacific and early in their radiation some crossed the Equator to establish a stock in South America. The South American fur seal, *Arctocephalus australis*, is probably very close to this ancestral stock, which gave rise to a number of closely similar species. One of the most numerous of these is the Antarctic fur seal, *Arctocephalus gazella*, which is found on islands lying to the south of the Antarctic Convergence, though generally only on those which are not regularly invested with sea-ice.

## 5a.3. THE CRABEATER SEAL

The crabeater seal (Fig. 5a.1) is one of the most remarkable, though least known, of the mammals of the world. Its population probably numbers between fifteen and forty million animals, making it one of the most abundant large mammals in the world. More than one in every two seals in the world is a crabeater seal and the population biomass of crabeaters is about four times that of all other pinnipeds put together.

Crabeaters are large seals. Females reach a length of about 200cm and a weight of 227 kg. Males are slightly smaller (Bertram, 1940). When freshly moulted in January or February the coat is mainly dark brown above and fawn below. The pattern is generally blotched, particularly in young animals, with

Fig. 5a.1. Crabeater seal. The crabeater seal is by far the most abundant seal species, not only in the Antarctic, but in the world.

patches of darker colour on a lighter ground. The pattern is strongest on the flanks behind the fore flippers and around the posterior end of the body. As the hair ages it fades to a pale blond, causing Edward Wilson (1907) to refer to the animal as the 'white Antarctic seal'. The new born young are uniformly clad in cafe-au-lait coloured woolly hair. Nearly all crabeater seals are marked with numerous sets of pairs of parallel scars (Fig. 5a.2). These will be referred to later (page 212).

Crabeater seals, like the other Antarctic phocids, occur circumpolarly. Probably the uniformity of oceanic conditions arising from the lack of physical barriers, has prevented speciation taking place. Crabeaters are characteristically seals of the drifting pack ice and are rarely seen ashore. When the ice breaks up in the summer the seals mostly retreat to the areas of residual pack. Some, however, enter open water, for example, the Bay of Whales where they have been seen at 79°S (Lindsey, 1937). Crabeater seals are readily recognized at sea, where groups of a few tens to several hundreds can sometimes be seen swimming along in a manner very reminiscent of a school of porpoises.

For such characteristically Antarctic seals, crabeaters are great wanderers. Occasional specimens have been recorded from New Zealand, Tasmania and southern Australia. The most northerly sightings are from Rio de la Plata in South America and from the tip of South Africa (Ross et al., 1978; Vaz Ferreira, 1965).

Not only do crabeaters wander far to the north of their normal range, they also make unexplained excursions inland on the Antarctic continent. Wilson recorded a mummified crabeater carcase on the Ferrar Glacier at an altitude of 1100 m (probably a record for any seal), and a living male weaned pup was found on the Crevasse Valley Glacier (76° 50'S, 145° 30'W) at an altitude of 920 m and 113 km from the open sea (Stirling and Rudolph, 1968). Perhaps these seals are ones that had been trapped by the freezing of the sea after they had penetrated to the head of inlets. In attempting to escape over the sea-ice, some may take the wrong direction and end up far from the sea, where they face inevitable starvation and death (Stirling and Kooyman, 1971).

Fig. 5a.2. A group of crabeater seals on an ice-floe in the South Orkney Islands. In the foreground is a young leopard seal. The parallel scars on the flank of the crabeater seal on the left were caused by a leopard seal's canine teeth.

Fig. 5a.3. A female leopard seal and her newly-born pup on an ice-floe in the Weddell Sea. The characteristic humped outline of the adult leopard seal is well shown in this picture.

## 5a.4. THE LEOPARD SEAL

The leopard seal is one of the largest of the phocids (Fig. 5a.3). As in the other lobodontine seals, the female is slightly larger than the male. The largest recorded was 358 cm long (Hamilton, 1939), but females average 291 cm and males 279 cm. Average weights are 367 kg for females and 324 kg for males but some very large female leopard seals estimated to weigh 500 kg have been seen hauled out at South

Georgia (Kooyman, 1981). Despite their size, leopard seals are slenderly built and appear sleek and rather reptile-like. This is accentuated by the large head with its enormous gape and massive lower jaw. The fore flippers are long and tapered and the whole animal has the appearance of having been built for speed.

The coloration on the back is dark grey to black which changes at the flanks to steely grey while the underparts are silver. On the ventral surface there is a scattering of dark spots while on the dorsal surface this pattern is reversed and the spots there are lighter. The new-born pup is patterned similar to the adult.

Leopard seals are another characteristic animal of the pack ice, but are not so restricted to it as the crabeater seal. They are often conspicuous in the sea around penguin rookeries, though this in no way represents their true distribution, and they are frequently seen hauled out ashore or on fast ice off the coast or offshore islands. Sub-Antarctic islands such as South Georgia, Macquarie and Heard Island also afford habitat for leopard seals. I have counted twenty-nine leopard seals, mostly young animals, hauled out on the beach of a small cove in South Georgia at the end of the winter. Even when present in such numbers there are no indications of social bonds between neighbouring animals. During the winter leopard seals migrate north (most of the sightings at sub-Antarctic islands are made at this season and in the spring) and occasionally individuals are reported from the coasts of Patagonia, South Africa, southern Australia, Tasmania and New Zealand. The most northerly record is from Raratonga in the Cook Islands at 21°S (Berry, 1961).

## 5a.5. THE ROSS SEAL

The Ross seal is probably the least abundant of all the Antarctic seals, though it is in no way a rare animal, and it is certainly the least well known. In the hundred years after its discovery by Sir James Clark Ross there were probably fewer than fifty sightings. However, this is largely to be explained by the fact that the Ross seal seems to be a denizen of heavy pack ice which even modern ice breakers can penetrate only with difficulty and which was quite impassable to earlier vessels.

Ross seals are the smallest of Antarctic phocids (Fig. 5a.4). Of the few that have been measured the largest female was 236 cm long and weighed 204 kg, and the largest male 208 cm and 216 kg (Ray, 1981). The profile of the Ross seal is instantly recognizable. It has an extraordinarily short snout set on a wide head. The eyes are large, up to 7 cm in diameter, but this is not readily discernible in the living animal, as the eye openings are no larger than in comparable sized phocids. The fore flippers have reduced claws and the terminal phalanges are greatly elongated, but are not themselves proportionately longer than in the crabeater or leopard seals. On the other hand, the rear flippers, which may reach 22% of the standard body length, are proportionately the longest for any phocid (Ray, 1981). As there are practically no observations on locomotion in Ross seals the reasons for these modifications of the limbs are unknown.

Ross seals are uniquely patterned (Fig. 5a.4). The body is dark brownish grey dorsally, shading into paler on the flanks, where there may be considerable spotting, and becoming nearly white on the ventral surface. Around the throat and on the sides of the head the dark colour is continued posteriorly as a pattern of streaks. This type of longitudinal streaking is found in no other seal. There seems to be a good deal of individual variation in the extent of the streaking, but few specimens have been examined.

Ross seals are distributed right around the Antarctic, but there is evidence that they are very much more abundant in some areas than in others. The King Haakon VII Sea is one such area (Condy, 1976) and they may be more abundant near Cape Adare than further south in the Ross Sea (Ray, 1981). Although ice distribution is clearly an important factor in determining the abundance of Ross seals, it is

Fig. 5a.4. A Ross seal in the typical position when disturbed. The dark streak from the angle of the mouth is characteristic of this species, as is the blunt head. (Photo M.W. Green)

Fig. 5a.5. Weddell seal. This seal feeds close to shore and is the seal most often seen ashore in the higher latitudes.

only part of the picture. Food distribution is also likely to be of importance. Nothing is known of seasonal movements in the Ross seal. The only sighting north of 55°S has been made at Heard Island (Ingham, 1960).

## 5a.6. THE WEDDELL SEAL

The Weddell seal is another very large phocid (Fig. 5a.5). A female measuring 329 cm in length and a male of 297 cm were recorded by the Scottish National Antarctic Expedition (Bruce, 1913). In early spring, when the seals are fat, both males and females may commonly weigh around 400 – 500 kg

Fig. 5a.6. Elephant seals at South Georgia. The disparity in size between the mating pair is apparent. A subordinate bull lies at the water's edge.

(Kooyman *et al.*, 1973). In contrast to the leopard seal, the head of the Weddell seal seems small for its body. The snout is short and the corners of the mouth tilt upwards, giving it a benign appearance.

Weddell seals are strongly spotted with irregularly shaped patches. The background colour is blue — black with silver grey spotting, the general colour being paler beneath. As the hair ages it fades to a yellowish hue, the darker areas of the back becoming rusty brown. The coat of the new born pup is a very light brown or silvery grey with no traces of the spotting that is characteristic of the adult until the moult begins at about nine to 21 days after birth.

In contrast to the Antarctic seals already described, the Weddell seal is an animal of the fast ice near-shore region. It is the most southerly naturally-occurring mammal in the world. It is rarely found far out in the pack but is the commonest seal seen ashore by visitors to the Antarctic. Weddell seals are fairly sedentary animals, though it has been suggested that there is a seasonal migration northward into the pack during the winter (Smith, 1965). There is certainly some seasonal dispersion and aggregation, perhaps in response to the scarcity of secure areas with recurring tide cracks suitable for breeding sites.

Although Weddell seals are normally associated with ice, there is a small relict northern population breeding at South Georgia. Here in most years the seals have no association with sea-ice in any form, but they show a definite predilection for lying on snow banks when ashore.

There are few observations of Weddell seals outside the Antarctic, though four specimens have been recorded from New Zealand, one from Australia and one from Uruguay. The Uruguayan record, at 35°S is the most northerly (Vaz Ferreira, 1956).

## 5a.7. THE SOUTHERN ELEPHANT SEAL

The southern elephant seal is the largest of all the pinnipeds, and one of the largest of all mammals, excluding the Cetacea. Fully grown males reach a length of 4.5 m and a weight of around 4000 kg. There is a striking sexual dimorphism, but in the opposite direction from that shown by the lobodontine seals (Fig. 5a.6). The female elephant seal is not larger, but very much smaller, than her mate, reaching around 2.8 m and 900 kg.

Elephant seals got their vernacular name partly from their size and partly from the development of a peculiar trunk-like proboscis in the males which plays an important part in antagonistic behaviour during the breeding season. The coat of the elephant seal is composed of short, very stiff recurved hairs rather sparsely distributed on the skin. In pattern it shows no trace of the spotting that is characteristic of the lobodontine seals. Elephant seals are a uniform brownish colour though there is considerable variation from very dark chocolate to pale sandy. Some specimens show a darker vertebral stripe. Old cows have a conspicuous lighter mask. Adults of both sexes are much scarred, the males about the chest and neck, as a result of wounds received during fighting in the breeding season, and the females about the back of the neck, where the bull holds them during copulation. The young are born in a coat of jet black woolly hair which is shed at about 3 − 4 weeks. The newly exposed coat is a beautiful steely-black above and silvery below.

Elephant seals are circumpolar but there seem to be three main breeding stocks. The first is centred on South Georgia and includes breeding populations in South America, the Falkland Islands and the islands of the Scotia arc. A second group breeds at Iles Kerguelen, Iles Crozet and Heard Island. The third stock is found at Macquarie Island and the New Zealand sub-Antarctic islands. These three stocks roughly correspond with three subspecies described by Lydekker in the early part of this century, but the evidence of genetical isolation and morphological differentiation is tenuous. Observed differences may be the result of environmental factors.

With established temperate populations northern records of elephant seals are perhaps of less interest than for lobodontine seals. The most northerly recent records are from Mauritius and Rodriguez Island at latitude 20°S. Perhaps of more significance is the fact that the southern distribution extends as far as open water permits, even to 77° 40′S in the Ross Sea (Bonner and Laws, 1964). However, elephant seals do not seem to be able to come to terms with ice as a breeding substrate, and the population at Signy Island in the South Orkneys, which from time to time has to breed on the ice, remains small.

## 5a.8. THE ANTARCTIC FUR SEAL

The Antarctic fur seal is a very different animal from the five phocid seals described so far. As an otariid it is very much more nimble on land than the phocids, being able to turn its hind limbs forward underneath its body and, by balancing its weight on its very large foreflippers, walk or even gallop about the beach.

The Antarctic fur seal is a medium-sized seal, adult males reaching a length of 200 cm and a weight of around 125 − 200 kg. As in the elephant seal, the female is smaller (Fig. 5a.7), up to 140 cm and about 50 kg. As their name implies, Antarctic fur seals are provided with fur. The coat is composed of two layers, an outer coarser layer of stout guard hairs, and an inner layer of very fine underfur fibres. It is this fur layer which provides the seal fur of commerce after the guard hairs have been removed. The guard hairs protect and support the underfur, preventing it from collapsing into a compacted mat. The layer of air trapped by the underfur provides the necessary thermal insulation and fur seals do not have such a well developed blubber layer as the phocid seals.

The colour of the fur seals is determined by the guard hairs, since these cover the underfur. The colour on the back and sides is grey to brownish, the throat and breast are creamy and the belly is a dark gingery colour. In adult males a heavy mane is developed round the neck and shoulders. The guard hairs forming this mane have long white tips, giving the bulls a grizzled appearance. As the tips of the hairs are worn away the animals appear darker. The underfur is dark fawn, but reddish in adult males. The new born pup is clad in a black coat. In all animals the exposed skin of the flippers is densely black. The naked surface serves to dissipate heat by radiation when the animal is active. When it is necessary to conserve heat the blood supply to the skin is cut down and the flippers are tucked under the body.

Fig. 5a.7. Antarctic fur seal family. There is considerable sexual dimorphism in this species.

A small proportion of Antarctic fur seals are white in colour, the guard hairs lacking all pigment. They are not albinos, for pigment is present in the underfur and in the eyes, but the skin pigmentation is much paler than in normal animals.

Antarctic fur seals occur on islands to the south of the Antarctic Convergence and north of about 65°S. The major breeding population is found at South Georgia. Smaller groups occur on the South Shetlands, South Orkneys, South Sandwich Islands, Bouvetøya, Heard and McDonald Islands. Perhaps a few breed on Kerguelen. The colonies disperse in autumn, but no definite migratory movements have been described. A small northern group occurs on Marion Island (Prince Edward Islands) where this species overlaps the range of another fur seal, *Arctocephalus tropicalis*. Although *A. tropicalis* and *A. gazella* are clearly differentiated species, some hybridization appears to occur.

## 5a.9. THE NUMBERS OF SEALS

To understand how seals fit into the general ecology of the Antarctic, it is necessary to know something about the size of their populations. The remoteness and inaccessibility of their habitat, coupled with the very large numbers involved, have made this information hard to come by.

No real estimate of the pack ice seals was possible until quite recently when a group of American biologists used helicopters operating from ice-breakers to make surveys of seals hauled out on the floes (Erickson *et al.*, 1971; Gilbert and Erickson, 1977; Siniff *et al.*, 1970). Obviously, it was necessary to cover very large areas, and the size of the sample was very small compared with the area surveyed. Counts obtained from the helicopters were adjusted to allow for the activity patterns of the seals, which spend most time hauled out around midday, and then converted into estimates of seal density.

From 4,257 km$^2$ sampled in the Amundsen and Bellingshausen Seas the mean density was found to be 1.76 seals per km$^2$, or 3.66 seal per km$^2$ when adjusted for activity pattern. Using estimates of the extent of the pack ice obtained from satellite photographs, the figures were extrapolated to yield the following population estimates: crabeaters, 14,858,000; leopards, 222,000; Ross, 220,000; and Weddells, 730,000.

These figures are likely to be underestimates, since only the seals on the ice are counted, and even at maximum haul out some will still be in the water. Furthermore, the estimates take no account of seals in areas of open water, or hauled out on shore. (The latter point would tend to underestimate Weddell seals seriously.)

Elephant seals, which aggregate on shore during the breeding season, can be counted directly. The easiest approach is to count the weaned pups (which are a relatively immobile part of the population) and multiply the count by a factor (3.5) derived from the age structure of the population. Using this method the South Georgia stock (including the seals in South America, the Falklands and the Scotia arc) was estimated at about 300,000. There are probably about 200,000 elephant seals at Kerguelen and Heard Island and about 100,000 at Macquarie Island and the New Zealand sub-Antarctic islands (Laws, 1960). This estimate is now rather out of date. In South Georgia the stock may have increased since commercial sealing ceased in the mid-1960s, while at Kerguelen there may have been a downward trend. However, it is the best available and gives an indication of the abundance of these huge seals.

The Antarctic fur seal is even more difficult to count. Counting pups on the beaches is the accepted method, but these are more difficult to find than the elephant seal pups, and the terrain occupied by fur seals is usually very broken, so that it is both difficult and dangerous to search for the pups. A further complication is that the breeding population at South Georgia has been increasing very rapidly recently. In the 1950s I was able to count the pups born there without too much difficulty, but by the mid-1970s it was necessary to resort to indirect methods of estimation. A mark-recapture method was employed. Pups were marked by shearing a patch of fur from the tops of their heads. After about 40% of the pups present in a study area containing about 10,000 pups had been marked, a week was allowed to elapse so that the marked pups could mingle with the others. Observers then counted the number of marked pups present in blocks of 25 pups. After making allowance for mortality that had occurred between marking and counting the total number present was estimated by assuming that the proportion of marked pups in the original population was the same as that in the samples counted later.

From this it was found that the best direct counts of pups needed to be increased by a factor of 1.54 to obtain the true number present (Payne, 1977). The number of fur seal pups born at South Georgia in the 1975 season was found to be about 90,000, from which a total population of 369,000 was calculated. However, this population had been increasing at a very high rate — 16.8% per annum — which was unlikely to continue indefinitely. Assuming the rate had fallen to 11.5% per year (which is what it was in the last year in which it could be calculated with accuracy) the South Georgia population in 1982 would have produced some 209,000 pups from a total population of around 856,000. Probably the total is not as high as this, but somewhere between 700,000 and one million animals is a likely estimate for the world population, including the very small breeding groups at other localities.

## 5a.10. FEEDING ECOLOGY OF ANTARCTIC SEALS

The seals of the Antarctic exploit the Southern Ocean's resources in a way that avoids undue competition between themselves. There are differences between the species, not only in diet, but also in geographical range.

The crabeater seal is a specialist feeder. Over 90% of its diet is made up of Antarctic krill, *Euphausia superba*, with only trivial amounts of fish and squid and other invertebrates. The crabeater's teeth are perhaps the most remarkable of any mammal. Each cheek tooth is divided into a complex series of cusps. The main backwardly-directed cusp is flanked by one anterior and two or three posterior cusps (Fig.

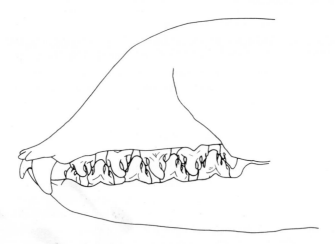

Fig. 5a.8. Cheek teeth of the crabeater seal. The lobed teeth fit together to form a strainer to separate the water from the krill that forms the food.

5a.8). The teeth of the upper and lower rows interlock to form a strainer which enables the seal to retain the krill while expelling the water taken in with the prey. It is these lobed teeth that have given the crabeater its generic name, *Lobodon*. We do not know whether crabeaters catch krill individually, selecting each shrimp from a dense swarm, or whether they take mouthfuls out of the swarm, hoping to secure several krill with each mouthful. From what we know of the density of krill within swarms, the former seems more likely. Whatever the method, it is highly successful. Crabeaters now consume at least 63 million tonnes of krill a year, more than the amount eaten by the surviving Antarctic baleen whales (Laws, 1977).

In contrast, leopard seals are opportunistic predators, taking a wide variety of prey. Leopards are perhaps best known for their activities when feeding on penguins. At the landing place at most Antarctic penguin rookeries there will be a leopard seal lying in wait to take a toll of the birds as they arrive or depart. Adult penguins are pursued in the water and seized at the end of a violent chase. Often the penguin can evade the seal and escape, perhaps bearing severe wounds. Young penguins, however, are easy prey and the leopard can often just swim up to them and take them without a chase.

This behaviour, well known as it is, is misleading. Only a very small proportion of leopard seals are found at penguin rookeries, and this way of feeding is clearly relatively unusual. Leopard seals may also be seen fishing in kelp banks just off-shore, but this again is not very typical of their feeding habits.

Much more important as a food for leopard seals is the flesh of another seal, the crabeater. Leopard seals are important predators on young crabeaters up to the end of their first year of life. As the crabeaters grow larger they are better able to avoid the leopards, though it seems that there are few that are completely successful in this. The spacing of the pairs of parallel scars seen on the flanks of nearly all crabeaters corresponds exactly to the distance between the massive canines of the leopard seal, and there is little doubt that it is the leopard that inflicts these wounds.

In comparison with this large active prey, it seems perhaps surprising that leopard seals take up to 37% of krill in their diet. Like the crabeater, the teeth of the leopard are intricately cusped. Each cheek tooth has a sharp principal cusp with a smaller, but equally sharply pointed cusp, before and behind (Fig. 5a.9). As in the crabeater, the teeth interlock to form a strainer through which water but not krill can escape.

Little is known about feeding in the Ross seal, but what observations there are indicate that it feeds largely on squid (about 57%) and fish (about 34%). The Ross seal shows the rather weak jaws and

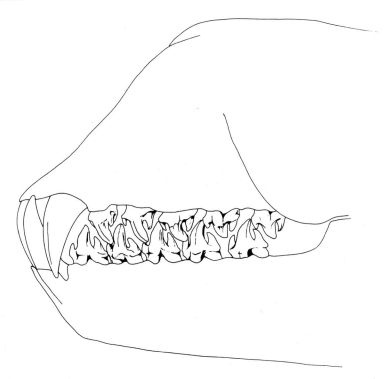

Fig. 5a.9. Cheek teeth of leopard seal. Though leopard seals are much more versatile feeders than crabeater seals, their pointed teeth, like the lobed teeth of the crabeaters, form an effective strainer when feeding on krill.

reduced dentition commonly seen in squid feeders. The teeth show no trace of the cusping so conspicuous in leopard and crabeater seals.

The Weddell seal eats mainly fish (58%) together with some crustacea and a small amount of squid. Weddell seals have the ability to dive deeply to find their food. Although they do not usually stay under for more than 15 minutes, the longest observed dive lasted for 73 minutes (Kooyman, 1981). Dives to 300 – 400 m are common and it seems that one of the favourite food fishes, *Dissostichus mawsoni*, is most commonly caught at these depths.

Elephant seals are mainly squid eaters. Like the Weddell seal they are capable of deep diving and the form of visual pigment in their eyes makes it likely that they find much of their food at depth. The cheek teeth of elephant seals are reduced to little more than widely-spaced pegs, a character probably associated with the squid diet. However, elephant seals do not feed only on squid. About a quarter of their diet is fish.

Antarctic fur seals are krill eaters. They take some fish and squid, perhaps each making up about 10% of the diet, but the remainder is krill. These are swallowed whole, and intact krill can be found in the stomach of a fur seal that has recently fed. In contrast to the crabeater seal, the cheek teeth of the Antarctic fur seal do not form a complicated strainer. In fact, they are greatly reduced, the posterior ones being little more than buttons of enamel. Exactly how the food is separated from the water in the mouth is not clear. Probably it is strained out through the lips. An unexplained mystery about the fur seal is the great amount of wear on the lingual surfaces of the teeth. Krill do not appear to have abrasive carapaces, yet the posterior teeth may be worn nearly level with the gums.

From this account we can see that the seals of the Antarctic feed on different proportions of the available resources, krill, fish and squid. Additionally, one species has become a successful predator on other warm-blooded animals. Antarctic fur seals do not compete with the very abundant crabeater seals for krill, which they both eat, for their ranges do not overlap. Similarly the two squid-eaters, the Ross seal and the elephant seal, are latitudinally separated. The Weddell seal has a mainly coastal distribution and in this region it can exploit fish stocks occurring at considerable depths. The leopard seal has the widest range, both of food and latitude. It is an opportunistic predator, able to take advantage of whatever food, krill, fish, birds or seals, that is available locally. It is perhaps surprising that the leopard seal, like that other versatile and successful predator, the killer whale, is not more abundant.

## 5a.11. REPRODUCTIVE BIOLOGY AND SOCIAL ORGANIZATION OF ANTARCTIC SEALS

Apart from the walrus, pinnipeds show little social structure except at the breeding season, and in most seals the duration of this has been reduced to a bare minimum. The reason for this is not hard to seek. Those same adaptations that have made seals so supremely agile in the water have rendered them clumsy and vulnerable on land. The ancestors of phocid seals in the Northern Hemisphere were liable to attack from terrestrial predators when ashore for breeding, and responded to this threat by seeking out remote secure breeding sites. Additionally they curtailed the time that the mother must spend suckling her offspring to a degree seen in no other comparable sized mammal. Although Antarctic seals have nothing to fear from terrestrial predators, the abbreviated breeding season characteristic of phocid seals has persisted.

The reproductive behaviour of the pack-ice seals is little known. Only in recent years have accounts become available (Bengtson and Siniff, 1981; Siniff et al., 1979). During the breeding season adult male and female crabeater seals form pairs, each to a separate floe in the pack-ice, where they remain for extended periods. The pups are born in September and October, with most births in early to middle October. Lactation lasts about 4 weeks, during which the pup increases its weight from about 20 to about 113 kg, the mother losing up to 50% of her weight in consequence. During this period the male defends an area with a radius of about 50 m around the female from other males, leopard seals and people. Despite the wide spacing of the pairs of seals there is frequent competition for reproductive rights, and the males bear wounds characteristic of interspecific encounters. Some of these wounds may have another origin, however, as the females are very aggressive towards the males. Any approach by the male is violently rejected until towards the end of lactation the female becomes less resistant. Copulation has not been observed, but the male makes great efforts to prevent the female entering the water, which suggests that it takes place on the ice. Perhaps this is because copulating seals would be vulnerable to predators (killer whales and leopard seals) in the water. Weaning of the pup takes place about 4 weeks after birth and is perhaps initiated by the male driving off the pup. Mating occurs soon after.

During the breeding season concentrations of 50 – 1000 crabeater seals, mostly immatures, are found on the seasonal fast ice. It is possible that the aggressive activities of the breeding males have driven them out of the pack-ice.

Very little is known of the breeding behaviour of the leopard and Ross seals. Both breed in the pack and there is no evidence of any social organization, other than that essential for reproduction to occur. Leopard seals probably pup sometime between October and November. The only birth recorded took place on Heard Island on 14 November, but the pup was stillborn, and its mother had been confined in a cage for a month. I have seen a pup that appeared not to be more than three weeks old on a snow-covered beach in South Georgia on 11 September, and at the other end of the scale, a female with a young pup on an ice floe in the Weddell Sea on 4 January. Males do not seem to accompany the females, as in the crabeater seal, and perhaps this may be because copulation occurs in the water.

The few observations on Ross seals indicate that birth takes place during November. It is likely that both Ross and leopard seals have an abbreviated lactation period, as in crabeater seals. Since pack-ice is a relatively unstable substrate, a short lactation period and early independence of the young would be favoured by natural selection.

Weddell seals form breeding assemblies along pressure cracks and tide cracks in fast ice. The seals can keep breathing holes open in the ice by gnawing with their procumbent upper canines and incisors, but prefer to use naturally-occurring open water for breeding. The shortage of suitable cracks causes the concentrations of seals around them.

During the pupping season, from September to November, several females may share a single breathing hole, giving birth on the ice around it. The mothers are active in defending their pups and mothers with young are more widely spaced than seals at other times of the year. The pups weigh about 25 kg at birth and by 10 days have doubled this, being weaned at around 6 − 7 weeks and 110 kg.

The breeding males spend most of their time in the water beneath the cracks, where they defend aquatic territories against other seals. This enables them to have exclusive access to all the females using the same breathing hole or length of crack. As the season wears on and competition for females becomes more intense, so the size of the territory declines. Mating takes place underwater, but has been observed only once. The female involved on this occasion had given birth 43 days previously and was still nursing her pup. Weaning occurs when the mother abandons her pup. Usually shortly after weaning the sea ice breaks out.

Elephant seals have a much more complicated breeding pattern. Early in the spring in September the first of the pregnant females begin to come ashore. The preferred beaches are formed of open sand or shingle and the use of particular beaches is traditional, cows returning again and again to the same site. Some breeding bulls arrive before the cows and take up positions on favoured beaches. The cows are gregarious and form groups, though individuals are aggressive to each other and there is no body contact. The bulls join these groups and drive off intruding males. Usually this is achieved by threat alone. A territorial bull, on sensing an intruder, will first roar at it, inflating the proboscis and producing a deep, grating roar that resonates through the grotesque nasal appendage, which can be seen to vibrate. If roaring is not sufficient to deter the intruder, the bull will rear up, raising the anterior half or two-thirds of its body 3 m or more vertically into the air, whilst continuing to roar. Very exceptionally, two bulls may come to blows, rearing vertically in front of each other and delivering massive swipes and slashes with their greatly enlarged canines to their opponents' neck and chest.

In this way a hierarchy of bulls is set up. The dominant bull, or beachmaster, is in the middle of the group of cows, whilst subordinate bulls are found around the group of cows and in the water just off the shore. As the season advances the groups of cows increase in size and coalesce, reaching a maximum in October (Fig. 5a.10). I have seen a single group of 4000 cows on a beach in South Georgia. As large a group as that would contain several beachmasters.

The pups are born about 8 days after their mothers haul out and weigh around 46 kg. This weight is doubled in 11 days and quadrupled by the end of lactation, which lasts for about three weeks. The cows come into oestrus about 19 days after birth and are receptive for about 4 days (McCann, 1980) during which time they mate several times with one or more bulls. Usually the dominant bull is the first to detect and mate with a cow in season.

Neither the females nor the males feed while ashore. For the females the fast lasts for about 30 days but for the males it is much longer, perhaps 90 days for some of the beachmasters. Because of the strain of fasting coupled with the exhausting activity of defending a place on the beach, beachmasters probably have only one or two seasons of dominance, but during this time they will pass on their genes to many females, perhaps as many as 80 a season. There is thus a strong selective advantage for those characters — size and aggressiveness — that enable a male to maintain his place on the breeding beach, and it is for this reason that there is such a great discrepancy in the sizes of the males and females.

Fig. 5a.10. A breeding beach of elephant seals at South Georgia at the height of the breeding season. The cows form a crowd on the beach. Note the dominant bull (left centre) and the subordinate bulls (foreground and at water's edge).

Adult male elephant seals reach sexual maturity at about 4 years, but do not achieve physical maturity or breeding status until at least 7 years old. Cows produce their first pup at age 4, though not all females are mature as early as this.

The Antarctic fur seal, like the elephant seal, has a complex social structure during the breeding season. The breeding grounds tend to be rocky coasts with very broken relief (Fig. 5a.11). The seals return to their breeding grounds year after year, showing strong site fidelity. The bulls arrive in late October and establish territories 2 − 3 weeks before the females haul out to pup. Territory size declines from about 60 m$^2$ at the beginning of the season to 22 m$^2$ when the beaches are fully occupied.

Bulls spend much time displaying at territory boundaries (usually some physical feature, like a prominent rock, or even a piece of driftwood). The boundary display consists of the male sitting upright with the chest out and head and neck vertical. The vocal threat is a surprisingly high-pitched whimper. If an actual encounter occurs, most commonly when a new bull arrives from the sea, the opponents slash at each other with their sharp canines. Most of the bites connect with the chest and sides of the neck, the regions protected by the bull's heavy mane, but severe wounds can be inflicted, often at the base of the foreflipper. Sometimes a bull will push at its opponent with its chest and drive it back by sheer strength. In prolonged fighting, the two contenders may pause with their heads resting on each others' shoulders.

Successful fighting and territory maintenance means that the bull is on site to mate with the oestrous cows in his territory. A cow produces her pup about 2 days after coming ashore. For eight days she remains with her pup, suckling it at about six-hourly intervals (Fig. 5a.12). At the end of this period she comes into oestrus and is mated by the bull in whose territory she occurs.

Thereafter, the female departs to sea to feed and synthesize more milk for her pup. She is away for 3 − 6 days, then returns to suckle her pup for three days or so. This pattern is repeated for the whole of the lactation period which lasts for 110 − 115 days until April.

Pups grow rapidly, though not so dramatically quickly as the phocid pups. Males gain weight at an average of 98 g per day from a birth weight of 5.9 kg. Females, born at about 5.4 kg, grow more slowly at about 84 g per day (Payne, 1979).

Fig. 5a.11. Antarctic fur seal breeding beach at Bird island, South Georgia. Breeding bulls hold tightly-defined territories, and there is no single dominant bull on the beach.

During the active part of the season, bulls never leave their territories and lose weight at about 2 kg per day. After most of the females have been mated, however, the territorial system begins to break down, with bulls leaving for short periods and their places being taken by newcomers. Non-breeding adult and juvenile males penetrate behind the beaches to take up places on the tussac grass slopes behind. They may attain considerable altitude. Occasionally a young female will produce her pup inland and the pair will soon be adopted by a juvenile male, who defends the cow as a harem. Juvenile females are rarely seen ashore, and perhaps spend most time at sea. At the end of the summer, in April, the seals disperse from the breeding beaches to adopt a pelagic existence until the following breeding season.

## 5a.12. EXPLOITATION AND RECOVERY

Historically, the first of the Antarctic's resources to be exploited were the fur seals. Captain Cook, returning from his voyage of discovery in 1775, reported that the beaches of South Georgia swarmed with fur seals. Unsurprisingly, the sealers were soon at work. By 1822 James Weddell, a Scottish sealer, calculated that no fewer than 1,200,000 skins had been taken from the island and the species was practically extinct there. The discovery of the South Shetland Islands in 1819 revealed another rich stock

Fig. 5a.12. Female fur seal suckling pup.

of fur seals. By 1822 some 320,000 had been taken and this population was nearly wiped out. Further discoveries at the South Orkneys, South Sandwich Islands, Bouvetøya, etc., were treated similarly. Fur sealing declined for want of fur seals. There were minor recrudescences of the industry in the 1850s and 1870s but the seals were given no real chance to recover.

This was partly because the fur sealers had turned to taking elephant seals for their blubber. This was less profitable than fur sealing, so fur seals were killed whenever they were found, preventing any recovery of the stocks.

By the beginning of this century the old fashioned sealers had all but disappeared. The last old-time sealer, the *Daisy*, visited South Georgia in 1913. The master of this brig, Benjamin Cleveland, had previously visited South Georgia in 1907, when he had taken 170 furs. The 1913 voyage yielded none and Captain Cleveland did not return to South Georgia.

Elephant seals were quicker to recover than fur seals, not having been hunted so remorselessly. By 1910 there was quite a good stock of elephant seals at South Georgia and licensed exploitation was permitted by the British administration there (Fig. 5a.13). In contrast to the practice of the earlier sealers, who killed everything, only adult bulls were allowed to be taken. The island was divided into four sealing divisions, only three of which were to be worked in any year, sealing reserves and close seasons were instituted, and a quota of no more than 6000 bulls a year was set.

Under this control sealing was profitably pursued at South Georgia for three decades. Unfortunately, in the 1940s an attempt was made to raise the quota to unjustifiable levels, with adverse effects on the stock. R.M. Laws, a British biologist who had worked on elephant seals at Signy Island, prepared a management plan for sealing at South Georgia. This reduced the quota to what could properly be supported, and allocated it between the various divisions in proportion to the stocks present. Laws' discovery that the age of seals could be determined by examining sections of their teeth (Laws, 1952) (Fig. 5a.14), made it possible to monitor the age structure of the catch, and adjust it accordingly,

Fig. 5a.13. Sealers flensing bull elephant seal at South Georgia. On average each seal would produce about 2 barrels (c. 670 kg) of oil.

Laws' plan was brilliantly successful. The elephant seals recovered and the rate of catch and oil production per seal increased. The elephant seal industry at South Georgia was one of the best examples of successful biological management of a natural resource. The sexual dimorphism and great degree of polygamy shown by the seals made it easy to identify and harvest males without harming the breeding stock. Indeed, the female component of the herd may positively have benefited from sealing. With the removal of some of the males, the females suffered less competition for food from the much larger males. The average age of sexual maturity of the females at South Georgia during exploitation was about 2 years, compared with more than three years in 1977 when seals had not been taken for 13 years. This indicates that growth was faster for the females during exploitation.

The South Georgia elephant sealing industry finally ceased in 1964, not because of any shortage of seals, but because the whaling industry, of which it was an adjunct, had collapsed as a result of the lack of whales around South Georgia.

The fur seal stock took longer to recover. After the last catch at South Georgia in 1907 no fur seals were seen until 1919, when a young male was shot at the eastern end of the island. In the 1930s a small breeding population with 12 pups was found at Bird Island at the north-west end of South Georgia.

I visited Bird Island again in 1956 and found a flourishing population there with a pup production I estimated at 3500 (Bonner, 1958). This population has since been monitored carefully and its astonishing recovery documented (Bonner, 1968; Payne, 1977). I have mentioned earlier the difficulty

Fig. 5a.14. Section of canine tooth of elephant seal. The repeated cycles of light and dark dentine can be clearly seen. This seal was in its ninth year.

of counting the seals, but careful analysis has shown that between 1958 and 1972 the population had been increasing at 16.8% per year. At that rate of growth, the doubling time for the number of pups born is only 4.5 years.

The seals showed strong site fidelity, so density on beaches at Bird Island increased greatly before new colonies were established on the mainland of South Georgia. By 1975/76 over half the total of pups was born on mainland beaches and the proportion has continued to rise. However, fur seals are still very much concentrated at the north-west end of the island and many important former localities have no breeding seals, or only negligible groups.

But it is not only at South Georgia that the fur seal is increasing. Small breeding groups are now present at the South Shetlands, South Orkneys and South Sandwich Islands which are almost certainly derived from the South Georgia stock. At the South Shetlands, for example, the seals have been increasing at 34% a year (Aguayo, 1978), a rate which is impossible unless their numbers are reinforced by immigration from elsewhere. The occurrence of white-coated animals at these other island groups as well as at South Georgia makes it almost certain that the latter island is the origin of the other colonies.

It is encouraging that this mammal, which so nearly was exterminated, seems now well set on the road to recovery. By the turn of the century we may hope to see it present in something like its original numbers. However, it is ingenuous to suppose that the previous system can be re-established exactly as it was. The extreme perturbation that has taken place has had permanent effects. The unnaturally high density of fur seals at Bird Island, for example, has resulted in much destruction of vegetation and the erosion of the lowest of the raised beach systems in one locality. Such changes may appear trivial, but they are permanent and not until many years have elapsed and the total fur seal population of South Georgia has come into equilibrium with its food supply can we expect to see a reduction of the population pressures at places such as Bird Island. Only then will the system have returned to normality for the seals, but by that time the landscape of parts of South Georgia may have changed as a result of the activities of the seals.

The lobodontine Antarctic seals represent a vast natural resource that has been virtually completely untapped. The only occasion on which any substantial harvest appears to have been taken was in 1892/3 and 1893/4, when four Dundee whalers and a Norwegian vessel, prospecting for right whales and finding none, took 32,558 seal skins and a great quantity of blubber from the pack-ice near the Antarctic Peninsula. We do not know which species was concerned, but the numbers involved makes the crabeater a likely candidate. The only other commercial venture was in 1964 when a Norwegian sealer, *Polarhav*, took 861 seals in an exploratory cruise between the South Orkneys and Elephant Island.

The seals are certainly there in plenty, but it seems to me unlikely that harvesting Antarctic seals will ever be commercially viable. Such work would be highly labour intensive, even more so than the sealing in the North-West Atlantic, since the Antarctic seals do not form the close associations on the ice that harp seals do. Fuel costs would be greater and the product itself might be less valuable than harp and hood seal skins. Fur seals would be relatively simple to exploit, and the industry would be easy to manage in a way that would not injure the stock. However, this again would be very laborious since, in contrast to conditions in the Pribilof Islands where northern fur seals are harvested, the commercially attractive seals, the juvenile males, do not conveniently gather together in groups on accessible beaches but spread out over the inland hills. Perhaps only elephant seals offer a prospect of an economically feasible industry for the future.

Should sealing start again in the Antarctic there are adequate measures to control it and prevent a repetition of earlier catastrophes. However, there is another very significant factor in the redevelopment of a southern sealing industry. A vast body of public opinion opposed to killing seals has been mustered in recent years, and administrations might be reluctant to expose themselves to opprobrium from these people by licensing sealing, even if the conduct of the industry could be guaranteed to be irreproachable. The aesthetic, rather than the rational, approach may govern the future development of Antarctic seal resources, and perhaps the region's other resources as well.

## REFERENCES

Aguayo, A. (1978) The present status of the antarctic fur seal *Arctocephalus gazella* at the South Shetland Islands. *Polar Rec.* 19(119), 167 – 76.

Bengtson, J.L. and Siniff, D.B. (1981) Reproductive aspects of female crabeater seals (*Lobodon carcinophagus*) along the Antarctic Peninsula. *Can. J. Zool.* 59, 92 – 102.

Berry, J.A. (1961) The occurrence of a leopard seal (*Hydrurga leptonyx*) in the tropics. *Ann. Mag. nat. Hist.* 13th ser. 1960 (1961), 591.

Bertram, G.C.L. (1940) The biology of the Weddell and crabeater seals, with a study of the comparative behaviour of the Pinnipedia. *Brit. Graham Land Exped., 1934 – 37. Sci. Rep.* 1(1), 1 – 139.

Bonner, W.N. (1958) Notes on the southern fur seal in South Georgia. *Proc. zool. Soc. Lond.* 130, 241 – 52.

Bonner, W.N. (1968) The fur seal of South Georgia. *Brit. Antarct. Surv. Sci. Rep.* 56, 81 pp.

Bonner, W.N. and Laws, R.M. (1964) Seals and sealing. In *Antarctic Research*, Eds. R. Priestley, R.J. Adie and G. de Q. Robin, pp. 163 – 190. Butterworths, London.

Bruce, W.S. (1913) Measurements and weights of Antarctic seals. *Trans. roy. Soc. Edin.* 49, 567 – 77.

Condy, P.R. (1976) Results of the third seal survey in the Haakon VII Sea, Antarctica. *S. Afr. J. Antarct. Res.* 6, 2 – 8.

Erickson, A.W., Siniff, D.B., Cline, D.R. and Hofman, R.J. (1971) Distributional ecology of Antarctic seals. In *Symposium on Antarctic Ice and Water Masses*, Ed. G. Deacon, pp. 55 – 76. SCAR, Cambridge.

Gilbert, J.R. and Erickson, A.W. (1977) Distribution. and abundance of seals in the pack ice of the Pacific sector of the Southern Ocean. In, *Adaptations within Antarctic Ecosystems*, Ed. G.A. Llano, pp. 703 – 40. Smithsonian Institution, Washington.

Hamilton, J.E. (1939) The leopard seal *Hydrurga leptonyx* (de Blainville). *Discov. Rep.* 18, 239 – 64.

Ingham, S. (1960) The status of seals (Pinnipedia) at Australian Antarctic stations. *Mammalia* 24, 422 – 30.

Kooyman, G.L. (1981) *Weddell Seal: Consummate Diver.* Cambridge University Press, Cambridge.

Kooyman, G.L., Keren, D.H., Campbell, W.B. and Wright, J.J. (1973) Pulmonary gas exchange in freely diving Weddell seals, *Leptonychotes weddelli. Respir. Physiol.* 17, 283 – 90.

Laws, R.M. (1952) A new method of age determination for mammals. *Nature, Lond.* 169, 972.

Laws, R.M. (1960) The southern elephant seal (*Mirounga leonina*, Linn.) at South Georgia. *Norsk Hvalfangsttid.* 10 & 11, 466 – 76; 520 – 42.

Laws, R.M. (1977) Seals and whales of the Southern Ocean. *Phil. Trans. R. Soc. Lond.* B 279, 81 – 96.

Lindsey, A.A. (1937) The Weddell seal in the Bay of Whales, Antarctica. *J. Mammal.* 18, 127 – 44.

McCann, T.S. (1980) Population structure and social organisation of the southern elephant seal *Mirounga leonina* (L.). *Biol. J. Linn. Soc. Lond.* 14, 133 – 50.

Payne, M.R. (1977) Growth of a fur seal population. *Phil. Trans. R. Soc. Lond.* B 279, 67 – 79.

Payne, M.R. (1979) Growth in the Antarctic fur seal *Arctocephalus gazella. J. Zool., Lond.* 187, 1 – 20.

Ray, G.C. (1981) Ross seal. In *Handbook of Marine Mammals*, Eds. S.H. Ridgway and R.J. Harrison, pp. 237 – 60. Academic Press, London.

Ross, G.J.B., Shaughnessy, P.D. and Best, P.B. (1978) New records of crabeater seals (*Lobodon carcinophagus*) from South Africa. *Ann. S. Afr. Mus.* 75, 153 – 8.

Siniff, D.B., Cline, D.R. and Erickson, A.W. (1970) Population densities of seals in the Weddell Sea, Antarctica, in 1968. In *Antarctic Ecology*, vol.1, Ed. M.W. Holdgate, pp. 377 – 94. Academic Press, London.

Siniff, D.B., Stirling, I., Bengtson, J.L. and Reichle, R.A. (1979) Social and reproductive behaviour of crabeater seals (*Lobodon carcinophaga*) during the austral spring. *Can. J. Zool.* 57, 2243 – 55.

Smith, M.S.R. (1965) Seasonal movements of the Weddell seal in McMurdo Sound, Antarctica. *J. Wildl. Manage.* 29, 464 – 70.

Stirling, I. and Kooyman, G.L. (1971) The crabeater seal (*Lobodon carcinophagus*) in McMurdo Sound, Antarctica, and the origin of mummified seals. *J. Mammal.* 52, 175 – 80.

Stirling, I. and Rudolph, E.D. (1968) Inland record of a live crabeater seal in Antarctica. *J. Mammal.* 49, 161 – 2.

Vaz Ferreira, R. (1956) Caracteristicas generales de las islas Uruguayas habitados por lobos marinos. In *Ministerio de Industrias y Trabajo. Servicio oceanograficio y Pesca*, Trabajos sobre las Islas de Lobos y lobos marinos. Montevideo.

Vaz Ferreira, R. (1965) Ecologia terrestre y marina de los pinnipedios del Atlantico Sudoccidental. *Acad. Brasil. Cienc. An. Suppl.* 37, 179 – 91.

Wilson, E.A. (1907) Mammalia (Seals and Whales). In *National Antarctic Expedition, 1901 – 1904, Natural History*, pp. 1 – 66, Trustees Br. Mus., London.

# CHAPTER 5b

# Birds and Mammals — Antarctic Whales

RAY GAMBELL

International Whaling Commission, The Red House, Station Road, Histon,
Cambridge CB4 4NP, U.K.

## CONTENTS

## 5b.1. INTRODUCTION

The whales found in the Antarctic, illustrated in Fig. 5b.1, are the blue, fin, sei, minke, humpback and right whales which are all baleen whales, and the sperm and killer whales which are odontocete or toothed whales. There are also a number of less well known smaller cetaceans including the southern bottlenose and beaked whales.

The baleen, or whalebone whales, are recognizable by the triangular plates of horny baleen which grow from the gums of the upper jaws and form the food-sieving mechanism of these animals. The body is more or less streamlined and the head relatively large, with a paired blowhole on the top.

The blue whale (*Balaenoptera musculus*) is the largest of all species of whales and probably the biggest animal which has ever existed. Lengths of up to 30 m and a weight exceeding 150 tonnes can be attained. The fin whale (*B. physalus*) grows to 25 m and a weight of 90 tonnes, and the sei whale (*B. borealis*) 18 m and 30 tonnes. These three species are all similar in their torpedo shaped form, as is the small minke whale (*B. acutorostrata*) which seldom exceeds 11 m in length and 19 tonnes in weight. The humpback whale (*Megaptera novaeangliae*) also has a dorsal fin and ventral throat grooves and is placed in

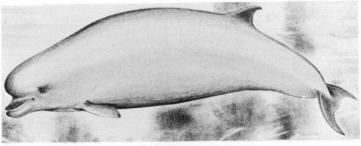

S.Bottlenose

Killer

Humpback

Fig. 5b.1. The Antarctic whale species.

Fin

Right

Sperm

Fig. 5b.1. *cont.*

the same family of rorquals, but it is rather plumper in form and the flippers are extraordinarily long, up to a third of the total length of the animal which may reach 16 m, with a maximum weight of 60 tonnes.

The right whale (*Balaena glacialis* or *australis*) is again a plump animal, reaching a weight of 90 tonnes for an 18 m specimen, with a large head making up about a quarter of the total length. There are no throat grooves or dorsal fin.

The sperm whale (*Physeter macrocephalus*) is the largest of the toothed whales and shows a marked sexual dimorphism. Only the larger males, which grow to 18 m and 70 tonnes, penetrate the Antarctic seas. The females, which are less than 11.5 m and 17 tonnes in size, juveniles and smaller whales remain

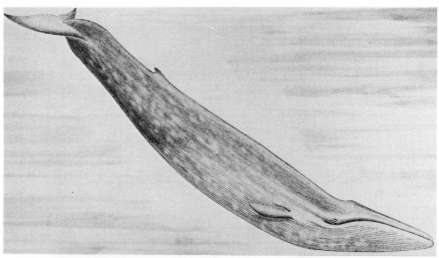

Blue

Sei

Minke

Fig. 5b.1. *cont.*

in temperate and tropical waters. Killer whales (*Orcinus orca*) show a smaller difference in size between the sexes, the males reaching 9 m and the females 8 m, with weights up to 7 or 8 tonnes. The dorsal fin in adult males can measure up to 1.8 m in height, twice the size found in the female.

Toothed whales differ from all other mammals in having only one nostril, normally set on the mid-line, although two nasal passages are present internally. The blowhole of the sperm whale is sigmoid in shape and set asymmetrically on the forward part of the left side of the head. The right naris (the tubular part of the nostril) has developed into the waxfilled 'case' set on top of the skull giving the sperm whale its characteristic blunt head, making up to one third of the total body length. It is thought that the spermaceti wax may have roles in both buoyancy regulation and sound transmission.

The southern bottlenose whale (*Hyperoodon planifrons*) has a somewhat cylindrical body tapering towards the tail, with a distinctive 'melon' giving rise to the common name from the bulbous head. The length reaches 7 m or a little more and the weight 3 − 4 tonnes.

The beaked whales are more spindle-shaped in form with Arnoux's beaked whale (*Berardius arnuxii*) reaching up to 10 m in length and 7 − 8 tonnes weight. Some of the smaller (5 m) beaked whales of the genus *Mesoplodon* may also penetrate into the Antarctic proper, although most species, including the strap-toothed whale *M. layardii*, appear to be more sub-Antarctic in their southern-most distribution.

## 5b.2. GENERAL LIFE HISTORY

Although the large baleen whales feed in the Antarctic waters during the summer months, they breed in the warm tropical and sub-tropical waters to the north. Their life cycle is very closely related to the pattern of seasonal migrations between the breeding and feeding areas. During the winter months the humpback, blue, fin and sei whales mate and conception occurs in many of the females. In the spring the whales migrate southwards towards the polar feeding grounds and as the ice retreats many of the whales enter the Antarctic and start to feed on the rich planktonic organisms which constitute their diet, and especially on Antarctic krill, *Euphausia superba*, which is abundant in these waters. After a period of three or four months of intensive feeding the whales migrate northwards once more in the autumn to the temperate waters where approximately a year after mating the pregnant females give birth, normally to a single calf. Rather little feeding appears to take place outside the Antarctic in these species, the animals living largely on the food reserves laid down in the blubber and body tissues during the summer.

The newborn calves accompany their mothers on the following spring migration towards the polar waters once again, living on their mothers' milk. Some six months after birth they are weaned in the rich Antarctic feeding grounds and can follow the normal cycle of migrations independently, although it seems that humpback calves may be dependent for up to a year on their mothers. The reproductive cycle therefore extends over 2 or 3 years in these species, and is also 3 years in the southern right whales observed calving in the protected shallow bay off Valdez, Argentina (Payne, 1983).

The migrations of minke whales do not seem to follow quite the same pattern as those of the larger rorquals, since some minke whales are present in temperate waters throughout the year, for example, in the south western Indian Ocean (Best, 1982). These are mainly small immature animals in summer and autumn, when the larger adults are in the Antarctic. It is only the adults which move into the Antarctic feeding grounds.

Sperm whales have a fundamentally different pattern of reproduction and associated seasonal migrations from the baleen whales. This is linked to a complex social organization in this polygynous species. The schools of females and young males which are found in the more temperate waters all the year round are joined by one or more breeding bulls during the summer when mating takes place. The exact relationship between a harem bull and the females is not known in terms of the length of time for which it persists, or if one male can associate with more than one group of females. The gestation period extends over nearly 15 months, and is followed by 2 years of lactation. The degree of dependence of the sperm whale calf on its mother appears to lessen towards the end of the lactation period, but much older animals have also been examined which contained both squid remains and milk (Best *et al.*, 1983). The reproductive cycle includes a resting period before the next pregnancy, giving a basic cycle of 4 or 5 years.

The reproductive behaviour of the small toothed whales found in the Antarctic is little known. The gestation period is 12 months in the killer whale, and is possibly of about the same duration in the bottlenose and beaked whales, but there is no firm evidence on the relationship of these southern animals with those which occur outside the Antarctic zone.

### 5b.2.1. Segregation

The sex ratio at birth and throughout the greater part of life is approximately 1:1 in all species, but because of the differential segregation of the sexes and sexual classes at various times in the seasonal migration of the larger species, the numbers of males and females present in an area at a particular time may show an imbalance. This is reflected in the catches which have been taken by whalers at land stations on the migration routes and by whaling operations at the extremities of the migration range. There is also segregation to some extent between the various baleen whale species as they follow rather similar migration paths.

Examination of the catches taken both at South Georgia and on the pelagic whaling grounds (Mackintosh, 1942) shows that blue whales arrive before the fin whales, the former being most abundant in the catches from November to February while the fin whales are more numerous in the later part of the season from January to March. Humpback whales are most frequent in December and January while the sei whales are the latest arrivals with the highest densities occurring in the period January to March (Gambell, 1968).

Not only do the species vary in their time of arrival, but the extent of their penetration into high latitudes also varies. The blue and minke whales go further into the Antarctic and right up to the ice edge, but the fin whales do not reach such high latitudes and the sei whales are rather more sub-Antarctic in their summer distribution. Blue whales were hunted up to the ice edge in the pre-war Antarctic pelagic fishery and the current minke whale fishery also extends into the pack-ice waters. In all these baleen whale species in the Southern Hemisphere the bigger and older animals tend to go further towards the polar regions than the younger whales.

Because humpback whales characteristically stay close to coastlines during their migrations between the summer and wintering grounds (Chittleborough, 1965), the composition of the migrating stream of animals has been more easily studied in this species. In the vanguard of the migration and arriving first in the Antarctic are a mixture of females in a sexually resting condition and others in early pregnancy carrying small foetuses. Following these whales at intervals of several days are successive groups comprising mainly the immature animals, the mature males, and females in early lactation accompanied by their young calves. These are the rearguard of the migration stream (Dawbin, 1966). On the Antarctic feeding grounds there appears to be a random aggregation of humpback whales once they have all arrived, but the pregnant females stay rather longer than the other classes, presumably to build up their food reserves before the main stresses of gestation and lactation. However, the lactating females stay the shortest time on the feeding grounds.

In fin whales there is not such firm evidence for segregation of the sexual classes because the animals do not concentrate so close to continental coasts. On the basis of catches made in the Antarctic it appears that the fin whale migration into the feeding areas is headed by the males while the females tend to come later. Older animals precede the younger, and Laws (1961) concluded that the older animals and pregnant females migrate first with the sexually immature whales towards the rear of the migration stream.

For the sei whales, Gambell (1968) found that the pregnant females are the first arrivals at South Georgia and that a greater proportion of older and bigger whales penetrate into the higher latitudes. Catches between 40°S − 45°S have a greater proportion of smaller and younger whales than those further south (Doi *et al.*, 1967).

Minke whale catch data have demonstrated that there is considerable temporal and spatial segregation of the sexes on the high latitude feeding grounds (Kasamatsu and Ohsumi, 1981). Females predominate nearest the ice edge while males are more abundant away from the edge of the pack-ice, even though they arrive earlier and stay longer on the feeding grounds than the females.

As male sperm whales grow they become increasingly independent. Adolescent males of 9 − 11 m length and aged 10 − 15 years become segregated from the female schools and form groups of medium-sized animals as their testes slowly mature over a period of a further 10 − 15 years. Although the young mature males seem to produce spermatozoa in small quantities it is the older bulls 14 m and longer and over 25 years in age which are socially acceptable as harem masters and which probably form the main breeding component of the male stock (Best, 1969; IWC, 1971). It is also only the older and bigger bulls which penetrate into the Antarctic waters.

Evidence of the migration of the larger male sperm whales out of the Antarctic to more temperate waters is provided by the occurrence of the Antarctic diatom *Cocconesis ceticola*. This organism forms a yellowish film on the skin of sperm whales in the Antarctic in about one month, but does not persist or develop in warmer waters. Medium and large sperm whales bearing the diatom found off the west coast of South Africa in the autumn (Best, 1969) and in early to mid-summer off Western Australia (Bannister, 1969) suggest strongly that these whales had only recently arrived from the Antarctic. Immigration into the Antarctic is implied by the absence or light development of the diatom film on sperm whales around South Georgia in the summer months, but its full blooming in the following months as it develops on the presumed new arrivals from further north (Matthews, 1938).

## 5b.2.2. Stocks

The humpback, blue, fin, sei and minke whales are thought to be divided into a number of stocks around the Southern Hemisphere. These stocks approximate to separate breeding groups, but there is a degree of interchange evident from the recovery of marked whales which indicates that the stocks recognized are not totally independent genetically.

Based mainly on evidence from the concentrations of whales reflected in the commercial catches and sightings in the Antarctic, Mackintosh (1942) proposed that blue whales can be divided into six areas extending through some 60° − 70° of longitude (shown in Fig. 5b.2). These feeding concentrations are considered as stocks at least for management purposes and are believed to correspond broadly to biological stocks. Fin whales are also considered to have similar stock divisions, with the addition of a sub-division of the fin whales in the Atlantic and western Indian Ocean sectors into eastern and western components, to give a total of eight stocks.

The evidence for comparable stock separation of sei and minke whales in the Southern Hemisphere is sparse at present. The traditional six Antarctic areas are therefore used for management purposes, although there is some evidence from the pattern of catches that sei whales in the Atlantic sector occur in eastern and western components (Gambell, 1976).

Catches of humpback whales in the Antarctic indicated a high degree of segregation of local stocks, and the five nearly separate concentrations on the feeding grounds were proposed as separate stocks by Mackintosh (1942) and are shown in Fig. 5b.2. Recoveries of marked whales have demonstrated that there is a direct connection between the humpback whales which migrate past New Zealand and East Australia to the breeding grounds to the north with those to the south on the feeding grounds. Similarly, marked whales have provided the link between the Western Australian humpback whales and those to the south, and between Madagascar and the area in the Antarctic to the south (Raynor, 1940; Dawbin, 1966).

Evidence for separation of the sperm whales in the Southern Hemisphere into separate stocks is sparse, although it is clear that the females occupying the oceanic basins and separated by the major continental land masses must constitute independent breeding units. It is possible that there may be some genetic exchange through movements of the large males into the higher latitudes and their return to a different female population. On the basis mainly of patterns of density in catches and sightings, together with the evidence from whale mark recoveries showing a broad north – south component of movement and discreteness in distribution, the nine stocks shown in Fig. 5b.3 have been recognized and are used for management and assessment purposes (IWC, 1973).

Nothing is known of the stock composition of the smaller cetacean species occurring in the Antarctic.

### 5b.2.3. Groupings

The numbers of whales commonly found together in the Antarctic vary considerably. Since the whales are largely occupied with eating as much as possible during the southern summer months their groupings may reflect more the abundance and distribution of the prey species, rather than the social tendencies of the whales themselves. Blue, fin, sei, minke and humpback whales are commonly found in groups of three to five, although solitary whales and much larger congregations are also not infrequent. Aggregations of up to 100 minke whales have been observed on the feeding grounds diving synchronously, and similarly large groups of fin whales have also been reported.

Right whales commonly form groups of less than six in the warmer waters, and this probably reflects a family unit. Little is known about their numbers when they penetrate south of the Antarctic

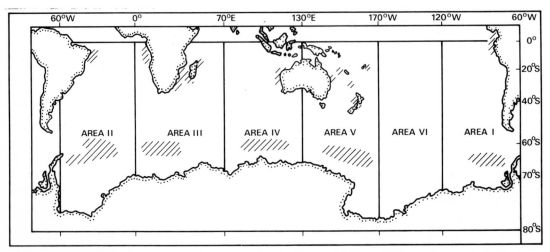

Fig. 5b.2. Areas of the Southern Hemisphere used for management of the baleen whale populations, based largely on the former concentrations of blue whales in the Antarctic. The shaded areas denote the concentrations of humpback whales on their feeding and breeding grounds.

Convergence, and there is a similar dearth of information on the infrequently seen southern bottlenose and beaked whales.

Sperm whales in the Antarctic tend to be solitary and evenly distributed, although short-term localized feeding groups may form on occasion. Killer whales by contrast exhibit a high degree of group co-operative activity, usually moving in an extended family pod of five to twenty animals but sometimes combining into a larger group of 100 or more whales.

### 5b.2.4. Feeding

Baleen whales feed in two main ways, either swallowing or skimming food from the water (Mitchell, 1975; Nemoto, 1959, 1970).

Swallowing-type feeding occurs when the whale engulfs a mouthful of food and water which is then sieved through the baleen plates by contraction of the previously distended ventral grooves beneath the chin and throat and by raising the tongue in the mouth. The food organisms are retained on the baleen filter and subsequently swallowed. Blue, fin, minke and humpback whales all feed in this way. It is claimed that the asymmetrical coloration of the fin whale lower jaw, being white on the right, is associated with the feeding behaviour of this whale which often turns on its right side as it swims through the swarms of krill and gulps the prey.

Humpback whales have been observed carrying out some rather exciting feeding behaviours. These include circling a swarm of krill then diving under the surface and coming up vertically with the mouth open to engulf the concentrated food organisms. A second method is for a humpback to swim in a circle below the surface loosing a trail of air bubbles which form a whorl rising to the surface. The whale then swims up through the centre of this bubble net to engulf the food concentrated in the centre once again.

Skimming feeders swim through the swarms of food organisms with the mouth half open and usually the head raised partly out of the water. The water and food are filtered through the baleen plates and when a quantity of food material has thus been accumulated on the filtering structure the mouth is closed and the food swallowed. This behaviour is typical of the right whale.

Sei whales can feed by either swallowing or skimming type activity.

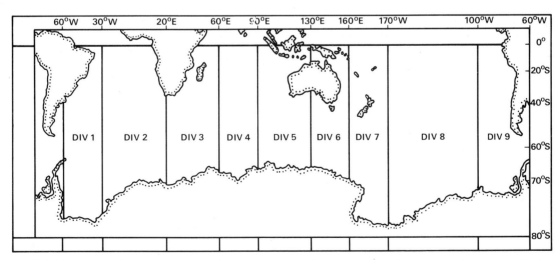

Fig. 5b.3. Divisions of the Southern Hemisphere used for the management of sperm whale populations, each comprising a region of higher density.

The baleen plates of the baleen whales have characteristic shapes and the fibres fringing the inside are different between the species in terms of their fineness of structure. The blue whale has a rather coarse bristle filter. The fin whale has somewhat finer fringes, while the sei whale and right whales have extremely fine fringe fibres which suggests that they are able to filter out much smaller food organisms. The baleen fringes in humpback and minke whales are intermediate in coarseness.

In the Antarctic the predominant whale food is Antarctic krill, *Euphausia superba*. It is commonly found in the stomachs of all the baleen whale species, but other euphausiids such as *Thysanoessa macrura* are sometimes taken by blue and fin whales, while the amphipod *Parathemisto gaudichaudi* is eaten by sei whales in some areas. Feeding in the Antarctic would appear to be determined by what is present rather than through specific selection or preference on the part of the whale. However, right and sei whales with their finer baleen filters are able to strain out smaller food organisms than the species with coarse baleen fibres, and they also consume copepods such as *Calanus* if and when they occur (Kawamura, 1980).

In areas outside the Antarctic these baleen whale species have much broader spectra of food organisms, so that fish are frequently eaten by humpback, minke and to some extent by fin whales; copepods are eaten especially by right and sei whales, and also by fin whales; whole squids also appear in the diet of these species. These organisms may also appear incidentally in the stomachs of the Antarctic whales from time to time.

Sperm whales feed almost exclusively on squid in the Antarctic, although fish are sometimes found in the stomachs too (Matthews, 1938). The cephalopods are normally of moderate size, about 1 m in body length, and are swallowed whole. Feeding appears to occur at great depths, because surface-living squid seldom occur in the stomachs. Just how the sperm whale with its apparently cumbersome long lower jaw manages to catch and consume its prey is unknown. Beale (1839) believed that the sperm whale when feeding remains still in the water and opens its mouth so that the cephalopods are attracted by the pale coloured lining of the mouth and lower jaw. However, the available light at the depths of down to 3000 m where sperm whales are known to feed is little or nothing, and a feeding method employing echo-location seems probable. Recently a sound-stunning technique has been proposed (Norris and Møhl, 1981), in which the sperm whale is thought to project a beam of sound onto its prey to render it immobile.

Killer whales feed on squid, fish, sea birds and marine mammals. They often appear to hunt in groups or packs, co-operating to attack, harry or distract the prey. The food items found in the stomachs of killer whales in the Antarctic include portions of a wide range of species, ranging from fish such as Antarctic cod, through squid, to Adélie penguins, crabeater, leopard and southern elephant seals, and minke whales (IWC, 1982). Feeding may be scavenging rather than direct predation for some of these organisms, but there are reports of groups attacking and killing minke whales and seals. It is likely that feeding is to a large extent opportunistic in the sense that any suitable prey, live or dead, will be taken. The smaller food items are swallowed whole, and the larger prey ripped apart by the powerful jaws armed with large and closely interlocking teeth.

As far as is known from the few bottlenose and beaked whales examined in southern waters, these whales feed almost exclusively on squid.

## 5b.3. REVIEW OF WHALING IN THE ANTARCTIC

Modern whaling dates from the year 1864 when the Norwegian Svend Foyn developed a new whaling technique. This was a combination of a steam-powered whale boat mounted with a cannon on the bow which fired a harpoon fitted with hinged barbs that opened out in the body of the whale. The tip of the harpoon consisted of a cast iron grenade full of explosive which detonated in the body of the animal. This harpoon was attached to a winch on the whaling vessel by a strong rope which passed over an accumulator, a series of springs which took up the strains and tensions caused by the different movements of the boat and the whale.

Whaling in the Antarctic dates from 1904 when the Norwegian whaler C.A. Larsen opened the whaling station at Grytviken in Cumberland Bay, South Georgia. On sealing and exploratory expeditions in earlier years, Larsen had noted the great number of whales in the area compared with the very reduced populations in the North Atlantic where he had previously operated. He therefore founded the Compania Argentina de Pesca which ran the Grytviken whaling operation but under a British licence.

The success of the new operation encouraged other whalers to move to the Antarctic and set up whaling stations, not only on South Georgia where a British company operated at Leith Harbour, but also in the South Orkney, South Shetland and Deception Islands.

The amount of space available on land suitable for setting up whaling factories was limited and so a number of floating factory ships also moored in suitable harbours and the whales caught were worked up in the water alongside. From the first year's operation, 1904, with one factory operating with a single catcher and taking 195 whales, the Antarctic whaling enterprise grew rapidly so that during the 1912/13 season six land stations and 21 floating factories were operating with 62 catchers. They took 10,760 whales, nearly all fin, humpback and blue whales.

Because all this early Antarctic whaling was carried out in the Falkland Islands Dependencies they were within territorial waters claimed by the Government of the United Kingdom. The Government could therefore control the number of land stations and floating factories at work, charge for licences and tax the oil produced, which was the main product from the industry at that time. This government regulation encouraged whalers to seek a way round the restrictions imposed on them and the idea of whaling from a factory ship on the open sea became increasingly attractive. A number of attempts were made to solve the main problem of finding a means for getting the whales up onto the deck of the factory ship so that they could be cut up on board instead of in the water alongside the vessel. In 1925 the factory ship *Lancing* fitted with a stern slipway operated successfully for the first season in the Antarctic. This development led to a further boom in the Antarctic whale fishery so that in the 1925/26 season there were two floating factories at work with nine catchers, while in 1930/31 forty-one factories operated with 205 catchers and caught some 37,500 whales. Together with the catches from the 6 land stations, over 40,000 whales were taken in that one season.

Over-production of oil and a desire to stabilize the market for whale products caused a temporary reduction in effort and catches for a few years, but both increased again until the Second World War intervened.

Species of whales caught during the Antarctic activities have changed over the years. Initially the shore whaling operations were able to take numbers of humpback whales which were close to land around the island bases. As the numbers of these animals were reduced the catchers tended to go further and further afield taking the largest whales which they came across, the blue and fin whales. These were the preferred species during the heyday of pelagic whaling activity from the mid-1920s onwards. The Norwegian and British fleets were joined by Japanese and German operations in the mid-1930s and whaling extended over the whole area from the Weddell Sea eastwards to the Ross Sea. Just before the Second World War factory ships from the United States of America and one registered in Panama were operating, while after the war a Dutch and a South African floating factory were involved in the Antarctic fishery.

Up to 21 factory ship expeditions and three land stations operated each season once the industry recovered from the war, with an average 256 catchers active every year in the 1950s. But this was the end of the era, and the Antarctic whale fishery has since declined steadily until it is now reduced to two factory ships with nine catchers.

Pressure on the whale stocks created by the enormous pre-war catching effort was more than the blue whale stocks could sustain and as the numbers declined greater emphasis was placed by the whaling fleets on capturing fin whales in order to maintain their productivity. The annual catches of whales by species in the Antarctic are shown in Fig. 5b.4. After the Second World War large catches of fin whales continued to be taken with more than 27,000 removed each season during the 1950s, so that these stocks also were much reduced. As the numbers of these larger whales declined the industry attempted to maintain its profitability by increasing the amount of meat produced from the carcases with a lesser emphasis on oil production. This led to greater numbers of sei whales being taken, but these stocks too were reduced rather rapidly when more than 60,000 sei whales were caught in four seasons in the mid-1960s. The current Antarctic pelagic fishery is based solely on an operation by Soviet and Japanese vessels hunting exclusively for minke whales for meat which is marketed for human and animal consumption. Catches have been limited to 7 – 8000 whales in recent seasons.

The catch of sperm whales in the Antarctic has always been somewhat secondary in importance to the industry, although nonetheless a significant component of the total operation. Because the oil from sperm whales, which is the prime product, is chemically very different from baleen whale oil, it has to be kept separate. The whaling fleets therefore tended to catch sperm whales before or after the main baleen whaling season in order that the decks and processing equipment in the factories would not need to be continually washed down as the different kinds of whales were processed. In more recent years when the available catch of baleen whales has been much reduced, a greater emphasis on the catching of sperm whales occurred, but this has been a relative change rather than a real increase in the numbers of sperm whales taken. In the ten seasons before 1979/80 when international regulation prohibited the capture of all whale species in the Antarctic except minke whales by factory ships, the catch of sperm whales averaged 3211. This compares with 4393 and 4622 sperm whales in the preceding two decades respectively.

Catches of killer whales in the Antarctic have generally been small. The USSR took 110 in 1958/59, but from 1969 to 1978 the average was 24 per season. However, when the catch of sperm whales in the Antarctic was reduced to zero in the 1979/80 season, the USSR caught 916 killer whales which have a somewhat similar oil. Subsequently the prohibition on catching any whales except minke by factory ship operations has been specifically extended to include killer whales.

Because of uncertainties in the scientific assessments of the whale stocks, and the past failure of international management through inadequate knowledge of the stocks, the International Whaling Commission has decided that there should be a pause in all commercial whaling from 1986 to 1990. It is uncertain if all the whaling nations will abide by this decision whilst the stocks are being reassessed.

## 5b.4. CHANGES IN BIOLOGICAL CHARACTERISTICS

It is characteristic of these species of whales that they reach sexual maturity at a given body size. As with many other animals, the critical factor is probably body mass, but in the case of the large whales the body size is most conveniently measured in terms of length. The mean lengths at sexual maturity are shown in Table 5b.1, together with the normal maximum sizes attained by these species in the Antarctic.

The age of baleen whales is usually determined by counts of the growth layers found in the earplug. This is a horny plug lying in the external auditory meatus, and its primary function is thought to be associated with the acute hearing capabilities of the whales. The size, shape and consistency of the earplug varies between the different baleen whale species, but it was found that when the plug is cut longitudinally a series of alternating light and dark laminae is revealed (Purves, 1955).

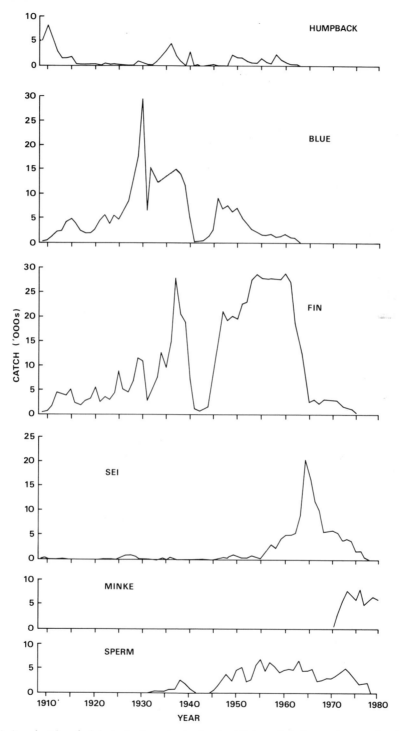

Fig. 5b.4. Annual catches of whales in the Antarctic, showing the shift in emphasis from species to species with time.

The original use of earplugs for age determination was based on the fin whale, in which it was shown by direct histological examination (Roe, 1967) that a growth layer comprising one light and one dark lamina is formed each year. Secondary confirmation of the rate of layer formation from the numbers of growth layers in recoveries of long-term marked fin whales is also available. Few earplugs have been collected from blue and humpback whales because of the small catches of these species since the usefulness of this technique was discovered, but sei and minke whales can also be successfully aged by this method.

The pattern of the growth layers is rather irregular in the early years but becomes more even and compact in the older whales. It has been shown in both fin and sei whales that the transition between these types of layer patterns is associated with the attainment of sexual maturity (Lockyer, 1972, 1974), and the same is also believed to be the case in the minke whale (Best, 1982).

The number of irregular growth layers formed in the early part of the sampling before the transition phase gives an estimate of the age of the whale when it became sexually mature. This has been correlated with the total growth layer count, equivalent to the age of the whale at capture. In the case of the fin, sei and minke whales in the Southern Hemisphere, all of these whales are now reaching the critical sizes at which they become sexually mature at younger ages than in earlier years. For the year classes of fin whales born up to 1930 the mean age at maturity was a little over 10 years. From the mid-1930s onwards this mean age fell and reached about 6 years for the 1950 year class. Direct study of the reproductive status of fin whales sampled in the mid-1960s also indicates an age of 6 to 7 years for both sexes (Ohsumi, 1972).

The reduction in age at sexual maturity for both sexes of sei whales from ear plug data shows that the year classes up to 1935 had a mean age at maturity of just over 11 years. This age fell to a little under 10 years in the 1945 year class. Direct analysis of the ovaries and ages of female sei whales sampled in the early 1970s shows that the average age at sexual maturity had decreased by then to 7 years in some areas (Masaki, 1978). Even greater reductions have been suggested in Antarctic minke whales, in which the age at sexual maturity is reported to have decreased from around 14 years in the year classes up to 1944 (the earliest year classes sampled) to 6 years in recent years (Masaki, 1979). However, minke whales in the southwest Indian Ocean off Durban are reported to show a smaller decline from 10 years in 1945 to just under 8 years in 1973 (Best, 1982).

There has been some doubt cast on the results described above arising from studies which have suggested that the trends observed may be artifacts of the data (Cooke and de la Mare, 1983, 1984). However, direct observations of the age at sexual maturity of fin whales sampled on board Japanese Antarctic fleets in 1957, 1961, 1964 and 1968 gave mean ages of 11.5, 10.6, 9.2 and 6.0 years respectively (Ohsumi, 1972). Whilst recognizing that both the precise values determined from the

TABLE 5b.1. Mean Lengths (m) at Sexual Maturity and Maximum Sizes Attained by Antarctic Whales

| Species | Sexual maturity | | Maximum size | |
|---|---|---|---|---|
| | Male | Female | Male | Female |
| Blue | 22.6 | 23.5 | 29 | 31 |
| Fin | 19.2 | 19.9 | 25 | 27 |
| Sei | 13.5 | 13.9 | 18 | 20 |
| Minke | 7.2 | 8.0 | 9 | 10 |
| Humpback | 11.6 | 11.9 | 17 | 18 |
| Right | ? | ? | 18 | 18 |
| Sperm | 11.9 | — | 20 | — |
| Killer | 5.8 | 4.9 | 9 | 8 |
| Southern | | | | |
| Bottlenose | ? | ? | 7 | 7.5 |
| Beaked | ? | ? | 9 | 10 |

earplug transition data and the direct observations of the catches in various seasons may have biasses due to the varying times and locations of the catches, is seems reasonable to conclude that these Antarctic baleen whales have exhibited real declines in age at maturity.

If this is the case, the implication is that the whales are growing faster now than in earlier years, to reach the same critical size at which they become sexually mature in a shorter span of time. For example, female fin whales born in 1930 grew to a mean length of nearly 20 m in 10 years, but their present day counterparts now reach this size in 6 years.

Such a dramatic increase in growth rate must involve considerably increased feeding, and the significance of the onset of the change in the 1930s in relation to the depletion of the blue and fin whale stocks at that time has been commented on as an illustration of a density dependent response by the whales (Gambell, 1973). As the numbers of blue and fin whales were reduced by over-catching, the remaining animals had the opportunity of consuming the 'surplus' food which was not being eaten by the whales that had been removed. They thus grew more quickly and began to mature at an earlier age. This is a direct reaction in the case of the fin whales, but the sei whales too have shown a comparable response even before their own stocks were directly exploited to a significant extent in the mid-1960s. Similarly with the minke whales, the age at maturity changed well in advance of the start of the direct intensification of catching effort which occurred on this species in the 1970s.

The pregnancy rates of blue, fin and sei whales in the Southern Hemisphere have also apparently changed in response to exploitation over the last half century. Whereas some 25% of the mature non-lactating blue and fin whales were pregnant in the Antarctic catches sampled in the early 1930s, the most recent samples indicate that now some 50% to 55% of this component in both these species are pregnant (Gambell, 1973; Laws, 1962). In the case of the sei whales a similar change has taken place since 1945. This reflects an approximate halving of the average interval between successive births in all three species during the sampling periods. In minke whales, up to 90% of the females sampled in the Antarctic fishery are pregnant (Masaki, 1979), but this may be the result of sampling bias due to the catching operations happening to fall on this component of the segregated populations.

The increased pregnancy rate in fin whales has been associated with an increase in the ovulation frequency during the normal breeding cycle (Gambell, 1973). Fin whales usually ovulate and conceive during the regular winter breeding period, with a few animals also ovulating again after the birth of the calf and/or at the end of the lactation period. A whale which ovulated and conceived soon after its calf was born would be simultaneously pregnant and lactating. Up to the end of the 1926/27 season such specimens had not been recorded at South Georgia. The first female that was both pregnant and lactating was found in January 1928, and fin whales in this condition later constituted up to 20% of all lactating females examined (Laws, 1961). Ovulating females are now also found in the Antarctic, whereas no lactating or resting females had been sampled at South Georgia before 1928 which showed signs of recent ovulations.

The conclusion to be drawn from this evidence is that since the reduction of the fin whale stocks, these whales are ovulating and calving more frequently, the latter being reflected in the increased pregnancy rate. However, there has been some criticism of the results described above (Mizroch and York, 1983) based on the suggestion that the apparently observed trends in age at maturity may have been produced by inappropriate treatment of the available data as statistical samples.

Surprisingly, although there is evidence that both the age at sexual maturity and the pregnancy rate have changed in these species since the major reductions have occurred in the blue and fin whale stocks, the expected corresponding increases in fin whale recruitment rates have not been confirmed by direct estimates. In fact, some analyses suggest a decline in recruitment (Clark, 1983). Nevertheless, normal population theory would suggest that density-dependent responses of the kinds indicated above would occur, since a species would otherwise very quickly be driven to extinction by any random decline in its numbers. Certainly in the case of the southern minke whales it is now generally accepted that there has been an increase in the abundance of this species prior to the onset of its exploitation. This is reinforced by evidence from the age distributions of the initial catches which suggested enhanced recruitment rates substantially exceeding the estimated natural mortality (IWC, 1979).

The interesting extension of this expected response is that reduction of the two major whale populations has apparently affected not only their own growth and reproductive parameters but also those of other whale species. Such inter-specific reactions have presumably been brought about because of the very heavy dependence of all the baleen whale species concerned on the single common food organism, Antarctic krill. Removal of a large proportion of the major consumers has permitted the remaining whales of the same and other species to take advantage of the relatively (to them) increased food supply.

Although the southern stocks of sperm whales are believed to have been depleted sufficiently for them currently to warrant complete protection from further hunting for a while, there is little direct evidence of any changes having occurred in their reproductive characteristics comparable with those found in the baleen whales. There is some evidence (Best, 1980) that the pregnancy rate in female sperm whales off Durban (Division 3) has increased from 17.1% in the years 1962 − 67 to 21.4% in 1973 − 75, which corresponds to the period of intensified exploitation for this sex. This response appears to have been mainly the result of higher pregnancy rates in the older females, which had a larger proportion of unsuccessful ovulations in the earlier period, although the mechanism for this change is not clear.

Pregnancy data from sperm whales off Western Australia (Division 5) show some evidence of a decline but there is no evidence from any sperm whale stock of a change in the age at sexual maturity in this species.

Analysis of the response of sperm whales to depletion of their stocks is confounded by their polygynous breeding behaviour. Reduction of the numbers of females can be expected to result in higher pregnancy and recruitment rates as suggested above in the southwest Indian Ocean. Reduction of the number of males may cause a decline in pregnancy rates and recruitment if the males remaining are too few to successfully impregnate all the available females. Reduction of both male and female components of the stock may have varying effects depending on the scale of the reductions and the balance of numbers between the sexes. Attempts to model these interactions and the ensuing predictions of the status of the southern sperm whale stocks have not been completely convincing (IWC, 1980).

## 5b.5. IMPLICATIONS FOR MANAGEMENT

In 1946 an International Convention for the Regulation of Whaling was signed in Washington, U.S.A., by 15 nations. These were mainly countries with whaling interests who saw the need to regulate the catches, particularly in the Antarctic, in order to provide for the proper conservation of whale stocks and thus make possible the orderly development of the whaling industry. This Convention followed a series of international conferences and agreements in the 1930s, as well as voluntary arrangements between the whaling companies, which gradually limited the catching operations. The measures adopted before the second world war included protecting right and humpback whales and females with calves, setting minimum size limits, restricting the operations of floating factory ships to the waters south of 40°S in the Southern Hemisphere, establishing a Sanctuary in the Pacific sector of the Antarctic where no whaling was permitted, and defining a relatively short catching season within set opening and closing dates, as well as establishing arrangements for the appointment of government inspectors on each factory ship, licensing of whaling vessels and the collection of catch statistics.

The Convention set up the International Whaling Commission as an executive body to formulate regulatory measures and with responsibility for their application. The Commission met for the first time in 1949, and now has 39 member governments. Limitation of the Antarctic catches was a prime topic for discussion in the early years, but it proved difficult to restrict the industry and the Antarctic stocks were successively reduced by over-catching.

Management of the whale stocks, as with most other marine resources, is based on the concept that each stock has a natural rate of increase through births and a rate of natural mortality which balances this to maintain the stock at an equilibrium level. Reduction in the numbers of whales in a particular stock leads to compensatory changes in the recruitment rate through earlier maturity and increased pregnancy rates, so that the stock can rebuild to its original level. Once the carrying capacity of the habitat niche occupied by the stock is reached, recruitment declines through such factors as food limitation so that the stock numbers stabilize once more.

The harvesting strategy of the International Whaling Commission since 1974 has been designed to bring all whale stocks to the level which maximizes the increased recruitment available from a stock held below its original abundance. This strategy based on the theory of Maximum Sustainable Yield (MSY) has not been entirely successful, since the scientific knowledge of the stocks and their yields has been less precise than has proved necessary to operate the policy. The result is that more and more stocks have been found to be already below their MSY level, so that they have been given total protection to allow the fastest possible increase to this optimum level.

This whole approach though presupposes that the numbers of animals in each whale stock are the direct and only cause of changes in their own vital parameters, through density-dependent responses to food availability for example. It has been demonstrated above that this is apparently not the case, with the early growth, age at sexual maturity and pregnancy rates all having changed in Antarctic sei whales well before any direct exploitation caused significant reduction in their own numbers. Similarly with the Antarctic minke whales, there has been an increase in recruitment associated with earlier maturation which has effectively allowed this species to increase in abundance as the larger species have been reduced by whaling. The inter-specific nature of the response in these baleen whales largely sharing a common food resource, Antarctic krill, therefore suggests that a more complex management approach is required, which takes account of these factors.

However, the Antarctic ecosystem includes a number of other animals which are primarily dependent on krill as their food, and some of these also show evidence of significant and comparable changes in recruitment parameters and numbers almost certainly brought about as responses to the declines in the large whale stocks.

Crabeater seals (*Lobodon carcinophagus*) in the area west of the Antarctic Peninsula show evidence of an advancement in the age at sexual maturity, based on the cementum layers in the teeth (Laws, 1977b). Up to 1955 the mean age at puberty was about 4 years but this declined following the opening of the whale sanctuary to 2½ years by 1970, an interpretation supported by direct examination of reproductive condition and age in recent years. It seems likely that the crabeater seal population has been increasing at least during the past two decades.

Clear evidence of increases in stocks of the Antarctic fur seal (*Arctocephalus gazella*) has been documented (Payne, 1976). These animals were hunted up to the beginning of this century and were much reduced. This species appears to be increasing since 1957 rather more rapidly than might be expected just in terms of recovery from over-exploitation, recolonizing beaches on South Georgia and the Scotia Arc as it experiences a population explosion.

There is also evidence that some Antarctic bird populations which feed on krill have been increasing in recent years. Increases in species such as the gentoo (*Pygoscelis papua*) and particularly the Adélie (*P. adelie*) and chinstrap penguins (*P. antarctica*) have been related to the decline in the major whale stocks (Sladen, 1964).

Rather more circumstantial evidence also suggests that some fish stocks in the Antarctic have increased following the decline in the whales. For example, the pelagic gadoid fish *Micromesistus* was not recorded from the Antarctic until 1963 (Merrett, 1963) although it is possible that it was present but undetected in earlier years. For other fish species the evidence is sparse but it seems likely that the pelagic krill-feeding fish stocks have enlarged in recent years, as revealed by the developing fisheries on them.

The evidence from the Antarctic is therefore of a complex of various animal populations all based on a common food supply, the krill. Management of the primary consumers, the baleen whales, which by their size and earlier abundance were the predominant predators on the krill, will therefore affect all these other components of the system. Now that the largest baleen whale stocks have been depleted, the smaller whale species, the seals, birds and fish have all had the opportunity to take up the large amount of food formerly consumed by the whales. The evidence outlined above suggests that they have done this and have increased in abundance as a result. The question then arises — can the whale stocks recover? It is not clear that the equilibrium of species which formerly held in the Antarctic can be restored, that the system is in fact reversible.

It has been suggested that since the minke whale stocks have increased in response to the reduction of the larger whales, reducing these stocks will aid the rebuilding of the blue whale stocks. But there is no guarantee that this would be the effect; it is quite possible that with fewer minke whales the seal, penguin or fish stocks would grow still further. We do not know the competitive aspects of the Antarctic ecosystem, so that management of one species or stock in isolation is neither predictable nor practical.

All the components of the whole food web, including the predators on the krill eating whales, seals, birds and fish, such as leopard seals and killer whales, as well as the squid which are consumed by a number of predators including sperm and killer whales, interact with one another directly and indirectly to confound any simplistic modelling in our present state of knowledge.

A further element which still further lessens certainty in any manipulation of the stocks of whales is the impact which the development of a major fishery for Antarctic krill might produce. There is a false notion that the reduced stocks of whales have left a *surplus* of krill which is available for direct harvesting. However, it is clear that this 'surplus' has already been taken up to a large extent if not completely by the other animals in the Antarctic (Laws, 1977b). Commercial scale catches of krill will therefore reduce the food available to the animals dependent on it. We do not know what effect this will have, in just the same way as we cannot yet predict the outcome of a change in the abundance of any one whale stock on all the other whale species and other competing organisms.

What is certain is that the large-scale reduction of the krill resource by direct fishing will increase the competition between the animals feeding on the krill. Further study may enable the outcome of any future manipulation of the whales, their food and the other inter-related component species of the Antarctic ecosystem to be predicted. In the meantime, all actions which may significantly change the abundance of any or all of these components of a complex system should be most carefully evaluated, because even the most cautious disturbance of the present relationships may have far-reaching and possibly irreversible effects.

## REFERENCES

Bannister, J.L. (1969) The biology and status of the sperm whale off Western Australia — an extended summary of results of recent work. *Rep. int. Commn Whal.* 19, 70 – 6.

Beale, T. (1839) *The Natural History of the Sperm Whale*, 1 – 393. John van Voorst, London.

Best, P.B. (1969) The sperm whale (*Physeter catodon*) off the west coast of South Africa. 3. Reproduction in the male. *Investl Rep. Div. Sea Fish. S. Afr.* No. 72, 1 – 20.

Best, P.B. (1980) Pregnancy rates in sperm whales off Durban. *Rep. int. Whal. Commn* (Special Issue 2), 93 – 7.

Best, P.B. (1982) Seasonal abundance, feeding, reproduction, age and growth in minke whales off Durban (with incidental observations from the Antarctic). *Rep. int. Whal. Commn* 32, 759 – 86.

Best, P.B., Canham, P.A.S. and Macleod, N. (1983) Patterns of reproduction in sperm whales, *Physeter macrocephalus. Rep. int. Whal. Commn* (Special Issue 6). 51 – 80.

Chittleborough, R.G. (1965) Dynamics of two populations of the humpback whale, *Megoptera novaeangliae* (Borowski). *Aust. J. mar. Freshw. Res.* 16(1), 33 – 128.

Clark, W.G. (1983) Antarctic fin whale, *Balaenoptera physalus*, recruitment rates inferred from cohort analysis. *Rep. int. Whal. Commn* (Special Issue).

Cooke, J.G., and de la Mare, W.K. (1983) The effects of variability in age data on the estimation of biological parameters of minke whales (*Balaenoptera acutorostrata*). *Rep. int. Whal. Commn* 33, 333 – 38.

Cooke, J.G. and de la Mare, W.K. (1984) A note on the estimation of time trends in the age at sexual maturity in baleen whales from transition layer data, with reference to the North Atlantic fin whale (*Balaenoptera physalus*). *Rep. int. Whal. Commn* 34, 701 – 9.

Dawbin, W.H. (1966) The seasonal migratory cycle of humpback whales. In *Whales, Dolphins, and Porpoises*, Ed. K.S. Norris, pp. 145 – 69. University of California Press, Berkeley and Los Angeles.

Doi, T., Ohsumi, S. and Nemoto, T. (1967) Population assessment of sei whales in the Antarctic. *Norsk Hvalfangsttid.* 56(2), 25 – 41.

Gambell, R. (1968) Seasonal cycles and reproduction in sei whales of the Southern Hemisphere. *Discovery Rep.* 35, 31 – 134.

Gambell, R. (1973) Some effects of exploitation on reproduction in whales. *J. Reprod. Fert.* Suppl.19, 531 – 51.

Gambell, R. (1976) Population biology and the management of whales. *Applied Biol.* 1, 247 – 343.

International Whaling Commission (1971) Report of the special meeting on sperm whale biology and stock assessments. *Rep. int Whal. Commn* 21, 40 – 50.

International Whaling Commission (1973) Sperm whale assessment meeting. *Rep. int. Whal. Commn* 23, 55 – 88.

International Whaling Commission (1979) Report of the special meeting on Southern Hemisphere minke whales. Seattle, May 1978. *Rep. int. Whal. Commn* 29, 349 – 58.

International Whaling Commission (1980) Report of the special meeting on sperm whale sssessments 27 November to 8 December 1978. *Rep. int. Whal. Commn* (Special Issue 2), 107 – 36.

International Whaling Commission (1982) Report of the Workshop on Identity, Structure and Vital Rates of Killer Whale Populations, Cambridge, England, June 23 – 25, 1981. *Rep. int. Whal. Commn* 32, 617 – 31.

Kasamatsu, F. and Ohsumi, S. (1981) Distribution pattern of minke whales in the Antarctic with special reference to the sex ratio in the catch. *Rep. int. Whal. Commn* 31, 345 – 8.

Kawamura, A. (1980) A review of food of Balaenopterid whales. *Sci. Rep. Whales Res. Inst., Tokyo*, No. 32, 155 – 97.

Laws, R.M. (1961) Reproduction, growth and age of southern fin whales. *Discovery Rep.* 31, 327 – 486.

Laws, R.M. (1962) Some effects of whaling on the southern stocks of baleen whales. In *The Exploitation of Natural Animal Populations*, Eds E.D. Le Cren and M.W. Holdgate, pp. 137 – 58. Blackwell, Oxford.

Laws, R.M. (1977a) Seals and whales of the Southern Ocean. *Phil. Trans. R. Soc. Lond. B*, 81 – 96.

Laws, R.M. (1977b) The significance of vertebrates in the Antarctic marine ecosystem. In *Adaptations Within Antarctic Ecosystems*. Third Symposium on Antarctic Biology, Ed. G.A. Llano. Gulf Publishing Co., Houston, Texas.

Lockyer, C. (1972) The age at sexual maturity of the southern fin whale (*Balaenoptera physalus*) using annual layer counts in the ear plug. *J. Cons. int. Explor. Mer.* 34(2), 276 – 94.

Lockyer, C. (1974) Investigation of the ear plug of the southern sei whale, *Balaenoptera borealis*, as a valid means of determining age. *J. Cons. int. Explor. Mer.* 36(1), 71 – 81.

Mackintosh, N.A. (1942) The southern stocks of whalebone whales. *Discovery Rep.* 22, 197 – 300.

Masaki, Y. (1978) Yearly change in the biological parameters of the Antarctic sei whale. *Rep. int. Whal. Commn* 28, 421 – 9.

Masaki, Y. (1979) Yearly change of the biological parameters for the Antarctic minke whale. *Rep. int. Whal. Commn* 29, 375 – 95.

Matthews, L.H. (1938) The sperm whale, *Physeter catoden. Discovery Rep.* 17, 93 – 168.

Merrett, N.R. (1963) Pelagic gadoid fish in the Antarctic. *Norsk Hvalfangsttid.* 52(9), 245 – 7.

Mitchell, E.D. (1975) Trophic relationships and competition for food in Northwest Atlantic whales. In *Can. Soc. Zool. Ann. Mtng. 1974*, pp. 123 – 33. Univ. New Brunswick, Fredericton.

Mizroch, S.A. and York, A.E. (1984) Have Southern Hemisphere baleen whale pregnancy rates increased? *Rep. int. Whal. Commn* (Special Issue 6), 401 – 10.

Nemoto, T. (1959) Food of baleen whales with reference to whale movements. *Sci. Rept. Whales Res. Inst., Tokyo*, No. 14, 149 – 290.

Nemoto, T. (1970) Feeding pattern of baleen whales in the ocean. In *Marine Food Chains*, Ed. J.H. Steele, pp. 241 – 52. Univiversity of California Press, Berkeley and Loss Angeles.

Norris, K.S. and Møhl, B. (1981) Do Odontocetes debilitate their prey acoustically? *Fourth Biennial Conference on the Biology of Marine Mammals*, San Francisco, Calif. 14 – 18 December 1981.

Ohsumi, S. (1972) Examination of the recruitment rate of the Antarctic fin whale stock by use of mathematical models. *Rep. int. Whal. Commn* 22, 69 – 90.

Payne, M.R. (1976) Growth of a fur seal population. *Phil. Trans. R. Soc. Lond. B*

Payne, R. (1984) Reproduction rates and breeding area occupancy in the Southern right whale, *Eubalaena australis. Rep. int. Whal. Commn* (Special Issue 6), 482.

Purves, P.E. (1955) The wax plug in the external auditory meatus of the Mysticeti. *Discovery Rep.* 27, 293 – 302.

Raynor, G.W. (1940) Whale marking: progress and results to December 1939. *Discovery Rep.* 19, 245 – 84.

Roe, H.S.J. (1967) Seasonal formation of laminae in the ear plug of the fin whale. *Discovery Rep.* 35, 1 – 30.

Sladen, W.J.L. (1964) The distribution of Adélie and chinstrap penguins. In *Biologie Antarctique*, Eds R. Carrick, M.W. Holdgate and J. Prevost, pp. 359 – 65. Hermann, Paris.

# CHAPTER 5c

# Birds and Mammals — Oceanic Birds of the Antarctic

## W. R. SIEGFRIED

FitzPatrick Institute, University of Cape Town, Rondebosch 7700, South Africa

## CONTENTS

## 5c.1. INTRODUCTION

There are many misconceptions about the circumpolar Southern Ocean and Antarctic seas. More particularly, it is widely believed, and often stated, even in the scientific literature, that the Antarctic marine ecosystem is enormously productive and, so, generally supports extraordinarily abundant and dense populations of a great variety of seabirds. The modern travel industry does its best to promote this fallacy, and its glossy magazines and brochures tell of spectacular arrays of myriads of seabirds to delight the tourist while he cruises in Antarctic waters. True, if he visits the Antarctic in high summer, the tourist will see seabirds, occasionally very abundant locally, but in the pelagic environment he will not regularly encounter anything like the variety or numbers that he would see while cruising in certain other parts of the world. If he visits the Antarctic in winter, the chances are that he will see very few seabirds. In short, contrary to popular opinion, the Southern Ocean generally does not support an exceptionally varied and dense avifauna. Indeed, both the species and density of birds tend to be greater in certain temperate seas at lower latitudes, where upwelling of nutrient-rich water occurs in the

Fig. 5c.1. Wandering albatross and chick (Photo: D.W.H. Walton).

vicinity of land. In the Southern Hemisphere, such more-or-less continuously highly-productive marine ecosystems are found along the west coasts of South America and southern Africa.

It is fair to ask how the myth of a super-abundance in the Antarctic arose. In the past, and to a lesser extent still today, most reports on Antarctica's birdlife were based on observations made during summer in the Southern Ocean where large numbers of seabirds congregate in certain areas. However, these areas, including islands on which the birds breed, account for a very small part of the vast oceanic region that stretches over some 40 million square kilometres between the edge of the permanent ice of the Antarctic continent and the Antarctic Convergence (see Chapter 2a) to the north. In winter, when reductions in daylight, temperature of water and air, and plant and animal productivity occur, many of the seabirds which fly migrate out of this region, as do (or formerly did) many whales.

The birds which remain behind, including the penguins which constitute approximately 90% of the region's avian biomass, prey largely on animals whose populations are built up rapidly during a short

Fig. 5c.2. Wandering albatross courting (Photo: W.N. Bonner).

period of concentrated reproduction and growth in summer. The important point here is that it is the rate of production, and not just the abundance or mass, of prey that determines the amount of food potentially available to consumers in an ecosystem. In this respect, when production is averaged over a whole year and over the whole region, the Antarctic oceanic region does not appear to be more productive than other oceans. What does set the Antarctic marine ecosystem apart from most others is the strong summer 'pulse' of solar energy which is converted rapidly into biomass. This is then concentrated and stored for relatively long unproductive periods in short food chains. The populations of seabirds and other predators at the tops of these food chains ultimately are limited by the stores of prey biomass. These stores contain surpluses large enough to ensure sufficient survival for viable growth and reproduction in the summer. Thus the relatively simple Antarctic marine ecosystem is characterized by a short seasonal pulse of high productivity, a high biomass (particularly of plankton), a slow rate of biomass 'turnover', short food chains, and a low number of species.

## 5c.2. TAXONOMIC COMPOSITION OF THE AVIFAUNA

A seabird is one which habitually obtains its food either directly or indirectly, as is the case in certain scavengers and predators, from the sea. There are approximately 250 seabird species in the world, of which only some 52, including seven penguins, breed south of a line corresponding very roughly to 50°S latitude, but including the Tristan – Gough group of temperate-climate islands situated to the north of that latitude. The majority of the Antarctic's flying seabirds consists of albatrosses and petrels (Procellariiformes), and skuas, gulls and terns (Charadriiformes). The albatrosses and petrels account for some 35 species of which most are truly oceanic, avoiding land except when breeding.

The order Procellariiformes is represented by members of all four of its families: six species of albatross (Diomedeidae), 23 typical petrels, prions and shearwaters (Procellariidae), four storm petrels

Fig. 5c.3. Grey-headed albatross feeding its chick. The grey-headed albatross feeds largely on squid (Photo: W.N. Bonner).

Fig. 5c.4. Black-browed albatross and chick. This species is a krill-feeder (Photo: British Antarctic Survey).

(Oceanitidae) and two diving petrels (Pelecanoidae). The balance of the flying birds consists of one species of gull and three terns (Laridae), two skuas (Stercorariidae), two sheathbills (Chionididae), which are almost exclusively land-based, and two cormorants (Phalacrocoracidae), now often regarded as a single species. This is not the place for detailed lists. For an introduction to the names of the species comprising the Antarctic breeding avifauna, their systematic arrangements and nesting localities, the reader is advised to consult Watson (1975).

These figures are approximate, because we are not yet sure, in quantitative terms, of the degree of genetic difference between the described subspecies and species in certain very variable taxa which include the small petrels, such as the prions or whalebirds (*Pachyptila* spp.). This problem, however, promises to be resolved with the modern development of biochemical techniques. A related hindrance to achieving a complete and comprehensive inventory of the Southern Ocean avifauna rests on our still fragmentary knowledge of where certain populations breed. Again, this applies particularly to species in the small-petrel category, which tend to nest in subterranean burrows on sub-Antarctic islands. Many of these islands have not yet been explored thoroughly, and even on the better explored islands hitherto unknown breeding populations continue to be discovered.

## 5c.3. WHY SO FEW SPECIES?

The number of species south of about 50°S latitude is, on balance, considerably lower than that north of about 50°N latitude. However, the Antarctic avifauna is proportionately richer in truly oceanic or pelagic species. It also is much more discrete than its boreal, or Northern Hemisphere, counterpart, in the sense that it is almost exclusively endemic. One notable exception to this is the kelp gull *Larus dominicanus* whose breeding populations are scattered widely in the Southern Hemisphere south of about

Fig. 5c.5. Light-mantled sooty albatross on nesting ledge (Photo: C.J. Gilbert B.A.S.).

10°S latitude. The Arctic tern *Sterna paradisaea* is the only boreal species migrating regularly to Antarctic waters. In contrast, many Antarctic breeding species migrate to the subtropics, and the sooty shearwater *Puffinus griseus*, greater shearwater *P. gravis*, Wilson's storm petrel *Oceanites oceanicus* and the south polar skua *Catharacta maccormicki* migrate across the equator to spend the austral winter in sub-Arctic waters.

The oceans cover about 70% of the surface of the earth, and nowhere is the proportion of sea to land greater than in the Southern Hemisphere between latitudes 50° and 70°S. Almost all of the islands in this area occur in the southern sector of the Atlantic and Indian oceans. The southern Pacific Ocean, exempting the New Zealand seas, contains very few islands. Many of the islands were created by volcanic action, as part of the continuing process of dispersal of the continental landmasses derived from the former super-continent of Gondwana. The ancestors of the penguins (Sphenisciformes) and the Procellariiformes probably evolved on the shores of these landmasses and subsequently dispersed to islands in the Southern Ocean, as the continents drifted apart and moved northwards. Today, the majority of the Antarctic region's avifauna depends on a suite of some 24 oceanic island-groups as breeding grounds. Only eight flying species breed on the Antarctic continent, and 11 on the Antarctic Peninsula.

In terms of geological time, the Southern Ocean island environment is a highly dynamic one, with new land appearing and old land disappearing. For instance, in the mid-1950s, tectonic forces created new land at Bouvetøya, facilitating the establishment of new colonies of breeding seabirds including species not known to have bred there previously. In addition to tectonic changes, repeated alternating advances and retreats of the Antarctic Convergence, or Polar Front, have been associated with expansion and contraction of ice-caps on islands, and rising and falling sea-levels, during the last few million years. Whole islands, or parts of islands, have been alternately created and destroyed, resulting in a constant state of flux in the avifauna as well. Varying levels of gene exchange and isolation have fluctuated spatially and temporally between avian populations with different degrees of faithfulness to particular breeding localities, and rates of breeding. This helps to explain why certain taxa apparently are more variable than others, and why there is still lack of agreement between taxonomists on the numbers of species and subspecies comprising the avifauna. It also helps to explain why the Antarctic avifauna, perhaps paradoxically, is characterized by a relatively low average number of species and a high degree of uniformity, in spite of the large area over which it is distributed.

The area is large, but it is also circumpolar and predominantly oceanic. It lacks any large-scale physical features arranged longitudinally, apart from the Antarctic Peninsula, which otherwise might interrupt and disrupt ocean currents, winds and other factors promoting regional differences in nutrient concentrations, and prey and predator diversity. There are no significant land barriers to the birds' dispersal. It is possible that a number of birds, such as the wandering albatross *Diomedea exulans*, spend many years repeatedly circumnavigating the globe by moving in an easterly direction with the assistance of the prevailing westerly winds. Although the avifauna tends to be uniform, across longitudes and between the Pacific, Indian and Atlantic oceans, it is distributed a great deal less evenly and uniformly in relation to latitude, as might be expected from the presence of the Antarctic Convergence and the Antarctic Divergence (see Chapter 2a). These two complex, oceanic, frontal systems mark major latitudinal changes in both the structure and the functioning of the Southern Ocean ecosystem. The frontal zones, being nutrient-rich and biologically productive, support enhanced avian diversity. The number of species tends to peak at or near the Convergence where, on average, between four and five species occur at a pelagic observation station. The average number of species decreases south of the Convergence to between two and three at the Divergence, and to about two south of the Divergence. These changes are accompanied by changes in species composition. For instance, the Antarctic petrel *Thalassoica antarctica* and the snow petrel *Pagodroma nivea* are largely confined to south of the Divergence, and the albatrosses tend to occur most commonly in the northern parts of the region.

It should be noted that the decrease in number of species southwards of the Antarctic Convergence is part of a trend which starts near the sub-Tropical Convergence at about 40°S latitude. The number of species also decreases to the north of the sub-Tropical Convergence. Unlike what has often been shown for continental faunas, a consistent trend of increasing species numbers from the Antarctic to the

Fig. 5c.6. Giant petrels at a fur seal carcass. This is the northern form, *Macronectes halli* (Photo: W.N. Bonner).

Fig. 5c.7. Antarctic fulmar in its nesting crevice on the Antarctic Peninsula (Photo: W.N. Bonner).

equator does not apply in the pelagic avifauna of the Southern Ocean. Given that the highly mobile Antarctic avifauna inhabits an extremely large and relatively uniform oceanic area, the interesting and important question to ask is not why are there so few species, but why are there so many endemic pelagic species and how do these species co-exist?

## 5c.4. HOW MANY BIRDS ARE THERE?

The short answer to this question is: we do not know precisely. What we have been learning in the last decade or so, however, is that the actual abundance of the Southern Ocean avifauna is well short of some of the gross over-estimates made earlier in this century.

Apart from penguins, our best current information is for the six species of albatrosses: wandering albatross, blackbrowed *Diomedea melanophris*, greyheaded *D. chrysostoma*, yellownosed *D. chlororhynchos*, sooty *Phoebetria fusca* and lightmantled sooty *P. palpebrata*. These large surface-nesting birds tend to congregate in groups at relatively few breeding localities and, thus, are easier to census than many other species. The wandering albatrosses' breeding population is about 20,000 pairs. Taken together, the breeding populations of all six albatrosses probably amount to 750,000 pairs. These figures help in debunking the still popular notion that the Southern Ocean, covering 50 – 100 million square kilometres, teems with millions of albatrosses.

Censuses for the balance of the surface-nesting avifauna are far less comprehensive and complete, although fair estimates of the numerical status of the giant petrels *Macronectes* spp., cormorants, skuas, kelp gull and terns are available. It is unlikely that their populations, in combination, account for more than 500,000 birds. Considerably less is known about the southern fulmar *Fulmarus glacialoides*, Antarctic petrel, Cape pigeon *Daption capense* and snow petrel which, according to one modern report (Mougin and Prévost, 1980), might total between four and five million birds.

Omitting the two sheathbills, we are left with the prions (possibly five species), gadfly petrels (six species), shearwaters (four species), storm petrels (four species), and diving petrels (two species). These birds tend to nest in burrows, in soil under vegetation or boulders, or in cracks and crevices in lava flows and screes. Normally, they visit land only at night. This, and the fact that many species are small, look alike and occur together in extremely rugged terrain, makes censusing of the birds very difficult. Some wild guesses have been made of the size of their populations, but modern methods are helping to refine the quality of our knowledge. Improved field techniques, involving special aerial photography, tape-recordings of the birds' calls, optical devices for night viewing, and electronically-operated automatic monitors of the birds' activities in burrows, have made possible for the first time, during the last five years or so, reasonably accurate assessments of the distribution and abundance of certain populations. However, this multiple-technique approach has been used at only a few of the birds' breeding grounds, for the work is arduous, time-consuming and tedious. Hence, many years are likely to elapse before confident, definitive pronouncements can be made on the total abundance of any of these species, let alone the combined numerical strength of the species as a group. Nevertheless, one recent series of estimates, given by Mougin and Prévost (1980), yields a grand total a little short of 150 million birds for these species.

So much for land-based estimates of seabird populations. What about the birds' abundance at sea? Here again, the quality of our knowledge has improved considerably during the last five years or so, with the development and practice of standardized methods for observing and counting birds from ships. We can now begin to make tentative comparisons, for the first time, between the abundance of birds counted simultaneously in different oceans and seas, or at different times of the year. We now know that large single concentrations, involving thousands of birds, are seldom encountered in the pelagic environment of the Southern Ocean.

Preliminary and provisional information indicates that, on average, minimum densities of six and eight birds per square kilometre prevail in summer between latitudes 50° and 70°S in the Indian and Atlantic oceans, respectively. Comparable figures are not yet available for the southern Pacific which,

Fig. 5c.8. A white-chinned petrel in its nesting burrow. Sudden thaws or heavy rain on frozen ground can create problems for burrow-nesting birds in the sub-Antarctic (Photo: J.C. Sinclair).

however, is suspected to encompass significantly lower densities, presumably because of its paucity of islands and, consequently, a relatively low abundance of resident breeding birds.

As in the case of the number of species, the latitudinal distribution of the at-sea abundance of the birds displays a clear general pattern. Abundance tends to be highest in the vicinity of the Antarctic Convergence, to decrease steeply towards the Antarctic Divergence and then to level off farther to the south. Unlike the trend for the number of species, however, avian abundance tends to decrease quite sharply north of the Convergence to levels roughly the same as those occurring near the Divergence, before rising again in the vicinity of the sub-Tropical Convergence. These changes are more pronounced in summer than in winter when a more evenly distributed abundance is approximately half of that observed in summer.

## 5c.5. WHAT DETERMINES THE DISPERSION OF THE AVIFAUNA?

Pelagic seabirds are highly mobile creatures, and foraging individuals probably move over greater areas per unit of time than most other animals. All marine birds, however, have to breed on a solid base. In essence, this means land, although the emperor penguin *Aptenodytes forsteri* breeds on ice. Moreover, these breeding sites should offer the birds a maximum degree of protection against predators. This, no doubt, prompted Murphy's (1936) perceptive remark: 'seabirds breed where they must and feed where they can'.

The total area of land suitable for breeding seabirds is very restricted in the Southern Ocean. Consequently, nesting birds tend to crowd together, often colonially, on small islands or at the relatively few safe, ice-free, sheltered localities on the coast of Antarctica. Visitors to these places

Fig. 5c.9. Wilson's storm petrel (Photo: A. Clarke).

Fig. 5c.10. Dove prion. This is one of the most abundant burrowing petrels in South Georgia (Photo: W.N. Bonner).

normally have been impressed by the abundance and tameness of the birds, the tameness reflecting a lack of large terrestrial predators. It is chiefly these visitors' anecdotal reports which have helped to reinforce the popular misconception that the Southern Ocean generally supports myriads of birds.

Does space for nesting sites limit the populations of marine birds in the Southern Ocean? No hard and fast general answer can be given. There is some evidence that nesting space might indeed be limiting for certain populations, but for others this does not appear to be so and for yet others the information available is equivocal. To what extent do the birds compete intraspecifically and/or interspecifically, including possibly with animals such as seals, for space on land? We do not have the answers. But, it is important that we should have the knowledge if we are to interpret sensibly possible changes in the birds' abundance when large-scale commercial exploitation of food resources, such as krill (see Chapter 4a), gets under way.

A seabird cannot breed successfully if it cannot obtain enough food of the right kind within an economical foraging range of its breeding site. Does this then mean, for instance, that the blackbrowed albatross which breeds mainly at sites in the Atlantic, does not breed at many islands in the Indian Ocean because that region is lacking in its food? The answer could be a qualified, Yes, the Indian Ocean is actually lacking in the food normally taken by the species or, alternatively, some other species, not necessarily avian, is exploiting the food to the exclusion of the blackbrowed albatross. The blackbrowed albatrosses' population conceivably could be limited by resources within its current breeding range, and/or in the species' wintering areas outside the compass of the Southern Ocean. Either way, this could restrict the possibility of it pioneering and colonizing a number of islands in the Indian Ocean, even should food be available there. These could be ultimate determinants of the species' current geographical distribution, and they would probably represent a dynamic consequence of events in the past (in the geological sense). It would be extremely difficult to reconstruct these events, nor would it be easy to unravel combinations of proximate factors which also affect the distribution of seabirds.

The dispersion of seabirds is neither random, nor even, at sea. Major factors, such as climatic regimes and the availability of breeding sites, obviously constrain the framework for any broad pattern of dispersion. One would, for instance, not expect to find wandering albatrosses, which rely on wind for their soaring flight, in the doldrums. Within a broad biogeographical framework, however, the abundance of any particular population varies from place to place. Aside from concentrating near their breeding sites, seabirds assemble elsewhere as well. These assemblages may or may not be indicative of places at which food is readily available. If concentrations of birds do represent responses to enhanced food availability, how does an individual bird discriminate between areas either rich or poor in food? Do the birds use one or more environmental cues, such as temperature, colour or salinity of water, either singly or in combination, to detect and assess indirectly the availability of suitable food, or do they respond directly to the presence or absence of food? Do the birds learn that certain areas are more dependable than others as sources of food? Do species differ amongst themselves in their assessments of feeding areas? Why is it that concentrations of birds are either absent or have lower densities in certain areas than those which might be predicted from the distribution of their food?

These are neither simple nor trivial problems, and it is only in the last decade or so that real progress has been made towards obtaining answers for some of them. There are two principal reasons for this: first, it is now accepted that seabirds can be useful indicators of the distribution and abundance of prey populations which are also actual or potential food resources for man; and, secondly, the study of marine birds is now an integral part of modern oceanographic research. These developments began to gain momentum in the 1960s. Since then there have been not only many more opportunities created for students of marine birds to work from oceanographic research vessels, but there has also been ever-increasing co-operation between scientists in multi-disciplinary research aimed at obtaining a better understanding of how marine ecosystems function. The days when organisms, not ecosystems, were the only subjects worthy of study by marine biologists are over. It is now possible, through the simultaneous collection of relevant information, to integrate the dispersion of pelagic birds with a wide range of environmental features in oceanic areas. This integration has allowed researchers to look for patterns in the co-occurrence of seabirds and certain environmental features, either independently or collectively, and to ask why there should be differences between patterns displayed by different species and between groups of species.

It is reasonable to assume that the dispersion of seabirds is associated with that of their prey in some determinable way mediated by their physical environment. Hence, statistical correlations have been sought between the incidence of seabirds and those physical features which help to define particular water masses, in attempts to predict the birds' occurrence and the occurrence of their prey. These attempts have met with limited success, other than on scales too broad to be really useful. There are

Fig. 5c.11. A pair of brown skuas with a gentoo penguin chick they have just killed (Photo: W.N. Bonner).

Fig. 5c.12. Kelp gull. In the northern parts of the Antarctic this bird lives largely by scavenging, but at the southern extremity of its range it feeds mainly on Antarctic limpets (Photo: W.N. Bonner).

many reasons for this but, in essence, it appears that either the assumption underpinning the correlations is spurious or, alternatively, both the intensity of sampling and the analytical methods involved are still too coarse. In this context, it should be stressed that statistically satisfactory associations need not as a matter of course represent cause-and-effect relationships. In other words, highly significant statistical associations between seabirds and particular water masses cannot be taken as proof that the birds are reacting directly to the physical parameters which we view as characteristic of the water types.

The fact that certain species and associations of species of seabirds have been found to be associated with particular water masses in the Southern Ocean, implies that they are components of particular biotic communities, since there is some evidence for their principal prey items being associated with the water masses as well. However, these links are still so diffuse and tentative as to offer little assistance in the accurate forecasting of particular systems, involving birds – prey – environment relationships, based on any one or more of the three components. Yet, this must remain a principal objective of the integrated, modern inter-disciplinary research, for without it we will continue to lack a proper understanding of the birds' pelagic ecology. More particularly, it is only through an appreciation of the structure and the functioning of the ecosystem that we can hope for the insights to answer the question: what determines the dispersion of the avifauna? Clearly, the advice of Ashmole (1981), as one of the pioneers of modern marine ornithology, is relevant: 'Ways must be found to focus more precisely on the interactions of the birds with their marine resources if seabird biology is not to languish in the doldrums'.

## 5c.6. WHAT DO THE BIRDS EAT, AND HOW DO THEY OBTAIN THEIR FOOD?

Very little is known about the diets of non-breeding seabirds in the Southern Ocean. The reader may find this somewhat surprising, given the many years that birds have been studied in the region. However, almost all of the studies have been land-based, and have focused on the stomach contents of breeding birds collected during the austral summer. There are good reasons for this. Few research ships visit the Southern Ocean in winter, and it is virtually impossible to collect good samples of pelagic seabirds from large ships. Small ships cannot operate safely in the Southern Ocean, especially in winter.

The lack of information on what the birds eat during the pelagic phase of their life-cycles is a major drawback to an understanding of their ecology, particularly since a presumably depressed availability of food in winter is likely to be an important factor in limiting and dispersing the populations of certain species. Unfortunately, no dramatic change in this state of affairs is likely until the seemingly insurmountable obstacle of collecting birds at sea can be overcome. In the interim, the only option open is to make full use of all opportunities for expanding the body of information that has been built up to date. In practice, this means more land-based diet studies, covering a wider range of breeding localities and species, and more collecting of birds at sea in pelagic areas in which collecting from ships is practicable.

To a lesser extent, similar logistical constraints apply to collecting information on the birds' foraging behaviour at sea. The information available on how the birds actually capture their prey is decidedly limited, and much of this is based on observations made outside the compass of the Southern Ocean. Indeed, Antarctic seabirds are seldom encountered actually feeding in the pelagic environment, reflecting, in part, the probability that many species feed at night when it is dark and their food is at the surface of the sea. Here, incidentally, modern research is providing exciting new insights into how albatrosses and petrels use their sense of smell, probably extensively, in locating food (Wenzel, 1980).

All of this means that at this stage it is dangerous to generalize too far but, on the other hand, generalizations have to be attempted in order to discern ecological patterns which prompt questions and hypotheses useful in guiding research in the future. Thus, based on existing information, the Southern Ocean avifauna can be ordered provisionally into trophic groups according to the species' principal diets and feeding methods. Although the individual species comprising these groups may be, to varying degrees, generalists or opportunists, or both, each group tends to rely on one principal food class. The species all use one, or at most two, feeding methods. This means that each group is reasonably distinct, and its dispersion should reflect the presumed dispersion of its prey.

Plankton-eating species account for approximately 55% of the total abundance of pelagic seabirds (excluding penguins) in the Southern Ocean. Most are prions and storm petrels which feed on the surface of the sea, taking crustaceans including krill. This group of birds tends to be most abundant south of the Antarctic Convergence. Squid-eating species, comprising approximately 20% of the pelagic avifauna, dominate north of the Antarctic Convergence. They are mainly albatrosses which apparently

Fig. 5c.13. Antarctic tern (Photo: J.C. Sinclair).

seize their prey at the surface of the sea. The third group is made up of fish-eating species, which dip or plunge into the surface water layer in pursuit of their prey. Except in the vicinity of land and the Antarctic ice-shelf, these account for an almost negligible proportion of the total avian abundance. Their scarcity apparently mirrors the general lack of surface-dwelling pelagic fish in the Antarctic region. The last of the four principal groups consists of species not readily classified by food type. Hence, these species comprise a mixed-diet group, which rely mainly on surface-seizing and scavenging in obtaining their food. Species in the mixed-diet group account for some 20% of the avian abundance, occurring most abundantly where both plankton- and squid-eaters tend to be least abundant near the Antarctic Convergence and in the pack-ice close to Antarctica. Relatively few species obtain their food by kleptoparasitism or piracy.

While it is clear that plankton-eaters predominate in the total avian abundance of the Antarctic region, the group probably accounts for less than 25% of the total avian biomass. The large albatrosses, in contrast, help to boost the squid-eating group to approximately 50% of the total avian biomass. This means that the southern part (but not the extreme south) of the Antarctic region supports many more relatively small-bodied birds than the northern part. Why are the plankton-eaters, on average, much smaller than the squid-eaters? Why do they tend to occur in the south and the squid-eaters in the north of the region?

Large birds generally have proportionately lower metabolic rates and take larger prey items than small birds. Small birds must feed more frequently than large ones and, therefore, they depend on readily-available concentrations of small food items. The pattering flight of the smallest pelagic seabirds, the storm petrels, appears to be an adaptation for almost continual feeding on abundant small prey items at the air — sea interface. Such food resources occur mainly south of the Antarctic Convergence where the density of zooplankton tends to be high in summer. Large seabirds are better equipped for coping with relatively sparsely dispersed, irregularly-occurring food resources. Squid might fall into this class of food. Little is known about the distribution and abundance of squid in the Southern Ocean (see Chapter 4d), but apparently there are large populations in the vicinity, and north, of the Antarctic Convergence. Many species occur at the surface of the sea only during periods of darkness. In high latitudes the short periods of darkness during the austral summer might reduce the availability of squid to surface-feeding species, including albatrosses. Other squid-eaters, such as certain penguins and whales, could be less affected, because of their ability to dive deeply. Finally, albatrosses and the other birds which rely mainly on dynamic soaring are at a disadvantage south of the Antarctic Divergence where the winds are far less predictable than farther north, and where the presence of icebergs and icefloes further helps to limit the avian community (penguins excluded) to those species which use active flapping flight for foraging. Flapping flight and flight manoeuvrability are advantageous for foraging in small areas of open water in sea-ice, as exemplified by snow and Antarctic petrels which tend to be predominant near Antarctica.

### 5c.7. HOW MUCH DO THE BIRDS EAT?

The preceding sections indicate that pelagic squid and krill constitute the two major food classes on which the bulk of the Southern Ocean avifauna depends. Both of these resources are viewed as being potentially important in the development of a large-scale commercial fishery in the Southern Ocean. Many of the squid eaten by the birds feed on krill as well, taking perhaps 100 million tonnes a year in Antarctic waters. It has been suggested by some experts that an equivalent amount of krill could be taken annually by a commercial fishery without depleting populations. How would such a fishery affect the birds? Clearly, it seems important to know what amounts of these resources are eaten by the birds.

Various estimates have been made of the overall food consumption of seabirds in the Southern Ocean, but all lack confidence because of the inadequacy of information on populations and poor knowledge of

Fig. 5c.14. Blue-eyed cormorant making a threat display from its nest (Photo: D.W.H. Walton).

the energetic costs associated with the birds' activities. It appears, however, that the birds consume, either directly or indirectly, a minimum of 100 million tonnes of krill annually, of which approximately 90% is taken by penguins. Once again, a figure of 100 million tonnes of krill emerges. What does it really tell us? Apart from simply emphasizing the importance of seabirds as one of the principal groups of top-order predators in the Southern Ocean ecosystem, it provides little that is helpful in indicating how a major commercial fishery would affect the birds. Indeed, no simple, straightforward answer can be given to the question, because the whole does not necessarily equal the sum of the parts. In other words, whole-ocean-style calculations and predictions tend to ignore or mask local conditions. Without knowing, for example, where and when intensive fishing is to occur and how intensive it will be, and what the rate of 'turn-over' of the birds' prey is, it is impossible to make confident forecasts. There are no easy shortcuts to solving complex ecological problems.

Commercial fishing for krill or squid, or both, in the Southern Ocean is likely to develop a great deal faster than the body of knowledge necessary to make accurate predictions of all the ecological consequences of such single- or multi-species fisheries. In the interim, scientific research, aimed at improving our understanding of the functioning of the Southern Ocean ecosystem, should be encouraged to proceed towards carefully-selected goals. One of these goals should be the identification of a suite of avian species about whose ecology enough can be learned in detail for their populations to serve as reliable agents for detecting and monitoring changes in the ecosystem. Obviously, a prerequisite for such a plan is adequate knowledge of the demographic characteristics of the avian candidates for monitoring environmental changes.

## 5c.8. DEMOGRAPHIC CHARACTERISTICS OF THE AVIFAUNA

The term 'demographic', as used here, refers essentially to the dynamic balance between increase and decrease of a population. Hence, this section concerns itself with rates of reproduction and mortality. In general, it is fair to say that seabirds reproduce slowly and live for a long time. It is also true that large-bodied species tend to live longer and breed more slowly than small species. The abundance of large species in the Southern Ocean is exceeded nowhere else. Hence, a significant part of the avifauna replaces itself relatively slowly. This holds obvious implications for using these species as monitoring agents. However, the question of moment is: why do so many members of the Southern Ocean avifauna breed slowly? In attempting to answer the question here, I have drawn on information synthesized by Croxall (1984).

Southern Ocean seabirds generally are restricted to breeding during a short summer season; the more southerly populations starting a couple of weeks later than the northern ones, due to the later onset of mild conditions in the south. Moreover, it appears that most species begin breeding as early in the season as is practicable, so that they can capitalize on abundantly-available food when their chicks grow rapidly. This is especially so for the large species. Climate and weather are important determinants of the birds' breeding seasons, acting either directly or indirectly, or both, as they affect the availability of food. The wandering albatross and certain petrels are winter breeders in the relatively mild sub-Antarctic. They, probably very importantly in this context, feed themselves and their chicks on squid, and not on crustaceans whose populations are much depressed in winter.

All the procellariiform species lay one egg, and both parents share incubation, brooding and chick-feeding duties. The eggs of small species tend to be larger in relation to female body-mass than those of larger species. Incubation periods tend to be unusually long compared to other birds, and this is believed to reflect overall slow development rates which evolved in relation to the difficulty that parents have in finding enough food for themselves and their chicks. It should be emphasized that it is not just the quantity of food that is important but also its quality. Thus, protein, for instance, could be limiting. Breeding success (i.e., the proportion of chicks which fledge from all eggs laid) is low; usually being less than 50%, due mainly to starvation of chicks. Predators on eggs, chicks or adult birds are normally not an important factor contributing to the low breeding success except at breeding sites where alien animals, such as cats and rats, have been introduced by man.

Fig. 5c.15. Sheathbill with macaroni penguin egg (Photo: W.N. Bonner).

Fig. 5c.16. The Antarctic's only passerine, the Antarctic pipit of South Georgia (Photo: W.N. Bonner).

Whereas mortality of chicks tends to be high, life-expectancy is greatly enhanced in seabirds once they become adult. For instance, there are records of wandering albatrosses living and breeding for 40 years and more in the wild. There is a tendency in these long-lived birds to delay first breeding until they are several years old. This deferment apparently is not a reflection of sexual immaturity, in that the young, pre-breeding adults are physiologically capable of producing sperms or eggs. The fact that they do not do so until they are actually ready to breed, probably is indicative of the high risk attending their survival if they attempt to breed too soon. In short, it is difficult to rear young successfully and, therefore, it is advantageous to wait until sufficient experience has been gained to allow a reasonable chance of success.

This is the general picture of the breeding biology of the procellariiforms. Considerable variation occurs within the group, but few of the details are known. For instance, demographic life-tables necessary for determining the proportion of pre-breeding birds in particular populations are available only for three albatross species. Progress in this field will be slow, and it can be furthered only through long-term studies spread over a number of decades and involving the regular monitoring of many individually-marked birds. The unravelling of the interplay of the host of adaptations involved in maximizing reproductive output and minimizing mortality in each species will take time. Nevertheless, the goal is a worthy one, as remarked by Croxall (1984): '... the ease with which these birds can be handled and manipulated in the Antarctic is such that the rewards of fully-integrated studies can be very considerable and one would expect many of the more significant future developments in the understanding of seabird ecology to come from work currently in progress in the Antarctic'.

The primary key to this complete understanding and, indeed, why breeding is difficult and why the procellariiforms breed slowly, is a thorough knowledge of the species' food requirements in relation to the availability of their food. It seems clear that the dispersion of food is the basis for a complex of breeding adaptations in an avifauna which has evolved in an environment relatively free of predators. As mentioned elsewhere in this chapter, very few detailed studies have been made of the diets of Antarctic seabirds. Even less is known about how far parents range in search of food and, therefore, what it costs them in energy and other vital resources to obtain food for themselves and their progeny. In this connection, new 'space-age' developments, involving sophisticated, miniature electronic devices, promise 'break-throughs' in the years ahead. Satellite-assisted radio-tracking of the birds' flight paths, and tiny digital integrators for recording wing-beats, heart-rates, time in and time out of water, distance moved while foraging, and so on, are, if not already in use, 'just around the corner'. Many exciting discoveries will be made by means of these devices, but I see no way of avoiding the long hard slog that is necessary towards complete and comprehensive inventories of the birds' diets, and knowledge of the dispersion of their prey.

## 5c.9. HOW DO THE BIRDS AFFECT TERRESTRIAL ECOSYSTEMS?

The high mortality of chicks and the relatively long-lasting incubation and chick-rearing phases of many procellariiforms contribute importantly to the structure of the avifauna while it is land-based, and the way in which it affects the functioning of terrestrial ecosystems on islands in the sub-Antarctic. In essence, the slow overall growth rate dictates fairly extended stays on land by both parents and chicks. This has three primary consequences: first, the birds are land-based for long enough to support populations of avian scavengers, such as the kelp gull and sheathbills, and predators, such as skuas; secondly, the birds deposit substantial amounts of organic products, in the form of excreta, eggs, carcasses and feathers; and, finally, they modify the physical nature of the terrain, chiefly by burrowing into the ground. So far only on Marion Island has much progress been made in quantifying these factors.

Typically, on a sub-Antarctic island breeding penguins occupy the coastal slopes, cormorants and smaller albatrosses (the mollymauks) the cliffs, larger albatrosses the higher flatter ground, and small petrels the coastal lowlands and inland slopes. Because the bulk of the avian biomass is concentrated in the coastal lowland, this area is affected most, and it appears that the nutrients introduced, recycled and redistributed by the birds are crucial for the maintenance of many vital processes in the lowland ecosystems. For instance, at Marion Island the soils are inherently deficient in nitrogen and the birds' contribution, mainly by way of manure, could be essential for the survival of the vegetation and its dependent terrestrial invertebrates. Most of this avian-derived nitrogen is contained in penguin guano which is washed straight into the sea, enriching inshore waters and promoting the production of food for animals which feed close to the coast. However, at least 100 tonnes does reach the vegetation, representing about 90% of all nitrogen introduced annually into Marion Island's vegetated areas. The total figure does not include nitrogen from other avian products, and contributions from burrow-nesting petrels. Thus, it is an underestimate.

The areas around albatross and giant petrel nests invariably have a more luxuriant vegetation than adjacent areas. On average, approximately three kilograms of guano (dry weight) are deposited annually at each wandering albatross nest. This rate of manuring is comparable to the rate of application of synthetic fertilizers to pastures and field crops in the U.S.A. Thus, it is very likely that the relatively high productivity of the vegetation at Marion Island is maintained largely by the continual introduction into the island of essential nutrients transported by birds from the sea. These findings were foreshadowed about 100 years ago by Moseley (1879) who, in describing aspects of the vegetation on Marion Island, commented: 'The plants are, no doubt, rendered especially luxuriant by the dung of the numerous seabirds'.

The amounts of avian products deposited annually by surface-nesting species in the 100 square kilometres of coastal lowland at Marion Island are impressive by any standard: 32,000 tonnes of fresh guano, 500 tonnes of feathers, 350 tonnes of dead birds and 200 tonnes of eggs. Moreover, these figures take no account of the contributions of hundreds of thousands of burrow-nesting petrels. (The eggs alone contain enough energy to drive a small motor car eight times around the world.) More importantly, however, the eggshells contain predominantly calcium which is released gradually into the ecosystem for take-up by plants. It is believed that calcium is an important element limiting primary production at the island. The calcium derived from eggshells is transported over much of the lowland by sheathbills, skuas and kelp gulls. These species play an important role as redistributers of nutrients within the terrestrial ecosystem, in that they either scavenge or prey on birds and eggs. For instance, skuas kill some 50,000 burrow-nesting petrels each year at Marion Island, in addition to feeding on chicks and eggs of penguins. The breeding seasons of the skuas and sheathbills are timed so as to allow the birds to take maximum advantage of the presence of nesting species, as providers of food. Indeed, without penguins to rob for food, obtained originally from the sea, sheathbills would not be able to live permanently on land. During the months that their penguin hosts are away at sea, the sheathbills switch to feeding on terrestrial invertebrates, such as earthworms and beetles, but this food is neither sufficient nor of the right quality to allow successful breeding. Even so, the impact of the sheathbills is considerable in that they consume annually about 8% of the standing crop of the macro-invertebrates in the coastal lowland at Marion Island, much of which is sustained indirectly by guano.

Seabirds are important agents of erosion at several sub-Antarctic islands. At Marion Island, the combined populations of the burrow-nesting petrels may have moved some two million cubic metres of soil in excavating their burrows. The effects of these birds are most evident where volcanic ash or glacial deposits occur on old lava slopes. Slumps and slides of unstable soil are created, which eventually are re-colonized by the birds after stabilization by new vegetation. The birds probably also play a second role in this dynamic process of continual renewal, by transporting seeds and other biotic propagules, such as moss spores, involved in the succession of vegetation. Indeed, the dispersal of certain plants between islands in the Southern Ocean is facilitated by seeds equipped with hooks and other structures for adhering to the plumage of birds.

All of this goes to show that marine birds constitute a major force affecting the growth, composition and succession of terrestrial vegetation on some sub-Antarctic islands, mainly through a combination of manuring and erosion. Moreover, there is a paucity of indigenous herbivores on these islands, and a detritus pathway is the normal route through which nutrients are recycled. The productivity of the micro-organisms is enhanced by energy and chemical elements contained in the birds' excreta. Macro-invertebrates, such as earthworms, snails, and mites, which also participate importantly in the formation of detritus, are important as food for sheathbills and kelp gulls which recycle nutrients for the detritivore community. Thus, while seabirds are major participants in almost all of the principal energy and chemical element fluxes in these terrestrial ecosystems, their primary role as transporters of nutrients from marine to terrestrial ecosystems is very important, if not crucial, at many islands in the Southern Ocean.

The penguins, at first sight, generally impress in this regard, because of the pre-eminence of their biomass. However, this impression may be misleading, since penguins normally concentrate in widely-separated, large colonies near landing beaches. Hence, a substantial majority of their potential contribution of nutrients to the terrestrial system is returned almost immediately to the sea. The relatively much smaller biomass of the balance of the avifauna tends, however, to be distributed more widely, more evenly, and over a longer time on islands. This means that these birds, especially the burrow-nesters, could be responsible for not only a greater input of nutrients into the terrestrial ecosystem but, most importantly, spread over a longer period of time. Studies dealing with spatial and temporal aspects of the birds' contribution of nutrients in relation to the plants' responses are still very new, but already they are indicating that big is not necessarily best.

## 5c.10 WHY SO MANY BIG SPECIES?

Visitors to the Southern Ocean often are impressed by the large size of albatrosses. A crude ordination of the Southern Ocean avifauna (excluding penguins) by body-mass shows that 25% weigh more than 2.0 kilograms, 15% 0.5 – 2.0 kilograms and 60% less than 0.5 kilogram. Comparable ratios for

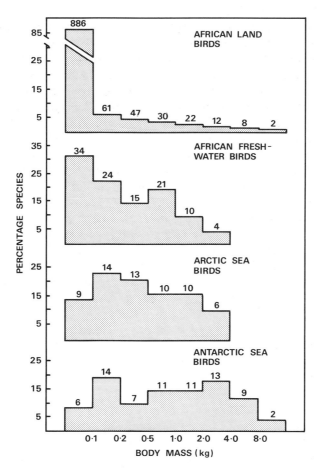

Fig. 5c.17. Distribution of body size of birds from different regions. The importance of large forms in the Antarctic is clearly shown. Numbers of species in each weight class are given above each histogram.

continental terrestrial avifaunas are biased much more heavily in favour of small birds, and major marine avifaunas elsewhere contain relatively few species exceeding 2.0 kilograms in body-mass (Fig. 5c.17). Continental freshwater avifaunas, despite large ibises, storks, swans, geese, pelicans, flamingos and the like contain proportionately fewer large birds. Why are there so many big species, and why so few medium-sized species in the Southern Ocean?

In attempting to answer these questions, it is best to begin by regarding them as parts of the broader problems of why are there so many different kinds of pelagic species, and how do they manage to co-exist? This invokes one of the central issues in modern ecology. The issue concerns the role that competition between species for common resources has played in selecting for the evolution of differences between co-existing species. The debate over interspecific competition contains primary assumptions which currently cannot be tested in the context of the Southern Ocean avifauna, because too little is known about the dispersion of the birds and their prey. Even at a most basic level, it is, for example, not known to what extent similar species in certain 'key' groups, such as the prions and diving petrels, overlap in their distribution at sea, because the birds can only be told apart accurately when examined in the hand. An additional complication arises in the form of wide variation in size between individuals in certain populations. In the snow petrel, for instance, very large and very small birds interbreed, and variability in size appears not to be correlated with either sex or geographical distribution.

Notwithstanding these drawbacks, it is reasonable to contend that there are different kinds of birds because there are different kinds of resources. This implies that there may be predominantly big and small sea birds in the Southern Ocean because there are predominantly big and small prey items, and not necessarily because there are different birds and/or other animals competing for common prey items. In other words, differences in size between species could be simply the result of specialization for foraging efficiency and prey capture. Thus, large food items could lead to selection for increased size in certain avian predators, independently of competitors.

There is no evidence either for or against the absence of interspecific competition affecting large flying birds in the Southern Ocean, but the premise that they feed on relatively large prey accords with their positions near the tops of short food chains in which there are dramatic increases in size between animals constituting successive trophic levels. Similarly, the small birds, as a group, feed mainly on small prey items, chiefly zooplankton, near the bottoms of the food chains. Apparently, intermediate-sized prey items either are too few or are not available to birds in the Southern Ocean.

In the Southern Ocean, pelagic food chains in their simplest form consist of four components: phytoplanktonic producers, primary consumers, and secondary and tertiary consumers. In terms of biomass, krill predominate amongst the primary consumers, and squid and marine mammals and birds predominate amongst the secondary consumers. Mammals and birds also predominate amongst the tertiary consumers. This arrangement differs markedly from that of most marine ecosystems in which fish normally predominate in pelagic environments. The endothermic, or 'warm-blooded', physiology of birds and mammals apparently gives them an inherent advantage over fish in exploiting the 'pulsed' nature of food supplies in the cold seas of the Antarctic. In fact, the birds and the mammals are utilizing representatives of food types which in other oceans would be exploited largely by fish. They presumably are able to do what fish cannot do, because of their superior body-temperature regulation, insulation against cold, ability to store substantial energy reserves, and ability to move rapidly over vast distances in search of food. Their exceptional mobility is, no doubt, a primary attribute for capitalizing on food resources whose abundance fluctuates in space and time. Many of the birds and mammals migrate out of the Antarctic region at the end of the austral summer.

There is an abundance of good circumstantial evidence favouring the view that selection for specialized foraging, and foraging efficiency, is important in Antarctic seabirds. Large avian predators typically hunt by means of sight, perching and waiting for prey to arrive or searching for prey while on the wing. Clearly, pelagic seabirds are largely restricted to the second, energetically more expensive, option and there is general agreement that their foraging behaviour has been selected to conserve energy and/or time, particularly by using the wind to glide instead of using body-energy to beat their wings. Thus, it can be expected that a seabird will tend to take larger prey as the distance over which it has to search for that prey, and transport it after capture, increases. Individually-marked wandering albatrosses, for example, have been observed at distances up to 1500 kilometres from their breeding stations at times when they had young chicks to feed. These birds are very probably at the 'cutting-edge' of evolution, involving intensive selection for characteristics which optimize foraging efficiency. In short, while not dismissing entirely interspecific competition as a factor in the ecology and evolution of the wandering albatross, this species, for one, could be what it is largely because of the nature of its prey and intensive selection for foraging efficiency.

This is fine as far as it goes, but it does not explain how two or more similar species potentially exposed to the same set of resources can co-exist. We know that certain closely-related, morphologically similar, co-existing species, such as the sooty albatross and the lightmantled sooty albatross, differ in their use of foraging areas. And, we assume that this is a form of two-dimensional spatial partitioning of resources. Indeed, in the pelagic environment of the Southern Ocean, co-existing seabirds (excluding penguins) have limited options for segregation, because their food resources largely are available only in a narrow horizontal plane — the surface layer of the sea. Thus, they either segregate spatially or they employ different foraging behaviour. In either event, we do not know whether the differences between the species reflect differences in the behaviour and the dispersion of the same prey, or whether the differences represent the consequences of competition for the same resources. To solve the problem we need to know whether prey is exploited fully by one species and, if so, when and where. This information will be very difficult to obtain, but it is essential if resource partitioning is to be interpreted in the context of avoiding or minimizing interspecific competition.

## 5c.11. CONSERVATION CONSIDERATIONS

Conservation of Antarctic ecosystems is treated fully in Chapter 8. Hence, this is not intended to be an exhaustive review of the conservation status of the Antarctic avifauna. Nor do I intend to provide a comprehensive and complete description of man's inimical past and present effects on the birds. Rather, I will attempt to highlight actual or potential man-induced influences on the functioning of Southern Ocean ecosystems, which affect or could affect seabirds. Some of these effects depend on subtle changes in ecological processes, not readily apparent to a concerned conservation-minded public. The focus, as throughout the rest of the chapter, is on flying birds in the pelagic environment.

There are two principal groups of human threats to Antarctic seabirds: direct disturbance or disruption of the birds' vital activities, such as breeding or feeding; and, indirect disruption of these and other activities. Direct disturbance is easiest to appreciate, and to control. In general, present controls are adequate in the Antarctic Treaty area, south of 60°S latitude, but less so in the sub-Antarctic where the total world populations of a number of species are small and restricted to breeding at relatively few localities. Some of these populations are threatened directly by man and introduced predators, such as cats and rats. Indeed, alien animals already occur at nearly 50% of the island groups in the Southern Ocean (see Chapter 6b). These and other direct threats to the birds at the islands on which they breed are discussed further by Watson (1975) and Williams et al. (1979).

Turning now to the indirect threats, it is convenient to examine them under two heads: pollution and food. Contamination of the Southern Ocean ecosystem by man-made waste is widespread and insidious. Synthetic chemical pesticides, their residues, and abnormal levels of heavy metals have been found in the tissues of Antarctic birds since the early 1960s. The existence of widespread pollution of the Southern Ocean by plastic was discovered in the 1970s. Polyethylene particles and pieces of polystyrene foam are now found in the stomachs of a number of species of Antarctic seabirds, especially surface-foragers which probably swallow these objects in mistake for food. No detrimental effects to the birds have been proven as yet, but preliminary research suggests that the plastic pellets interfere with their ability to digest food and to build up reserves of body-fat. Concentrations of chlorinated hydrocarbons, such as DDT, PCB and dieldrin, tend to be lower in the tissues of those Southern Ocean birds which spend more of their lives south of the Antarctic Convergence, and which feed essentially on plankton. Scavenging birds, and those which feed on squid or fish, contain higher levels of pesticide residues, especially if they frequent waters north of the Convergence.

In general, levels of residues of man-made chemicals are low in Antarctic birds, in comparison with what has been found in marine birds elsewhere and in large carnivorous birds in terrestrial and freshwater ecosystems. Apparently not all of these synthetic chemicals are toxic to birds. This does not, however, constitute grounds for complacency, because new chemicals are being introduced into the world's ecosystems all the time, and it is impossible to predict all their effects, especially over long periods. Moreover, for most of these chemicals which have been produced by man for many years, we still do not know fully either how they may affect different animals or how the animals' physiological mechanisms might detoxify certain of them. What we do know is that the Antarctic marine ecosystem 'turns-over' relatively slowly and that its animals tend to live for a long time. This, along with the cold environment which retards the rate of breakdown of certain pollutants to innocuous substances, probably means that various toxic chemicals will continue to accumulate in Antarctic organisms and concentrate in food chains leading to birds. Thus, it is necessary not only to monitor levels of as many potentially toxic chemical pollutants as possible in the tissues of Antarctic birds, but also to monitor the birds' physiological and behavioural functions which might be affected by pollution. This form of monitoring has implications for human health, because many of the organisms, such as krill and squid, on which the birds feed and thereby obtain pollutants, are potential resources for human consumption as well.

Man's intention increasingly to exploit the food resources of the Southern Ocean holds potentially serious consequences for the region's avifauna. Extensive fishing for krill, squid or fish, or very intensive local fishing for these resources in the birds' major breeding and wintering areas, are likely to be implemented in the foreseeable future, if not already in operation. As mentioned earlier, the potential effects of large-scale fisheries on the functioning of the Antarctic marine ecosystem cannot be predicted accurately at present. Some of man's stepped-up fishing activities might even benefit certain seabirds. For instance, it appears that the blackbrowed albatross and certain other sub-Antarctic breeding birds which winter regularly off the west coast of southern Africa, are able to increase greatly their rate of food consumption by scavenging offal provided by a recently expanded bottom-trawl fishing industry in the area. On balance, however, it is safe to say that the avifauna may be affected adversely, since both man and birds operate as efficiently as possible in exploiting common resources. In effect, this means that success in both parties is determined largely by the rates at which they encounter and capture prey. If the encounter rate for birds is reduced by man's fishing activities, then, it is likely that their foraging efficiency will decrease. For some species, the margin of safety between efficient and less than efficient foraging, and consequently success or failure, may be very narrow. Failure might manifest itself in the form of the birds' inability to nourish either themselves or their young adequately, or both. In other words, reproductive output could decrease and mortality could increase in the wake of even a slight reduction in foraging success.

Failure to capture sufficient food economically, and the likely consequent reduction in abundance of certain avian species, could have a 'ripple' effect culminating in the extinction of local populations. Once again, however, it is impossible to make a definitive prediction in the absence of information such as: to what extent are seabirds drawn to feed in favourable areas by the presence of prey alone or by associated environmental features, including feeding birds of the same or other species? If birds are attracted by the presence of other birds to good foraging sites, then there probably is a minimum avian population threshold below which such a system cannot work properly. This kind of arrangement could be especially important in certain species, such as prions, for which local depletion of prey populations could contribute to an increase in the distance travelled by individual birds in searching for food if they did not form flocks. In short, it is possible, on theoretical grounds, that prions exploit patchily-renewable food resources more efficiently by foraging over areas in flocks, so minimizing revisiting places where they have fed recently. Individuals foraging alone or in very small flocks would expend energy and time covering previously-exploited places, with uneconomical rewards. Unfortunately, too little is known about the spatial and temporal dynamics of the prey of prions to test this speculation at present.

Notwithstanding the preceding theoretical considerations, changes may have occurred already in the rate of encounter between certain birds and their food, following man's past exploitation of animals in the Southern Ocean. Whereas populations of certain krill-eating penguins have increased, apparently in the wake of commercial whaling, other birds may have been affected deleteriously by the decrease of whales. For instance, it seems possible that certain species, including the wandering albatross, may rely to a considerable extent on an association with whales for obtaining food. These birds appear to be adapted for opportunistic scavenging of squid from the vomit of sperm whales *Physeter catodon* which periodically regurgitate their stomach contents as part of their normal behaviour. Such relationships between animals and their food supplies underline the need for a soundly-based approach to regulations for the proper management and conservation of the living resources of the Antarctic marine ecosystem.

## REFERENCES

Ashmole, N.P. (1981) *A Review of Behaviour of Marine Animals, Current Perspectives in Research*, Vol.4. *Marine Birds*, Eds J. Burger, B.L. Olla and H.E. Winn. Plenum Press, New York. *Science* 212, 156.

Croxall, J.P. (1984) Seabirds. In *Antarctic Ecology*, Ed. R.M. Laws, pp. 533–620. Academic Press, London.

Moseley, H.N. (1879) *Notes by a Naturalist Made During the Voyage of H.M.S. Challenger*. Macmillan, London.

Mougin, J.L. and Prévost, J. (1980) Evolution annuelle des effectifs et des biomasses des oiseaux Antarctiques. *Rev. Ecol. (Terre Vie)* 34, 101–33.

Murphy, R.C. (1936) *Oceanic Birds of South America*. American Museum of Natural History, New York.

Watson, G.E. (1975) *Birds of the Antarctic and Sub-Antarctic*. American Geophysical Union, Washington, D.C.

Wenzel, B.M. (1980) Chemoreception in seabirds. In *Behaviour of Marine Animals, Current Perspective in Research*, Vol. 4. *Marine Birds*, Eds J. Burger, B.L. Olla and H.E. Winn, pp. 41–67. Plenum Press, New York.

Williams, A.J., Siegfried, W.R., Burger, A.E. and Berruti, A. (1979) The Prince Edward Islands: a sanctuary for seabirds in the Southern Ocean. *Biol. Conserv.* 15, 59–71.

# CHAPTER 5d

# Birds and Mammals — Penguins

## BERNARD STONEHOUSE

Scott Polar Research Institute, Lensfield Road, Cambridge, U.K.

## CONTENTS

## 5d.1. INTRODUCTION

Penguins are flightless sea-birds of the southern oceans, widely distributed in tropical, temperate and polar waters and well represented in the Antarctic avifauna. The distribution of the eighteen living species and their general biology has been reviewed elsewhere (Simpson, 1976; Stonehouse, 1967, 1975); here attention is focused on the seven polar species that breed on the Antarctic continent and islands south of the Antarctic Convergence. The polar species, their distribution and breeding data, are summarized in Table 5d.1.

Antarctic species represent three of the six general species into which living penguins (Sphenisciformes) are grouped. They include the two largest living species and five within a medium-size range; the smallest penguins live only in temperate, sub-tropical and tropical waters. The largest

TABLE 5d.1. Weights, Body Lengths and Breeding Localities of Polar Penguins.
Breeding localities are defined as follows: Continental; Antarctic continent and offshore islands to 65°S, Peter I Øy and Balleny Islands: maritime; Antarctic peninsula south to 65°S, and islands within the pack-ice zone: peripheral islands; South Georgia, Heard and Macdonald Islands and Archipel de Kerguelen: sub-Antarctic islands lie north of the Antarctic Convergence and within the Subtropical Convergence, separated by the 10°C summer isothern (Stonehouse, 1967a, 1970a, 1982).

| | Mean body-Weight (kg) | Length (cm) | Antarctic | | | Sub-antarctic | |
| --- | --- | --- | --- | --- | --- | --- | --- |
| | | | Continental | Maritime | Peripheral | Cold temperate | Warm temperate |
| Emperor *Aptenodytes forsteri* | 30.0 | 115 | + | − | − | − | − |
| King *Aptenodytes patagonica* | 15.0 | 95 | − | − | + | + | − |
| Adélie *Pygoscelis adeliae* | 5.0 | 70 | + | + | − | − | − |
| Chinstrap *Pygoscelis antarctica* | 4.5 | 68 | − | + | + | − | − |
| Northern Gentoo *Pygoscelis papua papua* | 6.2 | 81 | − | − | + | + | − |
| Southern Gentoo *Pygoscelis papua ellsworthii* | 5.5 | 71 | − | + | − | − | − |
| Macaroni *Eudyptes chrysolophus* | 4.2 | 70 | − | + | + | + | − |
| Rockhopper *Eudyptes crestatus* | 2.5 | 55 | − | − | + | + | + |

polar penguins (emperors) live furthest south and seldom appear even in sub-polar waters, but size and latitude are not otherwise closely correlated, either among the cold-water species, or among penguins as a whole. Though the greatest variety of species occurs in middle latitudes of the southern hemisphere (for example in the Falkland Islands and among the southern cold-temperate islands of New Zealand, where five species may breed in close company with each other), the greatest numbers and greatest total biomass of penguins occur further south in Antarctic waters.

The popularity of penguins with early Antarctic explorers, based partly on their quaintness but also on their ease of capture and palatability, gave them particular prominence in expedition narratives and reports, and established their popularity with the reading public. The popular accounts of penguin life by Murray Levick, coupled with Ponting's splendid photographs and Edward Wilson's art, established penguins as polar characters; in popular view they became the archetypal species, if not the only birds, of the high Antarctic. Huge numbers and relative tameness also made them relatively easy to study, so penguins came to figure largely in polar scientific literature. Adélie, emperor and gentoo penguins were among the first polar birds to be studied in detail, and penguins still claim a large — some would say disproportionate — share of Antarctic ornithologists' attention.

Prominent and numerous though they are in high latitudes, penguins in fact are a minority in the Antarctic avifauna as a whole; there are as many species of charadriiform birds (gull, skuas, terns, sheathbills) as of penguins breeding south of the Antarctic Convergence, and both are eclipsed by the two dozen or more species of procellariiforms (petrels) that breed in the same region. As individuals too penguins are far out-numbered by petrels and many other smaller, less spectacular and generally lesser known flying birds of the Antarctic region, here dealt with in Chapter 5c.

Perhaps the main reason for their prominence on land is the fact that all polar penguins nest in open, tightly-packed colonies numbering hundreds, thousands or tens of thousands of breeding birds. Although early estimates of population size were often exaggerated, systematic counts have shown many colonies of hundreds of thousands, and several that may be numbered in millions. Forming traditionally in the same areas each year, individual colonies may cover several hectares of coast, spreading both inland and upwards onto high ground. Penguins returning to their nests from the sea, heavily loaded with food for their young, may walk several kilometres over rough ground and climb more than a hundred metres in some of the largest colonies — a considerable achievement for flightless birds of short stature with stumpy legs and a restricted circle of vision.

On South Georgia, Iles Kerguelen and other islands close to the Antarctic Convergence, king, gentoo and macaroni penguin colonies are usually found among or behind the belt of tussock grass lining the shore. These are usually on flat or rising ground, often raised beaches and the scree slopes and cliffs behind them. Old-established colonies are generally marked by dead or dying vegetation. Gentoos especially favour tussock-clad slopes, nesting in the crowns of individual tussocks and colonizing high ground overlooking the sea. Rockhopper penguins prefer cliffs, nesting among the vegetation and boulders of steep screes.

The islands further south, surrounded by pack-ice in winter, are relatively bare of vegetation; the shores of the polar continent and its offshore islands, ringed with ice throughout the year, are colder and even emptier. Here southern stocks of penguins nest with only rocks, boulders and the slope of the ground to shelter them. Characteristically, the nests are clustered in groups of a few dozen or a few hundred, spaced two or three to a square metre with narrow alley-ways between. The clusters are separated by wider roads that take a constant, busy traffic of penguins passing to and fro during the breeding season. Penguin colonies can be identified from several kilometres off-shore, and stand out in aerial photographs, by pink or white droppings that persist for many years in the generally dry polar climates. In summer a large colony can be heard — and smelt — many kilometres down-wind. The presence of a big or well-established colony is a sure indication that there is a reasonable certainty of open water, with easy access to food, from mid-December onward (Stonehouse, 1967b).

Sea ice is critically important for several species of polar penguins. Adélies and many chinstraps ride the floes in winter, and may cross kilometres of fast ice (seasonal ice that forms close to land) to return to their nesting grounds in early spring. Emperor penguins live and breed on the sea ice; almost all their known colonies form on fast ice close inshore early in winter, and the birds court, mate, incubate and rear their chicks through the late winter, spring and early summer. Parents and juveniles disperse with the ice as it breaks up in mid to late summer, completing their moult on the drifting floes (Prévost, 1961).

## 5d.2. PENGUIN ORIGINS, FORM AND ADAPTATIONS

The order Sphenisciformes, with its single family Spheniscidae, is almost entirely restricted to the southern hemisphere; one species nesting on the Galapagos Islands breeds on the equator and extends a few kilometres north of it. A sparse fossil record from New Zealand, Australia, South Africa, South America and Seymour Island (Lesser Antarctica) shows that recognizable penguins have lived in southern oceans at least since early Eocene times — some forty to fifty million years, long pre-dating the present pattern of land and ocean that exists in the southern hemisphere. No fossil penguin material has so far been discovered beyond the geographic range of living species. The total record is meagre, but virtually every specimen known has been examined by the distinguished palaeontologist George Gaylord Simpson. It is doubtful if any other avian order has so well-documented a fossil history. Simpson's research papers of 1946, 1972 and 1975, and his popular account of 1976 provide reviews and extensive bibliographies.

Despite this record the origins and affinities of the penguins — in particular of the polar species — remain obscure. Even the earliest fossil material throws little light on phylogeny. Evidence from comparative anatomy (Zusi, 1975) and from serum or egg-white protein analysis (Baker and Manwell, 1975; Feeney and Osuga 1976) is as yet weak. Tentatively, the penguins appear to have originated from stocks of gull-like or petrel-like flying birds of Eocene age or earlier. Skeletal structures and musculature of the flippers and limb girdles indicate that proto-penguins passed through an auk-like stage, in which they could both swim and fly. Weight considerations make it clear that proto-penguins on the verge of losing their powers of flight cannot have been much larger than the largest modern auks — that is, roughly the size of the smallest living penguins (the eudyptulid or little blue penguins of Australia and New Zealand), standing about 30 cm high and weighing a kilogram or less.

No very small penguins or proto-penguins have yet appeared in the fossil record. Most fossil species fall within the size range of medium-to-large modern species; a few were considerably larger, standing 1.5 metres tall and with estimated weights of about 130 kg — the size of a small seal or dolphin. Throughout most of their history, from Eocene times at least until the mid-to-late Pliocene, penguins were necessarily birds of warm seas (Stonehouse, 1969a). Only with the relatively recent spread of the Antarctic ice cap, within the last ten to twelve million years, have high-latitude southern surface water temperatures fallen below 10°C; only in the last two to three million years have they been close to freezing point, with pack-ice a permanent feature of Antarctic seas (Kennett, 1980). Thus penguins appear to have evolved initially for life in warm temperate or even sub-tropical seas, and to have undergone their major radiations during the long period when southern oceans were cooling. They have given rise to polar species only within the last three million years.

The physical characteristics that distinguish penguins from other birds are briefly those that equip them for swimming and diving. Short-necked and squat on land, they become elongate and fusiform in the water, swimming like tiny torpedos with flipper-wings outstretched and vibrating. The legs, set well back on the body so that the birds walk upright, become steering organs at sea; the partly-webbed feet combine with triangular tail to form a flexible rudder. Huge wing muscles, covering a sternum that extends well down the body, provide both the motive power for swimming and the barrel-like shape; sleek feathers and subcutaneous fat combine to give both the streamlined form and the efficient insulation that helps them to maintain a high body temperature in cold water. This complex of characters enables penguins to swim efficiently, to spend long periods at sea without chilling or becoming waterlogged, and yet to remain mobile and efficient on land.

At sea they swim with the effortless ease of dolphins. Moving at speed they 'porpoise' in and out of the water, a method of swimming that allows them to snatch breath without loss of speed, and reduces friction by coating their plumage in a skin of air bubbles. Adélies, gentoos and other medium-sized penguins take the bulk of their food in surface waters, though they can feed deeper; gentoos have been caught in fishing nets set at 100 m (Conroy and Twelves, 1972) — well below the range of petrels and other species that compete with them for plankton. In winter when the plankton descends they may have to hunt deeper still. The larger species seem to feed mainly at depth; kings and emperors take squid and bottom-living fish that require them to hunt in a twilight zone well below the summer levels of plankton. While gentoos, Adélies and other small species normally surface after one or two minutes' diving, emperors feeding through sea ice habitually dive for longer and can stay down over a quarter of an hour. Emperors carrying depth recorders have dived well over 200 m deep and kings over 240 m (Kooyman, 1975).

Speed and efficiency in the water allow penguins to hunt far from their breeding colonies. It has been estimated that the five million Adélie penguins distributed among the colonies of a single island in the South Orkneys require up to nine thousand tonnes of krill and larval fish per day to feed their young at the height of the breeding season. This is probably available to them within a radius of 16 km; a contractor supplying sea food at a comparable rate would need a fleet of seventy medium-sized trawlers and port facilities ten times bigger than Aberdeen (Stonehouse, 1968).

Penguins spend much of their lives in the sea on hunting excursions and migration. Individuals of any species may be immersed for several days at a time; migrating penguins remain at sea for several weeks, and the crested eudyptid penguins (including macaronis and rockhoppers) are seldom found ashore during the four to six months between breeding seasons. The warmest seas that penguins experience — between 13°C and 23°C in the tropics and sub-tropics — are far cooler than a penguin's normal body temperature of 38°C to 40°C. Thus even tropical species require efficient and waterproof insulation. Much greater is the need of polar species, living in seas where temperatures seldom rise above 5°C and fall as low as −1.8°C when there is ice about.

The problem for penguins is partly one of size, for they are the smallest warm-blooded vertebrates that attempt to spend long periods in the water. The largest living penguins are 15 − 20% shorter than the smallest seals and 25 − 30% shorter than the smallest dolphins; their surface-to-volume ratios and consequent heat losses are correspondingly greater. Every penguin faces this problem, but the colder the sea and the smaller the penguin, the greater is its difficulty of conserving heat. Here is one possible reason why the smallest penguins are sub-tropical, not polar (Stonehouse, 1967a).

On land polar penguins experience a wider range of temperatures, though living close to the sea exempts them from the extremes of cold encountered inland. On the continent and many of the southern islands coastal temperatures in winter fall to −40°C and below. The only species likely to experience these very low temperatures consistently is the emperor penguin, incubating on the sea ice close inshore. The smaller southern species, wintering on the pack-ice, gain the benefit of a warmer maritime environment provided by leads of open water: they may well experience −10° − −20°C in winter, though seldom much lower. Coastal air temperatures in summer rise toward freezing point; sometimes they exceed it by 5 − 10°C on warm days, dropping back to 0°C and below at night. Sun temperatures — the surface temperatures experienced by birds basking in bright sunshine — may be very much higher, especially on calm days. Adélie penguins incubating in still air in their colonies often become overheated, exhibiting gular fluttering — the avian equivalent of panting — when the sun shines directly upon them. Even through a haze of thin cloud the sun can warm them to the point of discomfort; under a stronger insolation an incubating bird can suffer considerable heat stress, which it relieves by eating copious billfulls of snow as soon as its partner returns to relieve it at the nest.

On islands closer to the Antarctic Convergence maritime conditions prevail throughout the year. Air temperatures fall below freezing point frequently in winter and occasionally in summer, but strong winds from the sea keep mean monthly temperatures remarkably steady all the year round. Summer sunshine can be intense on these islands, which lie in southern latitudes equivalent to those of Britain in the north, but the combination of bright sun and still air is rare.

Winds blow strongly throughout the Antarctic region but hardly affect penguins. Living close to the ground they gain shelter from rocks and vegetation, and their plumage provides effective insulation (see below) which winds cannot disturb. Even blizzards have little effect on adult penguins, except those that are constrained by incubation. Blizzards over a colony during the breeding period may result in many hundreds of abandoned eggs. Chicks in down too may be devastated by drifting snow, which melts in their plumage and chills them to death (Sladen, 1958; Yeates, 1971, 1975).

Penguin insulation takes two forms — plumage and subdermal fat. In both tropical and polar species feathers cover the entire body (except for a seasonal brood patch) at a uniform density of 11 − 12 per cm. Each individual feather is short, with flattened shaft and slightly curved blade; a mass of filamentous

Fig. 5d.1a. Gentoo penguin at nest with incomplete clutch. (Photo: W.N. Bonner)

down forms the after-shaft close to the base. The shafts of the feathers grow outward at an angle from the body, leaving an airspace beneath that which is partly occupied by the down; the depth of airspace can be varied by raising or sleeking the feathers. The feather tips overlap like the tiles of a roof. Slightly oiled, they are water-repellent and combine to form a windproof outer shell.

Dorsal feathers vary in length from about 2 cm in small tropical species to over 4 cm in the largest penguins. Feather length generally increases in proportion with body length, but the dorsal feathers of emperors, Adélies, rockhoppers and southern gentoos are significantly longer than might be expected from body size alone. Kings, chinstraps, northern stocks of gentoos and macaronis have normal plumage length for their size; possible reasons for these differences and consequences are discussed under individual species. The actual thickness of plumage resulting varies from about 0.5 cm in tropical species to 1.5 cm in emperors. Plumage provides effective but only slightly variable insulation; in extreme cold inactive penguins fluff out their feathers, which may increase their insulating effect. Feathers account for about 80% of a penguin's insulation in air, but probably less in water (Drent and Stonehouse, 1971); much of the insulating effect is lost at depths, when the air contained between the feathers and down is compressed by the water pressure (Kooyman, 1975).

Subcutaneous fat provides a more variable insulation, and one that is no less effective at depths. Fat beneath the skin forms a layer up to 2 cm thick in emperors, 1 to 2 cm in Adélies and slightly less in gentoos; the amount varies considerably with time of year and feeding regime. Skin and fat are well supplied with blood vessels that dilate under the control of the autonomic nervous system. When blood flows freely through them the insulating effects of the tissues are negligible; when the vessels constrict, the fat becomes an insulating layer similar in efficiency to the plumage outside.

Apart from being an insulator, fat is an energy store that penguins need to see them through lean periods — in winter, for example, when food is scarce in surface waters, and during the two or three

Fig. 5d.1b. Gentoo penguin nesting site in tussock grassland at South Georgia. Note the pale (isabelline) penguin in the centre. (Photo: D.W.H. Walton)

weeks of annual moult when all the feathers are replaced simultaneously. Moulting penguins cannot go to sea. In preparation for the moult they fatten, adding some 50% to their normal weight; all of this surplus fat and sometimes more is lost as the feathers grow. Polar penguins fatten again before the start of their breeding season, which in all species involves stress and loss of feeding time, and in some involves fasting for several weeks.

The dual insulation of feathers and fat is remarkably efficient, allowing incubating or otherwise inactive birds to retain their full body temperature in conditions of extreme cold and strong winds. Light snow settling on Adélie penguins tends to stay there rather than melt; gentoos coccooned in snow during a blizzard emerge dry and warm, having lost very little heat through their down jacket of waterproof feathers.

Fig. 5d.1c. Adélie penguin. (Photo: W.N. Bonner)

Fig. 5d.1d. Chinstrap penguin and brood. The chick on the left is regurgitating some krill. (Photo: W.N. Bonner)

Such insulation brings its own problems for active birds on warm days. Most important of their heat-shedding surfaces are the flippers and feet. The flippers, covered with a mosaic of tiny, scale-like feathers, are bony and almost devoid of muscle or fat: the feet are unfeathered and bound in tough, scaly skin. Both feet and flippers are well supplied with superficial blood vessels but have little sub-dermal insulation. Penguins running, fighting or standing in the sun have an enhanced flow of blood through these surfaces, which thus become warm and act as radiators. Birds coming in from the sea, breathless and hot after the final dash through the surf, often show a pink under-surface to their flippers. The vascularized skin glows clearly through the wet feathers, losing its colour as the bird cools and returns to normal (Stonehouse, 1967a).

Some tropical species have unfeathered patches on their cheeks and around the base of the bill as well as the lower legs free of feathers which serve the same heat-shedding purpose. Polar species by contrast have well-feathered cheeks; in emperors and Adélies feathers grow over the base of the bill and well down the legs too, leaving only the feet exposed. Adélies, gentoos and other medium-sized polar penguins manage to be lively, active birds, well able to maintain a balance of body temperature in most circumstances. Kings and emperors are more liable to overheating. Kings move deliberately and cannot run far without stopping to pant. Emperors quite simply refuse to be hurried; their colonies are quiet, orderly assemblies that contrast sharply with noisy, bustling colonies of the smaller penguins.

## 5d.3. STUDYING PENGUINS

Most of what is known about polar penguins comes from land-based field and laboratory studies; we know little of their life at sea. Field studies on a broad scale include surveys and inventories of colonies (now done mainly by aerial photography), recording changes in colony size and distribution and attempting to assess the total biomass of penguins of all species throughout the Antarctic region. On a smaller scale long-term field studies have now been undertaken on most of the polar species, though the coverage is curiously uneven. Best known by far is the Adélie penguin, closely followed by the emperor and king; far less is known of gentoos, chinstraps, macaronis and rockhoppers, though the last two have to some degree been studied in sub-polar colonies north of the Antarctic Convergence.

Techniques for detailed studies of penguins are common for all species. The first essential is a banding programme to distinguish individuals whose life-style (at least on land) can be followed in detail for many years. Numbered leg bands are now commonplace in avian studies, but penguins wear flipper bands, which are less liable to damage them and much easier to spot from a distance. Fully grown chicks and young birds in distinctive yearling plumage are especially useful, for they form cohorts of known age. Mortality is high in young birds and many must be banded to ensure a few significant returns. Penguins that survive their juvenile years can be long-lived: the mean life expectation of emperors is 20 years, but of Adélies probably less than ten years. One practical problem of penguin study has been to find bands sufficiently durable to stand up to the wear-and-tear of these longer life spans.

With a marked breeding population, including a proportion of birds of known age, details of life history — survival, breeding ecology and behaviour, nesting success, significance of age and experience in breeding, stability of pair bonds, growth rates of chicks and other important facets of penguin life — can gradually be worked out (Croxall, 1982). Costly long-term studies involving several breeding seasons yield increasingly rewarding returns, and contribute much to our understanding not only of penguins, but of avian biology in general.

Emperor, king, Adélie and gentoo penguins have now been studied in some detail at their breeding colonies; we know less of chinstrap and macaroni penguins, but enough to make interesting comparisons with the better-known species (Table 5d.2). Pygoscelid and eudyptid penguins show many

TABLE 5d.2. Comparative Data on Breeding Cycles of Polar Penguin Species.

| | Nest | Number of eggs | Usual date of first laying | Parents with first incubation watch | Usual feeding interval (days) | Total incubation period (days) |
|---|---|---|---|---|---|---|
| Gentoo | Yes | 2 | June – Nov. | either | less than 1 | 35 |
| Chinstrap | Yes | 2 | late Nov. | female | 2 – 3 | 35 – 38 |
| Adélie | Yes | 2 | Nov. – Dec. | male | 2 – 3 | 35 |
| Rockhopper | Yes | 2 (unequal) | late Oct. – early Nov. | male (female present) | 1 – 2 | 33 – 34 |
| Macaroni | Yes | 2 (unequal) | Nov. | male (female present) | 1 – 2 | 35 – 37 |
| King | No | 1 | late Nov. – early March | male | 3 – 4 | 54 – 55 |
| Emperor | No | 1 | May – June | male | variable | 65 |

common features of breeding behaviour, differing only in detail from temperate and tropical species. Abandoning their colonies during the winter, they return to breed during a relatively short span of a few days or weeks in early spring. Experienced breeders usually settle to their breeding first, often mating with their partner of the previous year on the same site as before. Initially both partners are involved in nest building and in incubating and brooding their one or two chicks. The half-grown chicks leave their nests at similar stages to band together in crèches, leaving both parents free to hunt for food. Differences between the species — choice of nesting sites, precise timing of breeding seasons and details of family life, for example — can now generally be explained in terms of ecology and environmental adaptation.

Kings and emperors differ more radically in their breeding patterns, both from the smaller species of all latitudes and from each other. Neither builds a nest; both produce a single large egg which they incubate on their feet, but there the similarities end. Kings have evolved a remarkable pattern of family life that requires them to maintain their chicks throughout the winter, and allows them the chance to breed only twice in every three years. Emperors breed annually, but are unique in laying their eggs on newly-formed sea ice in autumn, incubating through the bitter cold of winter, and freeing themselves of parental responsibilities long before the chicks are fully grown (p.287).

Colony size is difficult to determine, and estimating changes in colony size over a number of years demands special techniques. Counting penguins on the ground is almost impossible, for penguins never stand still long enough to be counted. The population of a polar colony, large or small, increases from zero in winter and early spring to a series of peaks during the breeding season, fluctuating untidily toward the end of summer when juveniles, immature and mature birds throng the colonies. With penguins coming and going all the time, how can colony size be estimated? The most useful index is the number of nests at peak of breeding — a stage reached two to three weeks after the start of incubation, when the maximum numbers of birds have occupied their nests, laid eggs and begun to incubate. Counting nests is more reliable than counting heads, and counting from photographs — including aerial photographs — is the most reliable method of all for estimating and comparing colony size (Stonehouse, 1969b; Taylor and Wilson, 1982). Photographs taken from the air are often the only records available of remote colonies far from expedition bases: their value increases with time, for they allow us to detect changes over a number of years — to show how colonies increase or decline in size with changes in food availability, climate, sea ice and other environmental factors.

There still remain huge gaps in our knowledge of even the best-known species. Many of these gaps would be filled if we could follow our birds to sea. Few penguins (other than chicks) die on land, for example; what are the main causes of mortality? Emlen and Penney (1964) and others have studied their navigation ashore; how do they navigate at sea, and what are the stimuli that bring them back to the colonies after a long winter far from land? We would be glad to know more of their feeding behaviour and ecology. From shore-based studies we have some idea of what most of the polar species feed on (Croxall and Prince, 1980): it is not difficult to sample the food that a penguin brings back to the colony in the crop, though the half-digested mess is sometimes difficult to analyse and identify. But how do penguins catch their food, in competition with the huge flocks of petrels and other species that feed alongside them in surface waters? How do they find and identify shoals of food? What strategems do they adopt in the open ocean against predatory seals, and what other predators affect them? Does experience count in hunting and being hunted, as it seems to count in successful breeding? We know something of swimming and diving physiology from laboratory studies in temperate regions (Scholander, 1964); what special problems affect polar species in icy waters far from land? Our shore-side knowledge of penguins in general and of polar species in particular has increased enormously in the past three or four decades; perhaps the next few decades will bring answers to some of these equally interesting problems of penguins at sea.

## 5d.4. THE BRUSH-TAILED PENGUINS

The three species of the genus *Pygoscelis* form a closely-matched group. Similar in size, shape and general behaviour, they are all predominantly black and white birds, notable especially for their long, stiff tail feathers that stick out behind as they walk. No other penguins have so prominent a tail, and from its sweeping action comes the name *Pygoscelis* ('brush-tailed').

Gentoos, largest and most colourful of the pygoscelids, have the broadest latitudinal range. They live mainly on the peripheral Antarctic islands — South Georgia and Heard Island — that lie just south of the Antarctic Convergence, and on the cool-temperate sub-Antarctic islands — Iles Crozet and Kerguelen, the Falkland Islands and Marion, Prince Edward and Macquarie Islands — that lie to the north. Only a few thousand pairs live among the pack-ice in the far south. Chinstrap penguins are the smallest of the three species with the narrowest geographical range. They breed almost entirely in the maritime Antarctic region, including the South Orkney, South Shetland and South Sandwich Islands and the north-western coasts of Antarctic Peninsula. Adélie penguins are the southernmost species, confined entirely to islands and coasts within the northern limit of the pack-ice. Their northernmost stocks overlap with chinstraps and gentoos in the maritime Antarctic, but the bulk of Adélie populations live and breed far to the south on the shores of continental Antarctica, where no other pygoscelids penetrate. Comparative studies of the three species of pygoscelids have been made on the South Shetland Islands (Trivelpiece and Volkman, 1979; Volkman and Trivelpiece, 1980; Volkman *et al.*, 1980).

Characters that distinguish the species are borne mainly on the head. Gentoos have orange or coral red plates covering the lower mandibles and sides of the upper mandibles; head and face are black or brown, speckled with a scattering of white feathers and crossed by a narrow band of white that broadens over each eye. Chinstraps have a black bill, a solidly black cap, and white cheeks and throat; their distinguishing marks are narrow bands of black feathers that cross the cheeks and meet below the chin like the strap of a guardsman's helmet. The eyes are dark brown, rimmed with black, giving them a benign expression that belies their vitriolic temperament. Adélie penguins have a black head and face with a distinctive circle of snow-white skin about each eye. The feathers at the back of the head are erectile; agitated Adélies raise their hackles and roll their eyes in a striking threat display — one of a repertoire of visual signals that are significant in courtship and social life in the colony (Ainley, 1975; Spurr, 1975).

## 5d.4.1. Gentoo penguins

Gentoos are a heterogeneous species with a tendency to diversify and form recognizable subspecies over their wide geographical range. Southern gentoos, of the high Antarctic, all seem to belong to a small race with relatively short bill, flippers and feet. They also have the longest plumage for their size (Murphy, 1948; Stonehouse, 1967a, 1970a). Northern gentoos are larger birds, heavier and with longer bill and limbs. The total world population of gentoos is estimated to lie between 300,000 and 350,000, over 80% of which live south of the Antarctic Convergence.

About two thirds of this widely scattered population live on South Georgia. Norwegian whalers called them 'tussock penguins' — a good name, for gentoos more than any other penguins make use of the broad zone of coastal tussock grass. Gentoos are present on the beaches for most of the year. In winter their numbers are small, amounting to less than a quarter of the breeding population, and suggesting that a high proportion spend the coldest months at sea. Further south, on the Antarctic Peninsula for example, gentoos disappear altogether in May and return to the breeding beaches in August, September and October (Bagshawe, 1938).

October on South Georgia sees the main bulk of the population back in the colony areas. They settle about the traditional nesting sites, usually among the tussock grass, gradually treading down and melting the snow until the old nests reappear. Not every one of their colony areas is re-occupied each year. Often, during the first few days of re-occupation, a colony of several dozen birds shifts a few dozen metres away, perhaps to a site that is emerging more rapidly through the snow. However, as more of the ground clears, the birds become firmly attached to their sites and nesting begins in earnest. The gentoos scrape hollows in the ground with their sharp claws and collect stones, bones, and scraps of moss, tussock and other vegetation, carrying them industriously in their bills to the chosen sites.

Males, slightly larger and heavier than females, are usually the first to settle and defend their sites against intruders with ritualized threat postures and calls (Van Zinderen Bakker, 1971a). Fights are not uncommon, though gentoos are generally the least belligerent of pygoscelids. Females move into the colonies more gradually. Attracted primarily to the site of the previous year, each female stands a strong chance of re-mating with her previous partner, who may already be in residence. If he is not, she is likely to find another unattached male, move into his nesting area, and form a partnership with him instead. New or old-established, the two partners co-operate in nest building and defence. Both return to sea for short periods between bouts of courtship and nest-building, but never at the same time; a nest once established is rarely left unguarded. After preliminary courtship the birds copulate repeatedly. Here the most telling difference between the sexes appears, for males virtually always take the initiative in copulating behaviour. It is the male who stands on the partner's back during treading, and females at this stage are usually distinguishable by their muddy plumage, both back and front.

Two eggs are laid, bluish white, chalky, near-spherical and similar in size, the second following two to three days after the first. On South Georgia the first eggs appear during the second and third weeks of October. First eggs appear in mid- to late June on Marion Island, mid-September on Macquarie Island, late September to mid-October on the Falkland Islands, early November on the South Orkneys, and late November on Antarctic Peninsula. The partners take turns to incubate, lying close to protect the eggs from rain, sleet and snow for a total of about 35 days. Hatching within a few hours of each other, the chicks are at first covered with a plumage of fine, silky grey down; during the rapid growth of the first week this thickens to a complete covering of dense woolly down, grey on the head and dorsal surface and white underneath. Both parents brood the chicks closely, disappearing alternately to sea and returning after a few hours with cropfuls of krill or small fish. This they feed virtually undigested to the chicks in response to importuning calls.

After two to three weeks of almost constant brooding and feeding the chicks begin to wander, and at four to five weeks they join their peers to form crèches or nursery groups — gatherings of 20 to over

Fig. 5d.2a. Adélie penguin rookery at Avian Island. (Photo: W.N. Bonner)

100 chicks that offer company and a degree of protection from the predations of brown skuas (*Catharacta skua*). Crèches free the parents to hunt simultaneously; they effectively double the rate at which food can be brought back to the colony, and help to keep pace with the chicks' ever-increasing appetites. In their tenth to twelfth weeks the chicks, now almost the size of their parents, begin to shed down and take on the appearance of adults. Fully moulted, they leave the colonies in groups and move to the water's edge, swimming with adults and eventually leaving with them for the open sea.

Freed of responsibilities for their offspring, the parents fatten and undergo a complete moult — a process taking three to four weeks in late April and May. Then large numbers of gentoos leave the breeding areas altogether; only a few hundred remain through the early and middle months of winter where there were thousands before. They do not go far; a warm spell in August or September may bring droves of them back to the beaches and even to the nest sites, where the first stages of courtship may begin. Serious courtship and nesting, however, are unlikely to start before the fall of snow-level in spring. Gentoos wander widely during their juvenile years, appearing occasionally in New Zealand, South America and South Africa, and on Antarctic islands well south of their normal range. We do not know when young birds start to breed; by analogy with Adélies, many will probably court and nest for the first time in their third or fourth years (p.280). Gentoos survive well in zoos, where they may start to breed (though seldom successfully) in their second or third years.

## 5d.4.2. Adélie penguins

For Adélie penguins of the high Antarctic two needs are paramount — bare ground to nest on, and open water within easy reach in January and February. Bare ground is scarce at the southern end of their range, for over two thirds of the Antarctic continental coast is over-ridden by land ice, and heavy winter snowfall on many of the maritime Antarctic islands makes huge areas of the beaches and screes

Fig. 5d.2b. Macaroni penguin rookery at South Georgia. (Photo: D.W.H. Walton)

inaccessible. Open water — polynyas within the fast ice and wide leads, that form regularly every year — favours breeding from the moment that nesting begins in October or November; it becomes essential from January onward, for Adélies cannot feed their chicks adequately if they have to walk several miles across rough ice to reach the sea (Ainley and LeResche, 1973; Stonehouse, 1963; Taylor, 1962).

These two factors come together at many scattered points along the Antarctic continental coast and throughout the maritime sector. The southernmost colony of Adélies — a small one of about 2000 nests — is at Cape Royds, in McMurdo Sound (77°33'S). Elsewhere they are found at over 50 continental localities and they are plentiful, though patchily distributed, along the western flank of Antarctic Peninsula, on the South Orkney, South Shetland and South Sandwich Islands, Charcot Island, Peter I Øy, and on Bouvetøya which is their northern outpost. World population is estimated as at between five and ten million.

On their southern breeding grounds Adélie penguins experience not only the coldest conditions but also the shortest summers. Courtship, incubation, chick-rearing and post-nuptial moult must all be compressed within the few brief weeks when temperatures rise to and hover about freezing point, when the sea inshore is relatively accessible and rich in their preferred foods — macroplankton and larval fish — and snow cover is at a minimum. The tightest schedule occurs furthest south. In McMurdo Sound the first breeding birds arrive in the last few days of October and lay their two eggs during the second to fourth weeks of November. After five weeks of incubation and seven to eight weeks of growth, the chicks are ready for the sea during the first half of February, and the adults complete their moult in late February and March (Taylor, 1962). Towards the northern edge of their range Adélies can begin their courtship slightly earlier, lay over a longer period and if necessary take an extra week or two over chick-rearing and moulting before the summer weather ends.

The Adélie breeding routine is similar to that of gentoos. Courtship and nesting, involving striking visual and vocal displays (Ainley, 1975), take one to three weeks, and the clutches of two eggs are laid during November and early December. Males are almost invariably the first to incubate after the clutches are completed, holding the eggs end-to-end against a brood patch among their abdominal feathers, supported on either side by the feet, for one to two weeks. During this time the females return to the sea. Thus males that had already starved during two to three weeks of courtship might endure a further fortnight of starvation before being relieved at the nest. First incubation watches by the females are usually shorter — considerably so if open water lies close at hand — and thereafter the birds can alternate at intervals of two to three days. Despite high winds and low temperatures, Adélies contrive to keep their eggs within a few degrees of their own body temperature throughout the incubation period of 35 days.

The newly-hatched chicks are covered with fine grey-brown down that thickens to a dense chestnut-brown 'wool' during the first two to three weeks. By the end of their second week they have attained control of their body temperature and can wander freely from their parents in all but the coldest weather; by the third week most have joined crèches, usually within a few metres of their nest sites, where they await the frequent visits of their parents with food. Throughout most of their range Adélie penguins are subject to predation by McCormick skuas (*C. maccormicki*) which are quick to pounce on unattended eggs and chicks. Weak and ailing chicks that stand on the edge of the crèche fall quickly to their attacks. The crèches are made safer by the presence of sub-adult and non-breeding birds, and of failed breeders which return at intervals to re-possess their nest sites. No adult or sub-adult tolerates the presence of skuas and any that land are likely to be chivvied and moved on.

In seasons when food is plentiful many parents manage to raise both their chicks. When conditions are poorer mortality among the chicks is high; often it is the second chicks of many clutches that succumb. The chicks begin to shed down shortly after joining the crèches, and within three to four weeks have moulted completely into their first feathers — a suit of grey-black and white similar to that of their parents, though with a smaller eye-ring and a distinctive white chin. This they retain throughout their first year. Fully moulted and ready for the sea, they are as yet only 75 – 80% as heavy as their parents, undersized and very much slimmer (Stonehouse, 1970a; Volkman and Trivelpiece, 1980). Normal chicks that leave the colonies during February and early March still have time to grow and fatten before the end of the season; late chicks whose departure is delayed have less chance of surviving, and many neglected by their parents during the last days of summer fail to reach the sea at all.

Long-term banding studies of Adélie penguins have indicated that young birds spend their first and second years at sea or in the pack-ice, returning to the colonies late in their second and third seasons to wander and perhaps make desultory attempts at nesting. In subsequent seasons they return earlier, usually with the breeding birds, often taking up territories; a few females breed in their third year, many more in their fourth, but most males breed for the first time in their fifth or later years (Ainley, 1975; LeResche and Sladen, 1970). A high proportion of young Adélies return to breed in the colony in which they were hatched, often within a few dozen metres of their nursery site.

## 5d.4.3. Chinstrap penguins

Chinstraps are concentrated in the South American sector of Antarctica and seldom seen outside it. Most occur on the southern islands of the Scotia Arc with well over a million pairs on the South Orkney and South Shetland Islands, and probably several million on the large but virtually uncounted colonies of some of the South Sandwich Islands. About 60,000 pairs are estimated to breed along the western shores and islands of Antarctic Peninsula, south to the Joubin Islands and Waterboat Point (Croxall and Kirkwood, 1979). There are small breeding populations on South Georgia, and smaller groups breeding or attempting to breed on Peter I Øy, Bouvetøya, Heard Island and the Balleny Islands. Total world population numbers at least three million and may exceed ten million, and seems at present to be expanding rapidly. This may be due to increasing availability of plankton following recent reductions in stocks of plankton-feeding whales; many other species of Antarctic seabirds seem also to be increasing in numbers, but chinstrap penguins most rapidly of all.

Notably smaller and more slender than gentoos or Adélies, chinstraps are no less noisy than their congeners and tend to be more aggressive at the nest sites. Where two or three species of pygoscelids nest in neighbouring colonies (for example on several of the South Orkney and South Shetland Islands) chinstraps are usually the last to arrive at the start of the breeding season. They show preference for rocky, sloping, well-drained ground, sometimes occupying ledges above gentoos and Adélies. However, they are capable of infiltrating the colonies of other species and displacing them. Male chinstraps will attack incubating Adélies with pecks and flipper-blows, eventually taking over their nests and installing their own mates (Trivelpiece and Volkman, 1979).

Chinstraps follow the basic pygoscelid pattern of breeding. On the South Orkney Islands the first birds arrive in late October or early November. Pairs re-occupy their old nest sites, rebuilding with pebbles and bones. After brief courtship and mating the first eggs appear, usually about November 20; second eggs are laid two to four days after the first, and nearly all the clutches are completed by the first week of December (Conroy et al., 1975). At the southern end of their range chinstraps start to lay a few days later; on the warmer South Shetland Islands they begin two to three weeks earlier, but throughout their range practically all the eggs are laid within a span of three to four weeks. In contrast to Adélie and some gentoo penguins, chinstrap females take the first long incubation watch of eight to ten days, thus spending as long as 30 days ashore without feeding — in addition to producing two large, energy-demanding eggs.

On returning to their nests the males relieve their partners and then share incubation, alternating at intervals of two to three days for a total incubation period of 35 to 38 days. The chicks, clad in uniform grey — brown down, grow rapidly on a diet of regurgitated krill. Chinstraps feed almost exclusively on krill, which they are believed to catch much closer inshore than either gentoos or Adélies. At four to six weeks old the chicks gather in crèches. By their seventh week most chicks have reached about 80% of the weight of their parents and begun to lose their down; by their ninth week many have already left for the sea. Young birds are best diagnosed by their eyes, which remain pale and mottled for most of their first year. Late February sees the start of moult in the breeding adults; through March the colonies are reoccupied by birds in various stages of moult, some of which rebuild their nests and court while their new feathers are growing. By the end of March practically all have left for the sea, and the last remaining stragglers abandon the colonies in April and May.

## 5d.5. THE CRESTED PENGUINS

Macaroni and rockhopper penguins are the Antarctic representatives of the genus *Eudyptes*. Both wear the distinctive livery of the genus — blue — black dorsal feathers merging to coal-black neck and head, mahogany bill and vivid yellow crests that are shaken and flourished during courtship. In Macaronis the crests are orange — yellow, originating in a broad band across the forehead and sweeping both sideways and back above the dark brown eyes. Rockhoppers have golden to lemon-yellow crests, that start in narrow bands above the base of the bill and sweep back over the eyes, to droop in long filaments; the black feathers of the crown too are elongate, forming a coronet that gives the rockhopper head a distinctively angular shape. The eyes are bright red — much redder than those of any other penguin. As in all the eudyptid species, males are notably larger than females.

Macaronis, which are marginally smaller than the smallest pygoscelids, have a more southerly range than rockhoppers. Neither the thickness of their plumage nor their body proportions mark them out as polar species (Stonehouse, 1970a), yet macaronis at the southern edge of their range nest close to chinstraps and Adélies. Rockhoppers, the smallest of eudyptids, are seldom seen as far south as the fringes of the pack-ice, though many breed on islands south of the Antarctic Convergence. Macaronis appear to be a relatively homogeneous species, with little or no distinction between northern and southern forms. Rockhoppers are more variable; northern stocks of the warm temperate islands have markedly longer crests than southern stocks, and are sometimes given subspecific status (Warham, 1975).

### 5d.5.1. Macaroni penguins

Macaronis breed on South Georgia and throughout the Scotia Arc, reaching a southern limit on Deception Island in the South Shetlands (Croxall and Kirkwood, 1979). Further north they breed on Bouvetøya, Heard and Marion Islands, Iles Crozet and Kerguelen, and the Falkland Islands. Some authorities regard the royal penguin of Macquarie Island as a subspecies of macaroni (Warham, 1975). For breeding, macaronis favour flat or rough sloping ground, where boulders or tussock grass provide minimal shelter. Often this lies behind rough, weather-beaten coastline, for macaronis seem better equipped than other species for landing among reefs on storm-bound shores. The species is best known from early studies on Heard Island (Downes *et al.*, 1959), where several million pairs nest in over a

Fig. 5d.3a. Macaroni penguin and chick. (Photo: W.N. Bonner)

dozen huge colonies, and more recent work on Marion Island, where about half a million pairs are believed to nest (Williams *et al.*, 1979). Between five and ten million pairs (Croxall and Prince, 1979) nest on South Georgia, and many hundreds of thousands are estimated on other islands north of the pack-ice zone. Within the pack-ice numbers fall dramatically; groups of only a few dozen or hundreds nest on the southern islands, usually in colonies of pygoscelids. Macaronis feed mainly on krill in competition with their pygoscelid neighbours.

Macaronis disappear from their colonies completely during the winter, probably spending much of their time at sea. Throughout their range they return in October; males are the first to appear, taking up their nest sites and marking them with pebbles, and the females follow within a few days. Lively courtship follows, in which braying and violent displays of head-shaking figure prominently. Both partners scrape and furnish the meagre nests and take turns during the first few days of incubation. Then one bird — usually the male — leaves for a spell of ten to twelve days at sea; on his return the female disappears for a similar spell, usually returning with a crop full of food just as the chick is about to hatch.

Like other eudyptid penguins macaronis are remarkable in laying two eggs of different sizes. The first, a small though fertile one, is usually lost from the nest early in incubation; though capable of hatching, very few manage to survive. The second egg, some 50% heavier, stands a much better chance of surviving through the incubation period of 35 to 37 days. Reasons for this curious pattern of breeding are far from clear. Concentrating the main effort of breeding into one egg rather than two may favour the survival of chicks where the season of good weather and abundant food is short. Single chicks — especially those from large eggs — presumably benefit from a better endowment of yolk, and do not have to compete for food with siblings; they are thus able to grow faster, standing a better chance than paired chicks of attaining independence before the end of the season. This may well be one of the reasons that has allowed macaronis — least obviously adapted of all the polar penguins — to penetrate far south into the zone of short summers and compete with the better-adapted pygoscelids.

The chicks are guarded by the males for the first three weeks. At this stage only the females are hunting, returning from the sea with food in the crop at intervals that vary from a few hours to two or three days, depending on local availability of krill. From mid-January onwards the chicks — half-grown and wandering freely on the colonies — gather in crèches while both parents take up the hunt. By early February many have reached full adult weight and are shedding down freely; by early March most have left the colonies and begun independent life. Juvenile macaronis retain a distinctive plumage, with grey chin and only token crests of yellow feathers above the eyes, for their first year. Yearlings return to the colonies periodically through December and January and moult during February. Adults complete their moult in March, and by late April the colonies are deserted by all but a few stragglers. Most macaronis are believed to spend the non-breeding months at sea, probably in warm waters north of the breeding grounds.

Fig. 5d.3b. King penguin rookery at South Georgia in winter (Photo: W.N. Bonner)

## 5d.5.2. Rockhopper penguins

Shorter than macaronis and averaging little more than half their weight, rockhoppers are by far the smallest polar penguins. As their name suggests they are usually associated with rocky ground; cliffs and scree slopes are favoured breeding sites, often approached through heavy surf on exposed stretches of coast. Rockhoppers breed mainly on islands close to and north of the Antarctic Convergence. Hundreds of thousands have been reported from the Falkland Islands, Macquarie and Campbell Islands and there are large concentrations on Marion Island, Iles Crozet and Kerguelen, and many of the cool temperate islands south of New Zealand; a few dozen pairs nest on islands close to Cape Horn. South of the Convergence they are much rarer; the largest concentrations seem to occur on Heard Island, where a few hundred pairs nest mainly on the fringes of macaroni colonies.

Rockhoppers have been studied mainly on Heard and Macquarie Islands (Downes *et al.*, 1959; Warham, 1963), and on Campbell, Antipodes and Marion Islands (Warham, 1972; Williams *et al.*, 1979). Their breeding routine is similar to that of macaronis, in whose company they are most often found. Absent from the colonies in winter, they return to take up their nesting sites from October onward. On Heard Island, at the southern edge of their range, they are among the last to appear of all breeding birds, arriving in force during the first week of November. Males re-occupy their territories of the previous year and the first eggs are generally laid in late October and early November. On Heard Island they appear late in November, and most clutches are complete by the first week of December. Rockhoppers lay two eggs, the second some 50 − 60% larger than the first. A few pairs incubate both, but the majority manage to lose the smaller first egg in the course of incubation. Very few second chicks are ever reared. Incubation lasts about 34 days and is shared by both partners. As in macaronis, males tend the chicks continuously through a prolonged guard stage, while the females bring food back to the colonies every day. From the fifth to the tenth week both parents feed the chicks, bringing them up to full adult weight; most chicks leave the colonies when about 70 days old.

## 5d.6. APTENODYTID PENGUINS

Penguins of the genus *Aptenodytes* stand apart from other polar species in their size and brilliance of plumage, and in the apparent eccentricity of their nesting behaviour. Kings are two to three times, and emperors five to six times, the weight of pygoscelid penguins. Where lesser penguins are black (or at best blue − black) on their dorsal surfaces, kings and emperors are a paler blue − grey, with a capping of jet black. Kings have white ventral plumage, emperors pale lemon-yellow, and both have brilliant orange auricular patches and orange or lilac bill plates. Both king and emperor penguins lay only a single egg, which they incubate in a standing posture; though they incubate in large colonies which form each year on traditional sites, neither species builds nests.

Both king and emperor penguins appear to feed mainly on fish and squid, which they catch well below the surface at depths to which other species of penguin do not generally penetrate. Ability to dive deeply may be the most important single consequence of, and indeed the main reason for, their size, though other advantages accruing to large birds living in polar seas — their favourable volume-to-surface ratio, for example, and ability to store energy — cannot be overlooked. From size stems their apparent eccentricities in breeding, for both have long periods of incubation and chick-rearing which can be fitted only with difficulty into the short season of food abundance that characterizes the polar year. Their different ways of approaching the resulting ecological problems account for the differences in life style that distinguish kings from emperors in their widely separate ecological zones.

Fig. 5d.4a. King penguin positioning egg in brood pouch. (Photo: W.N. Bonner)

## 5d.6.1. King penguins

Kings are birds of the Antarctic fringe and cold temperate zone. They breed in greatest abundance on Marion Island (over 100,000 pairs have been estimated) and Iles Crozet (half a million pairs); tens of thousands breed on Macquarie Island, Iles Kerguelen and South Georgia, and there are small but slowly increasing populations on Heard Island and the Falkland Islands. Wandering individuals occur from time to time in New Zealand and South America, but are rarely seen south of their breeding localities and seem to avoid icy seas. World population, standing at about two million, seems still to be recovering from industrial exploitation in the nineteenth century, when king penguins were hunted for their oil (Conroy and White, 1973). Their life cycle has been studied in greatest detail on South Georgia (Stonehouse, 1960).

King penguin colonies are found typically on flat or shelving ground — often raised beaches backed by cliffs — with easy access to a gently shelving beach. The colonies are occupied throughout the year, dominated by adults in spring and summer and by brown woolly chicks, with a few adults in attendance, during the winter. The breeding cycle is complicated by two factors — the unusually long season in which laying occurs, and the long period of chick growth that extends from one summer to the next.

On South Georgia the breeding cycle starts in September when birds which have lost their eggs or chicks during the previous season undergo a prenuptial moult. In new plumage they return to sea for two to three weeks. Reappearing in November and December they court, mate and produce their single eggs. These are the 'early breeders'. From November onward they form tight knots of courting and incubating adults among the large chicks of the previous season, their numbers growing rapidly and coming to dominate as the chicks mature and leave for the sea. Courtship involves antiphonal calling and head-flagging, in which the brilliant orange auricular patches are displayed prominently. Courting pairs wander for several days before settling on sites which, despite the absence of a nest, they defend vigorously against intruders.

The first eggs of the season are laid late in November. Males take the first incubation watch of about fifteen days while their partners return to sea; females take the second watch of about the same length, and thereafter the partners alternate at intervals of three to four days. Hatching occurs in 54 to 55 days, the first wave of the season's new chicks appearing from mid-January onward. The parents share feeding responsibilities, taking turns to hold the chicks for spells of two to three days. After five to six weeks the chicks, well covered in brown down and already approaching half the adult weight, gather in crèches while both parents hunt for food. By late March and early April these early chicks, fed regularly by two busy parents, are sturdy and almost fully grown, weighing 12 – 14 kg.

Laying continues in the colony as the successful breeders of the previous season complete the care of their chicks, moult, and return to breed again. These 'late breeders' lay and incubate through January, February and March, and those that start early and tend their new chicks assiduously can almost catch up with the early breeders by the end of April. However, from late April courtship declines, incubation and brooding cease, and many eggs and small chicks are abandoned to die as food becomes limiting. Through May and June parental visits become less frequent, and more and more small chicks starve to

Fig. 5d.4b. King penguin chick in down. (Photo: W.N. Bonner)

Fig. 5d.5a. Emperor penguin rookery on Brunt Ice Shelf. (Photo: C. Johnson)

death, unable to survive in their meagre reserves of fat. Mortality increases in winter blizzards, and by late winter the population is reduced to those chicks, early and late, that had achieved a minimum weight of 7 − 8 kg in May.

The turning point of the year comes in October, when food becomes more readily available. Parental visits increase, and chicks whose weight had halved during the lean months begin to fatten visibly. By November and December the largest chicks, now up to full weight again, moult into juvenile plumage and leave for the sea. Their parents also moult, ready to become the late breeders of the new season; meanwhile the failed breeders of the past season are already back on the colony, many with eggs or chicks on their feet, as the early breeders of the new season. This curious régime implies that no king penguins on South Georgia can breed every year; some may breed in two years out of three, though the majority are probably successful only in alternate years. Biennial breeding may be the rule on Iles Crozet, and possibly elsewhere within the species' wide range.

### 5d.6.2. Emperor penguins

Emperors share with Adélie penguins the extreme cold and short summer seasons of the continental Antarctic zone. About thirty colonies are known, ranging in size from a few dozen birds to over 20,000 pairs. World population is difficult to assess as few colonies have been counted systematically; it is unlikely to exceed a quarter of a million. All but two of the known colonies form on winter fast ice lining the continental coast and its off-lying islands; colonies usually occur where the ice is stable from late autumn to late winter, but can be relied on to break away in spring. The exceptions are the species' northernmost colony, which forms on a raised beach on one of the Dion Islands (67°52'S) on the south western shore of Antarctic Peninsula and the Taylor Glacier colony (67° 28'S) on the Mawson Coast. Emperor penguins have been studied at several of their colonies, notably Cape Crozier, the Dion Islands and Pointe Géologie, Adélie Land: for a review of their ecology and physiology and an extensive bibliography see Le Maho (1977).

Unique among penguins, emperors lay their eggs in early winter and incubate through the coldest months of the year. The colonies assemble in April and May as soon as the newly-formed sea ice is strong enough to bear them. Heavy with stored fat, weighing 30 to 40 kg, the birds move in from the

Fig. 5d.5b. Emperor penguin brooding chick. (Photo: Ian Somerton)

open sea, often over several kilometres of ice. Courtship takes three to five weeks; the partners identify each other with trumpeting calls, bow and display their golden flashes, and follow each other about as the pair bonds strengthen. The single eggs are laid in May and early June. Male emperors take over the eggs within a few hours of laying and incubate them on their feet, protected from the weather by a feathered fold of abdominal skin. The females return to sea shortly after laying, and are not seen again on the colonies until early to mid July. Thus male emperors alone incubate for the full period of about 65 days.

This unique pattern of breeding is the only possible response of a large bird to polar winter conditions, when food is distant and hard to come by during the winter, and freely available for only a short season in summer. For the chicks to be ready for the sea by January and February, when the ice is open and food plentiful, laying must be set back as early as possible in the year, and winter incubation becomes a necessity. Food is not available close to the colonies so the incubating birds must depend on their own stores of fat. Laying and and incubating on sea ice rather than land represents a saving of energy, for the sea beneath the ice, though close to its freezing point, is considerably warmer than the land nearby, and provides the birds with their best possible thermal environment. While incubating the males huddle tightly together, especially in cold or windy weather, reducing their individual exposed surfaces by as much as five-sixths and helping to maintain their own body temperatures (only slightly reduced from normal) with minimal expenditure of energy. While incubating they are sleepy and inactive; there is little fighting among huddled emperors, and only in calm, relatively warm spells do they break up the huddles to wander and preen, still with the eggs on their feet.

During the courtship period emperors lose about a quarter of their initial weight and, by these energy-conserving strategems, the males contrive to pass their nine weeks of incubation with no more than a 15 % further loss in weight. Though still rotund at the end of their vigil, they are markedly thinner than the females that come up from the sea to relieve them. The return of the females coincides with hatching. Partners find each other by calling, and the females take over the eggs or newly hatched chicks without delay. Males whose partners are delayed feed their newly-hatched chicks on a secretion from their crop wall, tiding them over for a few days until the females arrive with fresh supplies. Within a few hours of being relieved the males depart for the sea, which in mid- to late July may still lie far beyond the horizon and take several days to reach.

For the first three to five weeks the chicks grow slowly on food which the females provide. Thereafter the two parents alternate, at intervals determined by the distance between the colony and open water. At five to six weeks the chicks gather in crèches. Normally by this time the ice edge has broken back to within sight of the colony, or a polynya of open water has formed nearby, giving the parents easy access to the sea and ensuring regular, frequent meals for the chicks. In hard seasons the sea ice persists locally; parents have a much longer walk to find open water and food, at a time when the growth rate of the chicks is most rapid and their demands for food are most pressing. Many chicks die of starvation in these circumstances; as many as 90% of the annual crop have been reported lost from the colony at Pointe Géologie due to persistent ice in late spring.

Chicks that survive the harsh régime begin to moult from grey woolly down to juvenile plumage, usually in December and January. Often the colonies break up about this time, and chicks in the last stages of moult may be carried out toward the open sea on floes. Fully feathered and independent, the young birds are still only about half the weight of their parents; like Adélie chicks they leave the colonies early and complete their growth in the pack-ice. Adults moult during January and early February, usually on the floating ice close to the shore, and are ready to return to their colonies for breeding by the following April. Emperors begin breeding during their fourth, fifth and sixth years, many of them in the colony of their birth (Jouventin, 1975).

## 5d.7. PENGUIN CONSERVATION

All the polar penguins are currently plentiful. Despite the harshness of their environment and its relative newness in evolutionary terms, all species appear well adapted. Even emperor penguins, once looked on as a rare and possibly a dying species, are now known to be abundant and flourishing. Though individual colonies have suffered reductions in numbers, mainly due to direct interference by man (see below), world populations of most polar species appear to have increased during the past few decades, possibly benefiting from surplus krill due to the removal of huge numbers of baleen whales (Chapter 5b).

Polar literature, human-oriented, tends to stress the problems of homeotherms living in polar environments and to overlook the advantages. In comparison with temperate or tropical species, polar penguins have few predators or parasites. Predators at the nest include only a few species of birds (skuas, dominican gulls, sheathbills, giant petrels) whose roles are mainly scavenging, removing eggs and chicks which are often already at risk from other causes. Predators at sea include leopard seals, fur seals and (north of the Convergence) sea lions, which are seldom present in large concentrations and are not generally considered to influence stock size seriously, even on a local scale. Ecto- and endo-parasites have been recovered from all species of polar penguins, but infestations are generally low and do not appear to affect the health of the birds materially. Pathogenic bacteria, viruses and antibodies have been detected in most species, though polar penguins in the wild seem remarkably free from diseases (Moore and Cameron, 1968; Morgan and Westbury, 1982; Sieburth, 1958; Sladen, 1962).

Long-term studies indicate that many colonies are subject to considerable fluctuations in size, for which climatic or other slowly-changing environmental factors are responsible. The Adélie colony at Cape Royds, for example, appears to have been considerably larger in the past few hundred years, with a neighbouring colony at Cape Barne which is now defunct. Both these colonies are and would in the past have been particularly sensitive to changes in the distribution of open water in McMurdo Sound. The Cape Crozier emperor penguin colony, which forms within enclaves of the moving Ross Ice Shelf, is from time to time exposed to the open sea. On these occasions it is particularly liable to destruction by unseasonable breaking up of the sea ice in late autumn — a condition in which it was almost certainly found during Edward Wilson's visit in July 1911. More recently the colony has been better protected by ice walls on its seaward side, and appears to have grown considerably.

Such long-term secular changes may be masked by more sudden changes brought about by man. Colonies that lie close to research bases are usually vulnerable to disturbance, especially in summer when both penguins and men are most active. A sudden and dramatic decline, for example, in numbers of breeding Adélies at Cape Royds during the 1950s and early 1960s can with some confidence be ascribed to repeated disturbance from activities associated with nearby bases in McMurdo Sound; numbers have more recently increased since steps have been taken to protect the colony (Thomson, 1977).

Disturbance may take many forms, with effects which may be readily apparent, or hidden by the apparent tameness and seeming indifference of breeding penguins. The effects of aircraft over-flying a colony during the breeding season are clear; birds run in panic from their nests, and eggs and chicks are exposed to cold and predation. Less obvious but possibly no less harmful are the effects of constant visits by sightseers during incubation, when lethargic birds are forced into activity and caused to draw more heavily than is necessary on their reserves of fat. Perhaps most serious of all is disturbance later in the season, when young birds may be discouraged from taking up nest sites and driven to breeding elsewhere; this is the most likely cause of the declining numbers of Cape Royds, a colony which could ill afford interference with natural processes of recruitment.

# REFERENCES

Ainley, D.G. (1975) Displays of Adélie penguins: a reinterpretation. In *The Biology of Penguins*, Ed. B. Stonehouse, pp. 503 – 34. Macmillan, London.

Ainley, D.G. and LeResche, R.E. (1973) The effects of weather and ice conditions on breeding in Adélie penguins. *Condor* 75, 235 – 9.

Bagshawe, T.W. (1938) Notes on the habits of the gentoo and ringed or Antarctic penguins. *Trans. Zool. Soc. Lond.* 24, 135 – 306.

Baker, C.M.A. and Manwell, C. (1975) Penguin proteins: biochemical contributions to classification and natural history. In *The Biology of Penguins*, Ed. B. Stonehouse, pp. 43 – 56. Macmillan, London.

Conroy, J.W.H., Darling, O.H.S. and Smith, H.G. (1975) The annual cycle of the chinstrap penguin *Pygoscelis antarctica* on Signy Island, South Orkney Islands. In *The Biology of Penguins*, Ed. B. Stonehouse, pp. 353 – 62. Macmillan, London.

Conroy, J.W.H. and Twelves, E.L. (1972) Diving depths of the gentoo penguin (*Pygoscelis papua*) and blue-eyed shag (*Phalacrocorax atriceps*) from the South Orkney Islands. *Bull. Br. Antarct. Surv.* 30, 106 – 8.

Conroy, J.W.H. and White, M.G. (1973) The breeding status of the king penguin *Aptenodytes patagonica*. *Bull. Br. Antarct. Surv.* 32, 31 – 40.

Croxall, J.P. (1982) Aspects of the population demography of Antarctic and subantarctic seabirds. *CNFRA* 51, 479 – 88.

Croxall, J.P. and Kirkwood, E.D. (1979) *The Distribution of Penguins on the Antarctic Peninsula and Islands of the Scotia Sea.* British Antarctic Survey, Cambridge.

Croxall, J.P. and Prince, P.A. (1979) Antarctic seabird and seal monitoring studies. *Polar Rec.* 19, 573 – 95.

Croxall, J.P. and Prince, P.A. (1980) The food of gentoo penguins *Pygoscelis papua* and macaroni penguins *Eudyptes chrysolophus* at South Georgia. *Ibis* 122, 245 – 53.

Downes, M.C., Ealey, E.H.M., Gwynn, A.M. and Young, P.S. (1959) The birds of Heard Island. *A.N.A.R.E. Rep. B* 1, 1 – 135.

Drent, R.H. and Stonehouse, B. (1971) Thermoregulatory responses of the Peruvian penguin *Spheniscus humboldti. Comp. Biochem. Physiol.* 40A, 689 – 710.

Emlen, J.T. and Penney, R.L. (1964) Distance navigation in the Adélie penguin. *Ibis* 106, 407 – 31.

Feeney, R.E. and Osuga, D.T. (1976) Comparative biochemistry of Antarctic proteins. *Comp. Biochem. Physiol.* 54A, 281 – 6.

Jouventin, P. (1975) Mortality parameters in emperor penguins, *Aptenodytes forsteri.* In *The Biology of Penguins*, Ed. B. Stonehouse, pp. 435 – 46. Macmillan, London.

Kennett, J.P. (1980) Paleoceanographic and biogeographic evolution of the Southern Ocean during the Cenozoic, and Cenozoic microfossil datums. *Palaeogeogr. Palaeoclimat. Palaeoecol.* 31, 123 – 52.

Kooyman, G.L. (1975) Behaviour and physiology of diving. In *The Biology of Penguins*, Ed. B. Stonehouse, pp. 115 – 37. Macmillan, London.

Le Maho, Y. (1977) The emperor penguin: a strategy to live and breed in the cold. *Am. Scient.* 65, 680 – 93.

LeResche, R.E. and Sladen, W.J.L. (1970) Establishment of pair and breeding site bonds by young known-age Adélie penguins (*Pygoscelis adeliae*). *Anim. Behav.* 18, 517 – 26.

Levick, G.M. (1914) *Antarctic Penguins: a Study of their Social Habits.* Heinemann, London.

Moore, B.W. and Cameron, A.S. (1968) Chlamydia antibodies in Antarctic fauna. *Avian Dis.* 13, 681 – 4.

Morgan, I.R. and Westbury, H.A. (1982) Virus diseases of Antarctic penguins. *A.N.A.R.E. News* 16, 14.

Murphy, R.C. (1947) A new zonal race of the gentoo penguin. *Auk* 64, 454 – 5.

Prévost, J. (1961) *Ecologie du Manchot Empereur.* Hermann, Paris

Scholander, P.F. (1964) Animals in aquatic environments: diving mammals and birds. In *Handbook of Physiology, Section W, Adaptation to the Environment*, Eds W.O. Fenn and H. Rahn, pp. 729 – 39. American Physiological Society, Washington, D.C.

Sieburth, H. McN. (1958) Respiratory flora and disease of Antarctic birds. *Avian Dis.* 2, 402 – 8.

Simpson, G.G. (1946) Fossil penguins. *Bull. Am. Mus. nat. Hist.* 87, 1 – 99.

Simpson, G.G. (1972) Pliocene penguins from North Canterbury, New Zealand. *Rec. Canterbury Mus.* 9, 159 – 82.

Simpson, G.G. (1975) Fossil penguins. In *The Biology of Penguins*, Ed. B. Stonehouse, pp. 18-41. Macmillan, London.

Simpson, G.G. (1976) *Penguins Past and Present, Here and There.* Yale University Press, New Haven and London.

Sladen, W.J.L. (1958) The pygoscelid penguins, Parts 1 and 2. *Sci. Rep. Falkld. Isl. Depend. Surv.* 17, 1 – 97.

Sladen, W.J.L. (1962) Studies on respiratory pathogens in Antarctica. *Polar Rec.* 11, 318.

Spurr, E.B. (1975) Communication on the Adélie penguin. In *The Biology of Penguins*, Ed. B. Stonehouse, pp. 449 – 502. Macmillan, London.

Stonehouse, B. (1960) The king penguin *Aptenodytes patagonica* of South Georgia. I. Breeding behaviour and development. *Sci. Rep. Falkld. Isl. Depend. Surv.* 23, 1 – 81.

Stonehouse, B. (1963) Observations on Adélie penguins *Pygoscelis adeliae* at Cape Royds, Antarctica. *Int. ornithol. Congr.* 13, 766 – 79.

Stonehouse, B. (1967a) The general ecology and thermal balance of penguins. *Adv. ecol. Res.* 4, 131 – 96.

Stonehouse, B. (1967b) Occurrence and effects of open water in McMurdo Sound, Antarctica, during winter and early spring. *Polar Rec.* 13, 775 – 8.

Stonehouse, B. (1968) *Penguins.* Arthur Baker, London; Golden Press, New York.

Stonehouse, B. (1969a) Environmental temperatures of Tertiary penguins. *Science, N.Y.* 163, 373 – 5.

Stonehouse, B. (1969b) Air censuses of two colonies of Adélie penguins (*Pygoscelis adeliae*) in Ross Dependency, Antarctica. *Polar Rec.* 14, 471 – 5.

Stonehouse, B. (1970) Adaptation in polar and subpolar penguins (Spheniscidae). In *Antarctic Ecology*, vol. 1, Ed. M.W. Holdgate, pp. 526 – 41. Academic Press, London and New York.

Stonehouse, B. (editor) (1975) *The Biology of Penguins.* Macmillan, London.

Stonehouse, B. (1982) La zonation écologique sur les hautes latitudes australes. *CNFRA* 51, 531 – 6.

Taylor, R.H. (1962) The Adélie penguin at Cape Royds. *Ibis* 104, 176 – 204.

Taylor, R.H. and Wilson, P. (1982) Counting penguins from the air. *Antarctic* 9, 366 – 8.

Thomson, R.H. (1977) Effects of human disturbance on an Adélie penguin rookery and measures of control. In *Adaptations within Antarctic Ecosystems*, Ed. G.A. Llano, pp. 1177 – 80. Smithsonian Institution, Washington, D.C.

Trivelpiece, W. and Volkman, N.J. (1979) Nest-site competition between Adélie and chinstrap penguins: an ecological interpretation. *Auk* 96, 675 – 81.

Van Zinderen Bakker, E.M., Jr. (1971) A behavioural analysis of the gentoo penguin *Pygoscelis papua* Forster. In *Marion and Prince Edward Islands*, Eds E.M. van Zinderen Bakker, J.M. Winterbottom and R.A. Dyer, pp. 251 – 72. Balkema, Cape Town.

Volkman, N.J., Presler, P. and Trivelpiece, W. (1980) Diets of pygoscelid penguins at King George Island, Antarctica. *Condor* 82, 373 – 8.

Volkman, N.J. and Trivelpiece, W. (1980) Growth in pygoscelid penguin chicks. *J. Zool., Lond.* 191, 521 – 30.

Warham, J. (1963) The rockhopper penguin *Eudyptes chrysocome* at Macquarie Island. *Auk* 80, 229 – 56.

Warham, J. (1972) Breeding seasons and sexual dimorphism in rockhopper penguins. *Auk* 89, 86 – 105.

Warham, J. (1975) The crested penguins. In *The Biology of Penguins*. Ed. B. Stonehouse, pp. 189 – 269. Macmillan, London.

Williams, A.J., Siegfried, W.R., Burger, A.E. and Berruti, A. (1979) The Prince Edward Islands: a sanctuary for seabirds in the Southern Ocean. *Biol. Conserv.* 15, 59 – 71.

Wilson, E.A. (1967) *Birds of the Antarctic*, Ed. B.B. Roberts. Blandford, London.
Yeates, G.W. (1971) Observations on orientation of penguins to wind and on colonization in the Adélie penguin rookery at Cape Royds, Antarctica. *N.Z. J. Sci.* 14, 901 – 6.
Yeates, G.W. (1975) Microclimate, climate and breeding success in Antarctic penguins. In *The Biology of Penguins*. Ed. B. Stonehouse, pp. 397 – 409. Macmillan, London.
Zusi, R.L. (1975) An interpretation of skull structure in Antarctic penguins. In *The Biology of Penguins*. Ed. B. Stonehouse, pp. 59 – 84. Macmillan, London.

# CHAPTER 6a

# The Sub-Antarctic Islands

## D. W. H. WALTON

British Antarctic Survey, High Cross, Madingley Road, Cambridge CB3 0ET, U.K.

## CONTENTS

## 6a.1. GEOGRAPHY AND CLIMATE

Outside the Antarctic Treaty area but well within the influence of the Antarctic Continent lies a ring of remote islands. They are mostly small, far from each other and from any continental land (Table 6a.1) and surrounded by cold seas. The Southern Ocean has a strong influence on their ecosystems, both

TABLE 6a.1.

| | Sovereignty | Total area km$^2$ | Snow-free area km$^2$ | Maximum elevation m | Latitude °S | Distance from nearest mainland km | Distance from nearest land km |
|---|---|---|---|---|---|---|---|
| South Georgia | UK | 3755 | 1500 | 2934 | 54 | 2210 | 1520 |
| South Sandwich Islands | UK | 618 | 85 | 1370 | 56 − 59 | 2670 | 730 |
| Bouvetøya | Norway | 50 | 4 | 780 | 54 | 2650 | 1600 |
| Prince Edward and Marion Islands | South Africa | 335 | 335 | 1230 | 46 | 1900 | 1900 |
| Iles Crozet | France | 233 | 233 | 934 | 46 | 2740 | 2740 |
| Iles Kerguelen | France | 3626 | 2900 | 1960 | 49 | 4110 | 1450 |
| Heard and Macdonald Islands | Australia | 380 + | 70 | 2745 | 53 | 4570 | 530 |
| Macquarie Island | Australia | 118 | 118 | 433 | 54 | 990 | 610 |

directly in terms of climate and indirectly in terms of biogeography. All the islands lie close to or within the oceanographic and biological boundary called the Antarctic Convergence (see Chapter 2a). Although in the short term the position of the Convergence is very stable, over geological time periods it has shifted considerably. Islands lying within the Convergence (South Georgia, Macdonald and Heard Islands, Bouvetøya, South Sandwich Islands) all have extensive ice caps and permanent snowfields (Table 6a.1). Iles Kerguelen lie on the Convergence and have only limited glaciers. Macquarie Island, Iles Crozet and Prince Edward and Marion Islands lie just outside the Convergence and have no glaciers or major permanent snowfields. All the islands appear to have been within the Convergence at some time and show signs of extensive glaciation. The most southerly islands — Bouvetøya and the South Sandwich Islands — possess such limited areas of snow-free ground and limited ecosystems that they have more in common with the Maritime Antarctic islands than those of the sub-Antarctic.

The origin of the island groups varies. South Georgia is a detached block of continental crust from the Upper Jurassic — Lower Cretaceous period. Its rocks are principally greywackes and tuffs but with intrusive pillow lavas in the south east corner. The fossil record is relatively poor. Kerguelen on the other hand has extensive fossil deposits. Geochemical data suggests that both it and Heard Island were formed by intense local volcanic activity up to 100 million years ago and so far no evidence for continental rocks has been found on either island. Macquarie Island is much younger, probably being formed during the Pliocene (c. 5 million years ago) by the raising of a block of oceanic crust lying along the submarine feature now called the Macquarie Ridge.

Iles Crozet are exclusively volcanic in origin with a possible Miocene date of formation. Prince Edward and Marion Islands are younger still and also volcanic in origin with an isotope date for the oldest lavas of only 276,000 years old. The most recently formed islands are those of the South Sandwich group which are still very active volcanically. This archipelago lies on the edge of a growing tectonic plate margin and most of the islands have marked fumarole activity.

During the glacial maxima the Antarctic ice sheet would have covered a much greater area than at present. One likely result of this was to push the Antarctic Convergence northwards, perhaps by up to 3° of latitude in eastern areas around Kerguelen and up to 7° in the Atlantic sector. This would have

Fig. 6a.1. The varied topography of sub-Antarctic islands: (a) South Georgia;

produced extensive glaciation on all the islands, and the consequent lowering of the sea level by massive sea ice formation is likely to have exposed areas of the Kerguelen shelf.

At the present time the extent of snow- and ice-free ground varies considerably (Fig. 6a.1). Iles Crozet, Macquarie Island, the Macdonald Islands and Prince Edward and Marion Islands have no glaciers and only rarely are able to support even limited persistent snow beds. Iles Kerguelen is about 80% free of permanent snow and ice whilst only 40% of South Georgia and 15% of Heard Island are free. Bouvetøya and the South Sandwich Islands are mainly ice covered (Walton, 1984).

Climatically the islands are also varied. The extent of snow and ice cover obviously influences the local climate but other features such as the size and height of the island and the origin of the air masses passing over it also have a major effect. Of the islands discussed in this chapter Bouvetøya, Peter I Island and Heard Island clearly have the most severe climates. Only Heard Island has any extensive meteorological data and these show that its summers are cold and short followed by long and severe winters. The growing season is limited and this feature, together with the very limited area of snow-free ground, must be of major significance in explaining the floristic and faunistic poverty of the island.

In contrast to this, the growing season on Macquarie Island is cool but long with relatively mild winter temperatures. Iles Crozet, Iles Kerguelen and Marion Island have slightly more rigorous climates with more definite winter periods and a greater frequency of snow falls. The South Georgian climate tends to be more similar to that of Heard Island but the island is sufficiently large and high for there to be significant differences between the north and south sides.

Radiation receipt on the islands is generally rather poor with high cloud cover throughout much of the summer. Macquarie and Marion Islands are particularly cloudy and both have frequent precipitation, mainly as rain. Snow showers are possible on any day of the year throughout the sub-Antarctic. All the islands lie in the path of the prevailing westerly winds and are subject to persistent winds, often reaching gale force. All the islands, except the South Sandwich archipelago and Bouvetøya, are beyond the pack-ice limit in normal winters but abnormal conditions may bring the pack close to South Georgia and Heard Island. In sheltered bays on South Georgia thin winter ice may form but it rarely lasts for long.

## 6a.2. BIOGEOGRAPHY

### 6a.2.1. Plants

As long ago as 1847 J.D. Hooker had noted the importance of the sub-Antarctic islands in any understanding of the distribution of plants around the southern continent. Conflicting theories developed as botanists struggled to reconcile the accumulating evidence of floristic relationships (Fig. 6a.2). It was not until the geological and fossil evidence was combined to produce our present theories of plate tectonics and continental drift that the connections began to assume explicable patterns. The existence of the widely distributed fossil floras, firstly dominated by *Glossopteris* and later by *Nothofagus* (see Chapter 2b), clearly supported the existence of the Gondwanaland supercontinent and with it long-lasting migration routes between the new disjunct continental lands of the southern hemisphere.

Fig. 6a.1. The varied topography of sub-Antarctic islands: (b) Marion Island. (Photo: V.R. Smith)

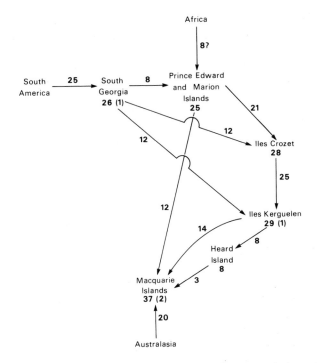

Fig. 6a.2. Biogeographical relationships of sub-Antarctic flowering plants. Numbers of species per island are in bold with numbers of endemic species in brackets; the arrows show the numbers of shared species between islands.

This may provide an explanation for the distribution patterns of some of the sub-Antarctic cryptogams which could perhaps have survived the ice ages in isolated refugia, but another explanation is needed for most of the vascular flora. Long distance dispersal of spores by wind was first suggested by Hooker in 1879 but seems suitable only for ferns. The obvious adaptations of many sub-Antarctic flowering plants to carriage by birds provides the obvious method of trans-ocean dispersal. For instance *Acaena* and *Uncinia* have hooked fruits, *Cotula* and *Stilbocarpa* have mucilaginous seed coatings, and some seeds may be eaten by birds. There is also the possibility of transport by sea with various reports of undiminished seed viability after weeks of immersion in sea-water. None of these methods provides a complete explanation for the origin of the flora and it seems likely that all have contributed to the present flora and fauna.

The remoteness of the islands, the often limited areas available for colonization and the cold summers have all tended to limit diversity in both the flora and fauna (Fig. 6a.3). The most recent tabulation of the flora (Smith, 1984) gives 24 species of grass, 32 forbs, 16 pteridophytes, 250 + mosses, 150 + liverworts, 300 + lichens and *c* 70 macro-fungi. Whilst the taxonomy of most of the vascular plants is now fairly well established the situation for cryptogams is much less satisfactory. Although many early accounts of the vegetation appeared to show large numbers of endemic species recent critical revision of several genera has shown this to be generally unfounded. Good phytogeographical accounts must wait on good taxonomy and until this is forthcoming critical discussion of biogeographical patterns must remain tentative.

The general affinities of the subantarctic flowering plants (Fig. 6a.2) are, not surprisingly, generally to the west. South Georgia has a strong Fuegian element which, although diminishing, is found in the flora of all the sub-Antarctic islands. Macquarie Island has a pronounced Australasian component whilst a very limited African component is found only on Prince Edward and Marion Islands and on Iles Crozet. Several taxa occur only in the sub-Antarctic, e.g. *Poa cookii, Pringlea antiscorbutica, Ranunculus moseleyi, Colobanthus kerguelensis,* but very few are endemic to a single island or archipelago. Examples of this group include the hybrid *Acaena magellanica x tenera* on South Georgia, *Lyllia kerguelensis* on Iles Kerguelen and *Poa hamiltonii* on Macquarie Island. Most of the islands have also had a number of alien species introduced (Walton, 1975), almost certainly accidentally, by man's activities. The most successful of these are *Poa annua, Cerastium holosteoides* and *Stellaria media* in terms of distribution but the most important in ecological terms differ between the islands. *Poa annua,* growing as a perennial, has proved a very successful replacement species on South Georgia where reindeer grazing has extensively damaged the natural vegetation (Fig. 6a.4).

Recent reviewers of Antarctic lichens (Lindsay, 1974) have stressed the importance of both the bipolar and cosmopolitan elements and the relatively minor endemic component. For instance amongst the macro-lichens on South Georgia the cosmopolitan and bipolar groups account for 60% of the species and no endemics have been recorded. A similar pattern seems to be emerging for the sub-Antarctic mosses, but for the liverworts the bipolar component is mainly absent with the majority of species having distinctly Fuegian — Australasian links. Schuster (1969) has concluded that the high proportion of endemic liverworts on the sub-Antarctic islands is mainly due to their relictual survival from a Gondwanaland period. He suggests some examples in support of long distance dispersal but concludes that this cannot explain the majority of the distributional data.

Fig. 6a.3. Examples of the diversity of sub-Antarctic flowering plants and ferns: (a) *Polystichum mohrioides;*

Fig. 6a.3. (cont.) Examples of the diversity of sub-Antarctic flowering plants and ferns: (b) *Ophioglossum crotalophorioides*;

Fig 6a.3. (cont.) Examples of the diversity of sub-Antarctic flowering plants and ferns: (c) *Uncinia meridensis*;

Fig 6a.3. (cont.) Examples of the diversity of sub-Antarctic flowering plants and ferns: (d) *Pringlea antiscorbutica* (Photo: Kevin Hall);

Fig 6a.3. (cont.) Examples of the diversity of sub-Antarctic flowering plants and ferns: (e) *Colobanthus quitensis*;

Fig 6a.3. (cont.) Examples of the diversity of sub-Antarctic flowering plants and ferns: (f) *Azorella selago* (Photo: Kevin Hall);

Fig 6a.3. (cont.) Examples of the diversity of sub-Antarctic flowering plants and ferns: (g) *Acaena magellanica*;

Fig 6a.3. (cont.) Examples of the diversity of sub-Antarctic flowering plants and ferns: (h) *Festuca contracta*.

## 6a.2.2. Animals

The native sub-Antarctic land fauna is composed entirely of invertebrates (Block, 1984). The history of the introduction and success of alien vertebrates on the island is dealt with in Chapter 6b. At the moment 387 invertebrate species (excluding Protozoa, Rotifera and Tardigrada) are known from the islands, distributed as follows — nematodes 22, worms 4, molluscs 3, and arthropods 358. Within the arthropods there are 144 spiders, 210 insects, 2 myriapods and 1 crustacean. Dispersal to the islands by wind and/or birds seems essential to account for present distributions. The lack of a good fossil record makes the historical biogeography of the group very conjectural. Aptery and flightlessness are common features of the sub-Antarctic fauna and endemism is well documented for some islands.

Some of the arthropods are associated with animals. Over 30 species of biting lice (Mallophaga) have been found on South Georgia and Macquarie Island, with fewer species on the other islands. Biting lice infestations have been recorded from 58 species of birds in the sub-Antarctic (Clay and Moreby, 1970). The sucking lice (Anoplura) on the other hand are limited to seals and appear to be species specific. Fleas (Siphonaptera) are known from sub-Antarctic seabirds, whilst three species of ticks (Ixodidae) have been found on 45 avian hosts.

A detailed examination of the distribution of some of the best known micro-arthropods — Collembola, Cryptostigmatid and Mesostigmatid mites (Wallwork, 1973) — has suggested that in these groups there is an important relictual component consistent with a Gondwanaland origin as well as more recent elements from outside the sub-Antarctic zone. The relatively high specific endemism is especially associated with the prostigmatid mites, whose ability to survive the glacial periods is implied by their present success on the Antarctic continent.

In a multiple regression analysis of the relationships between geographical parameters and plants, insects and birds on remote Southern Hemisphere islands (Abbott, 1974) the mean temperature of the coldest month and the island area accounted for 71% of the variation in number of plant species. Interestingly, the number of plant species accounted for 73% of the variation in number of insect species. Although the data base used for this study had serious defects, the results suggest that mathematical analysis may provide useful insight into biogeographical problems.

## 6a.3. SOILS

### 6a.3.1. Introduction

Despite the relatively short time that parts of the sub-Antarctic islands have been free of their ice cover there has been a substantial development of a range of soil types. They can be divided into four types, each encompassing a range of varieties related to the nature of the underlying rocks, the vegetation and to their local topographical position (Fig. 6a.5). The four types of soils are probably represented on all the sub-Antarctic islands but data from some islands are too limited to be certain of their extent and exact comparability.

All the soils fall into the classification, on a world scale, of tundra soils. They are generally cold, often wet and in many cases with sub-optimal levels of certain critical nutrients, usually nitrogen or phosphorus. Unlike most tundra soils in the Northern Hemisphere, very few of the sub-Antarctic soils have permafrost, although this may have been present in the early stages of soil formation thousands of years ago. The four categories are based mainly on a decreasing content of organic matter, from pure peat through to the skeletal mineral soils. It seems likely that all the islands have representative examples of the extreme soil types but the extent of the more highly developed intermediate soils is not currently known.

### 6a.3.2. South Georgia soils

*Peat soils*

Five major types of peat soil have been recognized on South Georgia (Smith, 1981) and several of these are known to occur on other islands. Two of the peats are clearly related to wetland peat found elsewhere in the world, one is a type of tundra ranker whilst the final two are limited to sub-Antarctic or Antarctic tundra by the nature of their constituents.

Mire peat, often over 2m deep, develops on the gentle seepage slopes on valleysides underneath communities of rushes and mosses. The water movement through the peat brings downwashed sand and clay with it resulting in the formation of narrow grey or blue — grey bands within the soil profile. If anaerobic conditions develop these bands may stain reddish brown with reduced iron. The mineral ion retention capacity of these peats must be high and calcium and magnesium often accumulate. This soil type is equivalent to valley or flush bogs described for Great Britain.

The other wetland peat, found on level ground where water movement is usually negligible and drainage is poor, closely resembles the ombrogenous blanket bogs of northern latitudes. These are the deepest peat deposits so far found on South Georgia and can be up to 3m deep. The soil is acid and much of the soil profile is permanently anaerobic. With *Rostkovia* or *Deschampsia* as the principal flowering plants the bogs are covered mainly by leafy liverworts and the two mosses *Drepanocladus uncinatus* and *Calliergon sarmentosum*. In a few places where the bogs are near major breeding populations of birds, there can be an unusually high nitrogen and phosphorus input to what are generally nutrient poor soils.

The coastal tussock (*Poa flabellata*) grassland community produces its own variety of peat on South Georgia. The peat soil may develop under the grass either on relatively well drained hillsides, where it remains relatively shallow, or on raised beaches or wet slopes where the peat depth might reach almost 2m. The association between seals and birds and the tussock stands ensures that many of them receive considerable nutrient input from guano. Localized erosion, usually due to the animal activities, often results in the loss of soil from between the plants, leaving them standing on peat pedestals (Fig. 6a.6). On Marion Island it appears that erosion by animal activity exceeds that of any other agent currently active on the island (Hall and Williams, 1981). Tussock peat tends to be fairly acid with little calcium accumulation and a more open and fibrous nature than is found in the other peats.

A tundra ranker peat generally develops on well drained hillsides beneath the dwarf shrub stands dominated by *Acaena magellanica*. The peat is not usually very deep although the productivity of the community is high. The high nitrogen content of the *Acaena* leaves and their deciduous nature make decomposition rapid compared to the other peat sites and the organic matter that is accumulated is derived mainly from the moss *Tortula robusta* and the woody stems of *Acaena*.

The final peat type is that found beneath the moss banks dominated by *Polytrichum alpestre* and *Chorisodontium aciphyllum*. These banks are very similar to those found further south on the maritime Antarctic islands, lacking only the permafrost and with a greater range of surface vegetation including scattered flowering plants. The peat is normally very acid with a low nutrient status and high water retention capacity.

*Meadow tundra soils*

This soil type is limited to seepage slopes and bryophyte marshes and consists of a shallow peaty layer overlying a brown or grey clay layer up to 50 cm deep. This soil seems fairly similar to the bog soil found below Alaskan tundra.

*Brown soils*

These are the most developed and structured soils on South Georgia. They are found beneath the dry *Festuca* grassland and are generally well drained. They have an upper litter layer overlying a partly humified horizon. Beneath this there is a layer of brown loam, often with grey clay inclusion, grading down into sandy till.

*Mineral soils*

Two types of mineral soil are developing on South Georgia. Those beneath the low altitude fellfield have scattered flowering plants and cryptogams contributing to the low percentage of organic matter (often less than 10%). In the high altitude fellfield areas there are few plants and very limited possibilities for nutrient input.

## 6a.3.3. Soils of other islands

Macquarie Island soils seem to correspond to several South Georgian types. The tussock peat is found beneath the *Poa foliosa* grassland and can accumulate up to 2m deep, whilst bog peats reach greater depths than on South Georgia with a maximum depth of 6m having been found. So-called fen peats (Jenkin, 1975) appear to be the equivalent of the Mire peats, although varying in mineral content from rich to poor. Finally, gravel soils with minimal organic matter can be found beneath the fellfield communities of the upland areas. The brown soil appears to be missing from Macquarie Island and this may be due to the high year-round precipitation.

The recent volcanic origin of Marion Island has resulted in a different mineral basis for soil development from South Georgia and Macquarie Island. The island is composed of two lava types of quite different ages — grey c. 276,000 years and black c. 15,000 years — together with large numbers of scoria cones. The black lava soils have a greater cation concentration than the grey soils and within each lava type slope soils tend to have higher amounts of cations than do mire peats and fellfield soils (Smith, 1978). Some of these slope soils seem analogous to the brown earth of South Georgia whilst the island also has mire peats overlying yellow clay, often with iron pans. Despite the high rainfall, low temperatures and high acidity, all of which favour podsolization, this does not seem to be well developed and they are perhaps more appropriately considered at the moment as more or less gleyed. The range of Marion Island soils does not seem to subdivide, at least on present information, into as many distinct categories of peat soil as on South Georgia and this may be attributable at least in part to the different dominant species in the various plant communities.

Little has been published on the soils of Iles Kerguelen and Iles Crozet. From the descriptions of the dominant vegetation types it is clear that both bog and mire peats are found, together with the ranker type peat developed beneath *Acaena* stands. Presumably a type of tussock peat has developed beneath the *Poa foliosa* grassland but there seems to be no evidence of the dry moss bank type peat. Fellfield soils are found on all the islands at the higher elevations and again permafrost is absent from all the soils.

## 6a.3.4. Soil development and climatic change

Whilst it is not known how quickly the Brown Earth type soils develop there has been extensive dating of the peat deposits on South Georgia, Iles Crozet and Iles Kerguelen. Coupled with this there has been some detailed palynological analyses to see if previous climatic changes have been reflected in floristic changes.

Radiocarbon dates for the South Georgian deposits show that some of the peat deposits have been accumulating for nearly 10,000 years. This corresponds well with the period since the last major glacial retreat began about 11 − 12,000 years ago. Thus all the soils on South Georgia are fairly young and indeed the same must apply to the other sub-Antarctic islands.

The oldest $^{14}$C date for any sub-Antarctic peat so far is $9700 \pm 150$ years for a depth of 3.5m on South Georgia. The mid north-eastern part of the South Georgian coast must have had significant vegetation established for nearly 10,000 years. The radiocarbon dates from a variety of depths, and profiles suggest that over this period there has been a relatively constant rate of peat accumulation and there does not appear to be any evidence for any major changes in climate. Pollen analyses from a number of sites suggest that fluctuations in the proportions of particular pollen types reflect minor variations in the climate which could have affected the extent of grassland or bog but there is no clear evidence that these changes correlate with geomorphological evidence of change during the last ten millenia, or that climatic change has been sufficient to alter the diversity of the flora.

## 6a.4. PLANT COMMUNITIES

### 6a.4.1. Vegetation types

Although attempts have been made to classify both sub-Antarctic and Antarctic vegetation in terms of categories developed for the Arctic, this has not been very successful. Although in initial appearance it seems similar to Arctic and alpine tundra (Bliss, 1979), its species poverty and the physiognomy of many of the most important species require a unique classification which recognizes these features. A recent synthesis by Smith (1984) goes a long way towards meeting this requirement.

Fig. 6a.4. The effects of reindeer grazing on South Georgian vegetation; the area inside the fence has been protected from grazing for eight years.

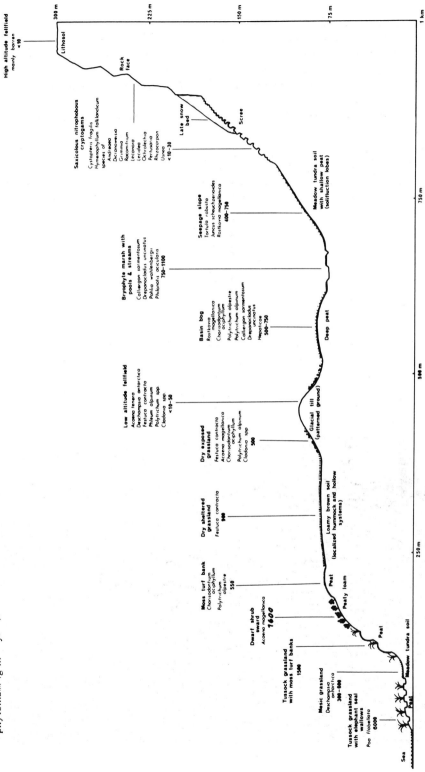

Fig. 6a.5. The altitudinal distribution of vegetation and soil types on South Georgia. (Smith and Walton, 1975)

. South Georgia IBP study area. Vegetation, soil types and estimated maximum net annual production for above and below ground living phytomass (g m$^{-2}$ yr$^{-1}$).

He defined ten categories to describe sub-Antarctic vegetation:

1. Tall tussock grassland — coastal vegetation dominated by *Poa* spp., usually biotically disturbed.
2. Short tussock grassland — usually closed stands on well drained soils dominated by short narrow leaved grasses e.g. *Festuca contracta*.
3. Meadow — closed vegetation mainly of broad-leaved grasses, e.g. *Agrostis magellanica* with herbs on moist sites.
4. Mire — usually rush- and bryophyte-dominated on seepage slopes with shallow peat.
5. Bog — rush- and bryophyte-dominated vegetation in very wet soils with deep peat.
6. Flush — mainly bryophyte vegetation on stream banks and below springs.
7. Swamp — submerged or floating vegetation at pool margins.
8. Herbfield — dense vegetation dominated by large leaved forbs, e.g. *Pleurophyllum hookerii*, *Acaena magellanica*, on moist sites.
9. Fernbrake — limited areas dominated by ferns, e.g. *Blechnum penna-marina*.
10. Fellfield — mainly patchy bryophytes and lichens with few phanerogams on dry exposed stony soils usually at high altitude.

These categories cannot be regarded as rigid and many examples exist of intergradations between them. Palynological data from South Georgia, Iles Kerguelen and Iles Crozet have shown that there has been no important change in species composition for several thousand years although minor climatic changes, perhaps associated with movement of the position of the Antarctic Convergence, have caused changes in the proportions of each species present. At a local level, minor changes in drainage patterns can initiate successional changes in particular vegetation stands whilst the introduction of alien herbivores has produced major changes in some communities.

Pattern is an obvious feature in stands dominated by some of the rhizomatous perennials. *Poa foliosa* on Macquarie Island often grows in parallel lines along stabilized landslips whilst *Rostkovia magellanica* on South Georgia produces two tiller types, clumped and pioneer. *Pleurophyllum* populations on Macquarie Island show clustering of rosettes which is probably due to a combination of limited seed dispersal and short axillary rhizomes, whilst *Phleum alpinum* in fellfield areas on South Georgia shows cyclic development with a break-up of the original colonizing plant at *c* 20 years old.

Fig. 6a.6. Erosion around stools of tussac grass (*Poa flabellata*) caused mainly by seal activity.

Some vegetation patterns are apparently associated with interference effects e.g. the spacing of *Poa flabellata* tussocks, whilst others are related to relictual periglacial features. Vegetated stripe hummocks on South Georgia are dominated by *Festuca* on the dry ridges and bryophytes in the damp hollows. Active frost sorting in small scale nets and polygons produces a moisture and disturbance gradient reflected in the circular patterns of vegetation around these features.

## 6a.4.2. Adaptations and life strategies

The rigorous climate experienced by plants of the Arctic tundra has selected for a number of important morphological and physiological features. Chief amongst these are preformation of flowers the season before they are needed, emphasis on vegetative reproduction, strong development of food storage organs and the ability to photosynthesize at low temperatures. Although the sub-Antarctic climate is less demanding in terms of temperature extremes, it is still sufficiently taxing for successful species to have some or all of these attributes.

These are no annual species in the sub-Antarctic flora. Even the alien grass *Poa annua* adopts a perennial life cycle. Preformation of flowers, although not widespread, has been found in a few species (*Poa flabellata*, *Festuca contracta*, *Deschampsia antarctica* and *Colobanthus quitensis*). Physiological ecotypes adapted to lowland and upland habitats have been described for *Phleum alpinum*, and there is some evidence that winter-green grasses can continue their photosynthesis below 5°C. Nearly all species have the ability to produce viable seed but most put the majority of dry matter production into storage materials and vegetative reproduction.

Many sub-Antarctic flowering plants can produce very large quantities of seed from certain populations but viability in the few species tested has been very variable. Seed from populations in exposed habitats appears usually to have a lower viability than that from sheltered populations. Despite the production of viable seed, seedlings of some taxa, e.g. *Rostkovia, Juncus*, have never been observed in the field. Some species retain their seeds in the inflorescence overwinter, e.g. *Azorella selago, Colobanthus quitensis, Ranunculus biternatus*, whilst others broadcast seed widely during the late summer − early autumn period. The South Georgian flora, which is probably typical in this respect of all the islands, shows two major reproductive specializations: wind pollination and self compatibility (Walton, 1982). Preformation of flowers in some species together with cleistogamy in others show that the island populations are adapted in various ways to ensure seed production is successful.

## 6a.4.3. Production and decomposition

Most estimates of production for sub-Antarctic plants and communities have been made by harvesting. Gas exchange data has recently been collected for some species and should in due course provide an independent assessment of the accuracy of the harvest estimates.

The highest values recorded have been for individual heavily manured *Poa flabellata* tussocks with above ground production of up to 10,000 g m$^{-2}$y$^{-1}$. Above and below ground community biomass in this species is also very high and can reach 34,000 g m$^{-2}$ on South Georgia. It must be said that these are exceptional values and more typical tussock communities have biomass and production levels of less than half those quoted. The smaller tussock grasses, *Poa foliosa* and *Poa cookii* have a range of typical total standing crop biomass of 2500 − 8000 g m$^{-1}$, with above ground production of less than 2000 g m$^{-2}$y$^{-1}$.

Annual above ground production rates for other types of communities are much less. The *Blechnum* fernbrake on Marion Island produces only 450 g m$^{-2}$y$^{-1}$, the *Roskovia* bog on South Georgia 200 g m$^{-2}$y$^{-1}$, *Acaena magellanica* stands 530 – 1100 g m$^{-2}$y$^{-1}$ and the *Azorella* dominated fellfield on Macquarie Island *c*. 300 g m$^{-2}$y$^{-1}$ (Smith, 1984). Estimates vary considerably, especially in community assessments, and depend both on the species proportions in the stands sampled and in the sampling methodology.

Decomposition of this biomass varies considerably between community types. The earlier descriptions of extensive peat soils has implied that in many cases decomposition does not match production and organic matter is accumulating. In *Acaena magellanica* stands the leaves decay quickly and there is an active fermentation layer in the soil. Peat accumulation here is of moss material and the woody rhizomes of *Acaena*. Standing dead is a major component of all the sub-Antarctic grasslands, with leaf material only entering the litter layer when several years old. Litter bag estimates of decomposition in the litter layer suggest *c*. 20% y$^{-1}$ loss is typical. Broad leaved herbs such as *Pleurophyllum* (36 – 66% y$^{-1}$) and *Stilbocarpa* (up to 90% y$^{-1}$) have much higher rates. Fellfield plants especially cushion species such as *Azorella*, lose their dead material by damage and abrasion rather than by decomposition. In the bogs and mires where anaerobic conditions can develop decomposition is very slow, allowing steady buildup of peat.

The major decomposers appear to be the fungi with bacteria playing a fairly minor role. Comminution of litter by the arthropod fauna appears to be limited and the lack of it may be the controlling process in decomposition. Of the phylloplane and litter fungi from South Georgia examined physiologically, none are psychrophilic but the dominant species, although having optima of *c*. 15°C, are able to grow even at 0°C, thus making it likely that decomposition continues throughout much of the winter. Some recent work on South Georgian sites (Smith, 1983) has shown that leakage of organic nutrients from rapidly-growing rhizomes and roots can have dramatic effects on local populations of fungi and bacteria, greatly increasing decomposition within localized microsites in the peat soils.

## 6a.4.4. Nutrient cycling

The sub-Antarctic ecosystems are principally reliant on marine sources for nearly all their nutrient input although some inorganic nutrients do come from weathering of the rock and soil. The most detailed investigations so far have been on Marion Island (Fig. 6a.7) and there the nutrient cycle comprises a major inorganic component (sea spray, atmospheric deposition, weathering) and a minor biological component (N fixation, decomposition, local biotic enrichment). The high rainfall on Marion Island results in rapid leaching, especially of N and P, except in areas frequented by seals and birds. Indeed, it is the activities of these animals that transfer most of the N and P into the terrestrial ecosystem. Local enrichment by guano around albatross nests is an obvious phenomenon on the island but less obvious is the volatilization of N as ammonia in the penguin rookeries and its subsequent deposition over the whole island by rain (Lindeboom, 1979). Most of the nitrogen fixation is probably by cyanobacteria epiphytic on bryophytes and is controlled by pH and water availability (Smith and Russell, 1982).

The rate of nutrient turnover is obviously linked to the rate of decomposition. The elemental composition of *Acaena magellanica* leaves is high (especially in N) and this is clearly instrumental in the rapid decomposition in this community. In the dry fescue grassland retention of significant quantities of elements in the standing dead and litter fractions coupled with slow decomposition greatly reduces the rate of nutrient cycling. This does, however, allow the dead component to act as a buffer store, allowing slow nutrient release into the soil system rather than a seasonally produced pulse. The long leaf life (up

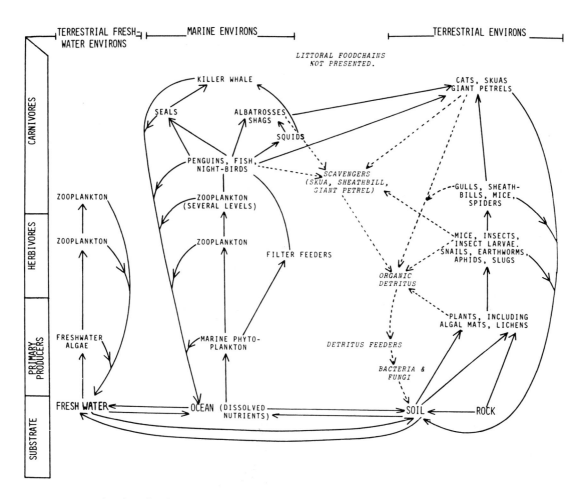

Fig. 6a.7. The relationships between the major nutrient pools for the Marion Island ecosystem (from V.R. Smith, 1977).

to 15 months) found in many sub-Antarctic species and the slow senescence, especially common in grasses, allows full withdrawal of useful nutrients before final leaf death. This husbanding of nutrient resources is typical of many Arctic species.

## 6a.5. MICROCLIMATES

### 6a.5.1. Plant and invertebrate microclimates

Whilst general meteorological data are useful for making climatic comparisons between the islands as descriptions of the local biological microclimates they are very misleading. From the very limited data so far published for the sub-Antarctic several surprising facts have emerged.

The most closely-studied parameter has been temperature. Measurements made in several Marion Island plant communities suggest that plant temperatures may reach 25°C or more in full sun (Huntley, 1971). In the fellfield temperature differences between north and south faces of boulders have reached over 18°C, but generally the high cloud cover over the island prevents the development of high temperature. Variations in wind speed appear to be of greater ecological significance than temperature gradients.

On South Georgia several communities have been studied including the fescue grassland, the tussock grassland, the moss banks and the dwarf shrub stands (Walton, 1984). The effective boundary layer for the moss bank is very shallow and for much of the time the temperature at the top of the vegetation is similar to the soil temperature. This is not the case in the grassland where the maximum temperature occurs within the standing dead leaves about 10 cm above the ground, with a sharp gradient above that to ambient air temperature at about 50 cm. In the dwarf shrub stands insulation by the moss layer accounts for the initial temperature rise from the ground to the 5 cm level. Above that the profile is almost isothermal to the top of the closed leaf canopy at about 25 cm.

Spot surface temperatures on leaves of *Poa flabellata* showed that standing dead leaves could easily exceed 30°C in full sun with green leaves $6-8$°C less. A differential of about 3°C existed between the sun and shade sides of these green leaves. Temperature changes at the leaf surface could be very rapid with a fall of 20°C in a few minutes when a cloud covered the sun.

Although there are no published data for the radiation microclimates in the sub-Antarctic, it is clear from the annual plant production data that light is not a limiting factor for the native species. In the more southerly islands where the period of winter snow cover is longer, radiation penetration of snow can be important in extending the growing season of cold-adapted plants. Measurements of attenuation of photosynthetically-active radiation by snow in the South Orkney Islands has suggested that up to 20% can penetrate 20 cm of snow. The formation of an ice layer and crystalline ice pipes by thawing and refreezing — a common phenomenon on South Georgia — is likely to improve the light penetration. Another important feature of the radiation régime for plants is daylength although, as yet, relatively little is known of the sub-Antarctic species in terms of photoinduction of floral development (Walton, 1982).

Polar microclimates are greatly affected by wind, the paucity of shelter making both plants and animals especially vulnerable to its effects. It has three important effects: disturbance of local temperature and humidity profiles, abrasion damage by transport of ice crystals and mineral particles, and transport of plant propagules and small animals. In exposed areas high winds can affect plant distribution. On Macquarie Island plants in the fellfield areas grow only on the lee side of terrace slopes whilst on Marion Island fellfield *Azorella* clumps always grow away from the wind. The relationship between wind speed and direction and local topography is of crucial importance in determining the position and depth of snow, whilst the protection offered by the snow from temperature and abrasion can be critical for plant and animal survival. Investigations of these complex inter-relationships have not yet begun.

## 6a.5.2. Vertebrate microclimates

Although, in some respects, the sub-Antarctic climates can be regarded as hostile or unpleasant, for some animals they are particularly suitable. Large animals have one major advantage over plants and most invertebrates — they can more quickly search out the most favourable microclimate.

Most of the information on the microclimate of surface nesting birds comes from the Maritime Antarctic and the Continent. The huddling behaviour of both king and emperor penguins in cold windy conditions is an interesting example of collective behaviour to modify the windchill factor affecting individuals (Le Maho, 1977), although it should be stressed that by ornithological standards the climates of most sub-Antarctic islands should not be regarded as especially rigorous. The use of nesting sites above ground level is a characteristic of several surface nesting species of seabirds and this protects the nests from flooding and the parents being buried by drift snow (Walton, 1984).

Many species of petrels live in burrows, crevices or caves. Most of the burrow temperature data are from work on the French sub-Antarctic islands (Mougin, 1975). For the parent bird the mean temperature inside the burrow is generally only slightly above the outside air temperature but the diurnal range is greatly reduced. Climatically, burrows and crevices have drawbacks. At snow melt they are prone to flooding and they can be blocked by drift snow during storms, perhaps resulting in the death of the occupants by starvation.

## 6a.6. THE ECOLOGY OF SUB-ANTARCTIC INVERTEBRATES

The limited diversity of the native land fauna and the numerical dominance of certain groups has resulted in a very uneven depth of knowledge across all the invertebrate groups present.

Information on Protozoa appears to be limited to South Georgia, Marion Island and Iles Kerguelen, and consists of limited population assessments and species lists. The South Georgian protozoan fauna shows a pattern with habitat type which can be related to the degree of soil development (H.G. Smith, 1982), i.e. vegetated habitats are richer than unvegetated ones, with angiosperm soils more species-rich than bryophyte peats. A similar relationship is apparent in the data from Kerguelen, whilst on Marion Island mire soils are the most species rich.

Nematodes are a mainly unknown group in the sub-Antarctic. There are no records from South Georgia, Prince Edward and Marion Islands, Heard and Macdonald Islands or Bouvetøya. Records from Iles Crozet, Iles Kerguelen and Macquarie Islands are inadequate (Maslen, 1980), as is the recent report of nematodes in fumarole soils on the South Sandwich Islands. Rotifers and Tardigrades appear not to have been investigated at all in the sub-Antarctic.

Earthworms are surprisingly widely distributed. Three of the four species on Macquarie Island are enchytraeids, and so are four of the five species on South Georgia. However, large earthworms (*Microscolex*) are also found and on Marion Island they comprise 75 − 89% of the biomass of macroinvertebrates in a wide range of vegetation types (Burger, 1978). These worms, at densities of over 1300 m$^{-2}$, form an important part of the diet of at least two bird species. Although snails and slugs have been collected on some islands they are always rare and nothing is known of their ecology.

Much of the ecological data for non-parasitic insects in the sub-Antarctic has come from detailed studies of springtails (Collembola) and beetles (Coleoptera) on South Georgia, beetles and other groups on Marion Island, and mites on Iles Kerguelen, with some general observations on arthropods from Iles Crozet. It is therefore difficult at the moment to draw any general conclusions about invertebrate ecology in the sub-Antarctic region.

Thirty-seven species of collembolan have been reported from the sub-Antarctic islands, but most available data relate to the more restricted fauna further South. In a recent detailed study on South Georgia, the population dynamics of six species were followed over two years at two sites. Maximum numbers occurred at the end of summer with minima in winter and spring. The fauna was similar for both the grassland and moss bank sites, with *Friesea grisea* dominant at the former and *Setocerura georgiana* dominant at the latter. The maximum density of any species was 4620 m$^{-2}$. Comparative studies on *Cryptopygus antarcticus* have shown that the South Georgian specimens are smaller, lighter in weight and sexually mature at a smaller size than those from the South Orkney Islands (Block and Tilbrook, 1978), although physiologically the two populations were not significantly different.

Over 50% of the beetle fauna of the sub-Antarctic islands belong to the weevil family (Curculionidae) and in the relatively simple ecosystems beetles may play an important role as herbivores. Beetle densities in *Festuca* grassland on South Georgia have reached 455 adults and 620 larvae m$^{-2}$. The species of *Perimylops* in particular have been observed to cause considerable damage to grass shoots as well as eating the leaves of *Acaena magellanica* and moss sporophytes. The weevils are important in fellfield areas on Iles Crozet and in lowland vegetation on Marion Island. In the only feeding study undertaken so far *Ectomnorrhinus similis* was found to ingest *c*.14% of body weight day$^{-1}$ when fed *Acaena magellanica* leaves and up to 37% of body weight day$^{-1}$ eating the moss *Brachythecium rutabulum*.

Ecological information on the Acari is patchy. Cryptostigmata or oribatid mites are an important soil group in the sub-Antarctic. On Kerguelen they are the predominant group in the Acari which were 67% of the total invertebrate fauna. Mite density was highest in halophytic mosses (12,652 l$^{-1}$) and lowest in degraded moss peat (572 l$^{-1}$). At South Georgia, maximum mite populations occurred in spring and summer and minimum in winter. The highest mite densities were found in the moss bank (156,750 m$^{-2}$).

## 6a.7. VERTEBRATE FAUNA

The larger animals of the sub-Antarctic islands are mainly dealt with elsewhere in this volume. They range in size from the elephant seal (Chapter 5a), through the introduced herbivores (Chapter 6b) to the oceanic birds (Chapters 5c and 5d). One small group of vertebrates, not dealt with elsewhere, are the land birds. There are only four native species of these comprising a pintail, a speckled teal and a pipit on South Georgia, and a pintail on Kerguelen (Watson, 1975). All the waterbirds feed on algae, small crustaceans and aquatic insects. The pipit appears to subsist mainly on collembola, beetles and flies although it may also eat some seeds. Whilst the waterbird populations appear to be adequate, rat predation of pipits on South Georgia has largely limited them to offshore islands for breeding.

## 6a.8. CONSERVATION AND SOVEREIGNTY

All of the islands fall outside the Antarctic Treaty area and are not therefore subject to any of its provisions. Each island or archipelago is subject to national sovereignty and legislation and this is exercized in different ways. South Georgia and the South Sandwich Islands are British, Prince Edward and Marion Island are South African, Macquarie, Heard and McDonald Islands are Australian, Bouvetøya is Norwegian, and Iles Kerguelen and Iles Crozet are French. The Auckland Islands and Campbell Islands, belonging to New Zealand, have important biogeographical connections but are outside the scope of the present article.

Bouvetøya is and always has been uninhabited. Its ice-free area is very small and it is of limited biological significance. Its conservation legislation is limited to decrees by King Haakon prohibiting interference with birds and marine mammals without permission. Macquarie Island is administered by the Tasmanian State Parks Department and is subject to Tasmanian State legislation on park use and protection. Although a manned station was run for a short time by Australia on Heard Island this was abandoned in 1955 and both Heard and the uninhabited Macdonald Islands are now very rarely visited. They are Australian Federal Government responsibilities and are subject to the same legislation as the Australian Capital Territory. The French islands are run through TAAF (Terres Australes et Antarctiques Français), as an overseas Department of France, whilst the South African sub-Antarctic islands are principally the responsibility of the Department of Transport, although legislated for separately in the national parliament. The British islands have always been the responsibility of the Falkland Islands Government, with legislation enacted by the High Commissioner of the Falkland Islands.

Historically South Georgia has had a resident population for longer than any of the other islands, and Man's industrial waste, in the form of abandoned whaling stations (Fig. 6a.8), still litters the island. The scientific importance of the island has, however, been recognized and new conservation legislation enacted by the Falkland Islands Government is now as stringent as that applying within the Antarctic Treaty area (Roberts, 1977). Specific conservation legislation also covers Prince Edward and Marion Islands, Macquarie Island, and Heard and Macdonald Islands. In general terms all this legislation attempts to restrict the introduction of alien species, prohibit the harvesting of any living resources except under special licence, and protect the native flora and fauna from interference and disturbance. The French appear, however, to see things differently. Their continued and deliberate introduction of a wide range of herbivores into Kerguelen, ostensibly to provide fresh meat for the scientific station, has resulted in the major degradation of such large areas of vegetation that there can be little on the main

Fig. 6a.8. The derelict remains of the old whaling station at Grytviken, South Georgia.

island sufficiently undamaged to be worth protecting. The reindeer on South Georgia and rabbits on Macquarie Island have also caused localized damage to native vegetation and allowed the spread of alien plant species within overgrazed areas.

Thus, although there are national provisions for conservation on the islands not all are comprehensive and their enforcement is decidedly patchy. Increasing interest in these islands, in particular South Georgia, from a tourists point of view will require more vigorous prosecution of the laws if they are to be of any use. The proliferation of private yachts in the Southern Ocean has added a new, and largely unwelcome element, to the tourist problem. The activities of these yachts seem, at the moment, to be beyond any general control.

It was apparently the intention of the Treaty nations to include all the living Antarctic marine resources within the jurisdiction of the Convention for the Conservation of Antarctic Marine Living Resources. Indeed, the Convention states explicitly that it will apply to the marine resources within an area bounded mainly by the Antarctic Convergence. This clearly impinges on potential national sovereignty around the sub-Antarctic islands and this has prompted two important exclusions from the Convention. Macquarie Island together with a 200 mile territorial limit is excluded, whilst France has reserved the right to decide on its own conservation measures for Iles Crozet and Iles Kerguelen and their surrounding waters. This has enabled France to continue to collect substantial licence fees from the USSR for the right to fish within these territorial waters, and to decide unilaterally on catch limits for these areas. In the waters around South Georgia, the other substantial sub-Antarctic fishing ground, no such system applies and there foreign trawlers are able to fish right up to the three mile limit with no effective national or international controls on catches.

Management of these islands from a conservation viewpoint has only just begun. For the larger herbivores, e.g. reindeer, population management plans are essential to control vegetation damage. Attempts to rid specific islands of unwanted introductions, e.g. cats on Marion, rabbits on Macquarie Island, have had only limited success. On South Georgia the use of designated Tourist Areas for limiting the effects of visiting tourist ships appears to have been useful. One major problem, seemingly intractable at some localities, is the growth of rubbish dumps. These provide a focus for foraging by some of the scavenging birds, are extremely unsightly and allow uncontrolled release of toxic chemicals and plastics into the ecosystem. In cold climates with slow rates of decomposition the creation of these mounds seems the very antithesis of conservation. Clearly, despite all the legislation, much remains to be done before both the letter and the spirit of the laws are applied fully throughout the sub-Antarctic.

## REFERENCES

Abbott, I. (1974) Numbers of plant, insect and land bird species on nineteen islands in the southern hemisphere. *Biol. J. Linn. Soc.* 6, 143 – 52.

Bliss, L.C. (1979) Vascular plant vegetation of the southern circumpolar region in relation to antarctic, alpine and arctic vegetation. *Can. J. Bot.* 57, 2167 – 78.

Block, W. (1984) Terrestrial microbiology, invertebrates and ecosystems. In *Antarctic Ecology*, Ed. R.M. Laws, pp.163 – 236. Academic Press, London.

Burger, A.E. (1978) Terrestrial invertebrates: a food resource for birds at Marion Island. *S. Afr. J. Antarct. Res.* 8, 87 – 98.

Hall, K.J. and Williams, A.J. (1981) Animals as agents of erosion at sub-Antarctic Marion Island. *S. Afr. J. Antarct. Res.* 10/11, 18 – 23.

Hooker, J.D. (1879) Observations on the botany of Kerguelen Island. *Phil. Trans. R. Soc.* 168, 9 – 16.

Huntley, B.J. (1971) Vegetation. In *Marion and Prince Edward Islands*, Eds E.M. van Zinderen Bakker, J.M. Winterbottom and R.A. Dyer, pp. 98 – 160. Balkema, Cape Town.

Jenkin, J.F. (1975) Macquarie Island, SubAntarctic. *Ecol. Bull. (Stockholm)* 20, 375 – 97.

Le Maho, Y. (1977) The Emperor Penguin: a strategy to live and breed in the cold. *Amer. Sci.* 65, 680 – 93.

Lindeboom, H.J. (1979) *Chemical and Microbiological Aspects of the Nitrogen Cycle on Marion Island (Sub-Antarctic)*. Ph.D. Thesis, Rijksuniversiteit te Groningen.

Lindsay, D.C. (1974) The macrolichens of South Georgia. *Br. Antarct. Surv. Sci. Rep.* 89, 1 – 91.

Maslen, N.R. (1980) Additions to the nematode fauna of the Antarctic region with keys to the taxa. *Bull. Br. Antarct. Surv.* 49, 207 – 29.

Mougin, J.L. (1975) Ecologie comparée des Procellariidae antarctiques et subantarctiques. *C.N.F.R.A.* 36, 1 – 195.

Roberts, B.B. (1977) Conservation in the Antarctic. *Phil. Trans. R. Soc. B* 279, 97 – 104.

Schuster, R. (1969) Problems of antipodal distribution in lower land plants. *Taxon* 18, 46 – 91.

Smith, H.G. (1982) The terrestrial protozoan fauna of South Georgia. *Polar Biol.* 1, 173 – 80.

Smith, M.J. (1983) *The Microbial Ecology of sub-Antarctic Tundra soils.* Ph.D. Thesis, University of Surrey.

Smith, R.I.L. (1981) Types of peat and peat-forming vegetation on South Georgia. *Bull. Br. Antarct. Surv.* 53, 119 – 39.

Smith, R.I.L. (1984) Terrestrial plant biology. In *Antarctic Ecology*, Ed. R.M. Laws, pp. 61 – 162. Academic Press, London.

Smith, R.I.L. and Walton, D.W.H. (1975) South Georgia, Subantarctic. *Ecol. Bull. (Stockholm)* 20, 399 – 423.

Smith, V.R. (1977) A quantitative description of energy flow and nutrient cycling in the Marion Island terrestrial ecosystem. *Polar Rec.* 18, 361 – 70.

Smith, V.R. (1978) Animal-plant-soil relationships on Marion Island (sub-Antarctic). *Oecologia* 32, 239 – 53.

Smith, V.R. and Russell, S. (1982) Acetylene reduction by Bryophyte- cyanobacteria associations on a subantarctic island. *Polar Biol.* 1, 153 – 57.

Wallwork, J.A. (1973) Zoogeography of some terrestrial micro-Arthropoda in Antarctica. *Biol. Rev.* 48, 233 – 59.

Walton, D.W.H. (1975) European weeds and other alien species in the Subantarctic. *Weed Res.* 15, 271 – 82.

Walton, D.W.H. (1982) Floral phenology in the South Georgian vascular flora. *Bull. Br. Antarct. Surv.* 55 11 – 25.

Walton, D.W.H. (1984) The terrestrial environment. In *Antarctic Ecology*, Ed. R.M. Laws, pp. 1 – 60. Academic Press, London.

Watson, G.E. (1975) *Birds of the Antarctic and Sub-Antarctic.* American Geophysical Union, Washington D.C.

CHAPTER 6b

# The Sub-Antarctic Islands — Introduced Mammals

N. LEADER-WILLIAMS

Large Animal Research Group, Department of Zoology, 34A Storey's Way,
Cambridge CB3 0DT, U.K.

## CONTENTS

## 6b.1. INTRODUCTION

The sub-Antarctic islands lack indigenous mammals. Since the nineteenth century, initially during sealing and whaling expeditions and latterly whilst manning observatories and bases, man has introduced at least 15 species of mammal (not including himself) to these islands. This chapter reviews both failed and successful introductions and identifies factors common to all of them, to provide a basis for the consideration of the control and management of introduced animals and their possible interest for man.

TABLE 6b.1.The Sub-Antarctic Island Groups, Their Geographical Location, and the Dates of Introduction and Present Extent of Distribution of Introduced Mammals that are Presently Established upon Them. L = Localised, W = Widespread, C19 = 19th century

| Location and introduced species | Island or island group | | | | | Source |
|---|---|---|---|---|---|---|
| | South Georgia | Macquarie | Kerguelen | Marion [and Prince Edward] | Crozet | |
| *Accidental introductions* | | | | | | |
| Norway rat (*Rattus norvegicus*) | *c.* 1800 W: several areas | | | | | Pye & Bonner (1980) |
| Black rat (*Rattus rattus*) | | C19 Widespread | *c.* 1950 L:1?island | | | Jenkin *et al.* (1979); Lesel & Derenne (1975) |
| House mouse (*Mus musculus*) | C19 L:1 area | C19 Widespread | C19 L:1?island | C19 W: Marion | C19 L: 1 island | Berry *et al.* (1979); Jenkin *et al.* (1982); Lesel & Derenne (1975); Skinner *et al.* (1978); Derenne & Mougin (1976) |
| *Deliberate introductions* | | | | | | |
| Rabbit (*Oryctolagus cuniculus*) | | 1879 Widespread | 1874 W:several islands | | *c.* 1850 L:1 island | Jenkin *et al.* (1981); Lesel & Derenne (1975); Derenne & Mougin (1976) |
| Sheep (*Ovis aries*) | | | 1909 L:1 island | | | Lesel & Derenne (1975) |
| Reindeer (*Rangifer tarandus*) | 1911 and 1926 L: 3 areas | | 1955 L: 2 islands | | (1980): | Leader-Williams (1980a) Lesel & Derenne (1975) |
| Mouflon (*Ovis ammon*) | | | 1957 L:1 island | | | Lesel & Derenne (1975) |
| Cat (*Felis catus*) | *c.*1820 | 1951 Widespread | 1951 L: 1 island | 1949 W:Marion | 1887 L:1 island | Jenkin *et al.* (1982); Pascal (1982); van Aarde (1979); Derenne & Mougin (1976) |

## 6b.2. INTRODUCED MAMMALS

Mammals have been introduced to sub-Antarctic islands both deliberately and accidentally. Some of these species have become established, but most have failed in the sense that they are no longer present on the islands.

Fig. 6b.1. Reindeer on South Georgia.

## 6b.2.1. Failed introductions

Several of man's husbanded and domesticated animals, such as cattle, goats and pigs; horses and donkeys (and mules); and dogs (Holdgate and Wace, 1961) have been introduced to provide both familiar sources of fresh meat and beasts of burden and companion animals. These species were confined mainly to the immediate vicinities of settlements, and where attempts were made to allow them to roam free, they were largely unsuccessful. Another failed introduction was that of mink, *Mustela lutreola*, to Iles Kerguelen in 1956 (Lesel and Derenne, 1975). The rigorous climate of the islands, relative to the usual conditions experienced by these various species without winter housing or supplementary feeding, appears the most likely reason for their failure to adapt. Indeed, introductions of sheep, rabbits and cats have failed on South Georgia which has almost the severest climate of any sub-Antarctic island, yet have succeeded on some of the other more equable islands.

## 6b.2.2. Established introductions

At the present time eight different species of mammal are known to occur in established populations on the six different islands or island groups in the sub-Antarctic (Table 6b.1). Only one island group (Heard and Macdonald) is free of introduced mammals. Rodents were introduced accidentally from ships and amongst foodstuffs to all five remaining groups from the nineteenth century onwards. Various ungulates and a lagomorph were brought to five groups as sources of fresh meat and in, the case of reindeer on South Georgia, as a sporting amenity (Fig. 6b.1). Cats were introduced as companions and to control rodent pests, and have become established on four groups.

The island groups fall into two distinct categories. The first includes those where a single introduction has become widespread across the whole group because there is only one main island and

TABLE 6b.2. The Main and Preferred Foods of Introduced Mammals on Sub-Antarctic Islands; where Quantitative Data are Available, Items Marked * Predominate in the Diet

| Island | Reindeer | Sheep/Mouflon | Rabbits | Rats | Mice | Cats | Source |
|---|---|---|---|---|---|---|---|
| South Georgia | Poa flabellata* Acaena magellanica Lichens | | | Poa flabellata* Invertebrates Carrion | Poa flabellata Invertebrates | | Leader-Williams et al. (1981) Pye & Bonner (1980) Berry et al. (1979) |
| Macquarie | | | Poa foliosa Stilbocarpa polaris | Poa foliosa* Invertebrates Carrion | Invertebrates* Stilbocarpa polaris | Rabbits* Prions & petrels* | Costin & Moore (1960) Jenkin et al. (1982) Berry & Peters (1975) Jones (1977) |
| Kerguelen | Acaena magellanica Azorella selago Pringlea antiscorbutica Lichens | Pringlea antiscorbutica Acaena magellanica | Poa cookii Azorella selago Pringlea antiscorbutica | ? | Azorella selago Acaena magellanica | Rabbits* Prions & petrels* | Lesel & Derenne (1975) Pascal (1982) Derenne (1976) |
| Marion | | | | | Poa cookii Acaena magellanica Agrostis magellanica | Petrels* Mice | Skinner et al. (1978) Van Aarde (1980) |
| Crozet | | | Acaena magellanica Pringlea antiscorbutica | | Plants* Carrion | Prions & petrels Penguins Mice | Derenne & Mougin (1976) |

TABLE 6b.3. Time of Birth of Young and Present Numbers [with Date of Census or Estimate] of Introduced Mammals on Sub-Antarctic Islands

| Island | Reindeer | Sheep/ Mouflon | Rabbits | Rats | Mice | Cats | Source |
|--------|----------|--------|---------|------|------|------|--------|
| South Georgia | November 2000 [1976] | | | November February ? | November to January ? | | Leader-Wiliams (1980a) Pye & Bonner (1980) Berry et al. (1979) |
| Macquarie | | | August to March ? | ? | Continuous ? | November to March 250 – 500 [1975] | Skira (1978) Berry & Peters (1975) Jones (1977) |
| Kerguelen | September to October 2000 [1974] | September to November 300 [1974] | October to February 177,000 [1968] | ? | ? | September to March 3500 [1975] | Lesel & Derenne (1975) Derenne (1976) Lesel (1968) Pascal (1982) |
| Marion | | | | | Continuous ? | September to March 2000 [1975] | Berry et al. (1978) Van Aarde (1979) |
| Crozet | | | December to March 2000 [1975] | | Summer 200,000 [1975] | October to May 100 – 400 [1975] | Derenne & Mougin (1976) |

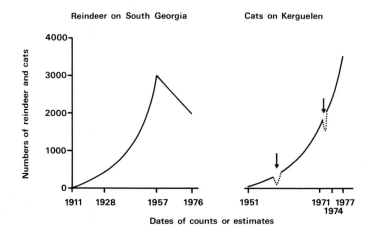

Fig. 6b.2. Simplified population trends in some introduced mammal populations on sub-Antarctic islands. ↓ = dates of shoot-outs of cats on Kerguelen. Data from Leader-Williams (1980a,b) (Barff herd only); Derenne (1976), Pascal (1982).

TABLE 6b.4. Examples of the Ways that Introduced Mammals have Affected the Floras and Faunas of Sub-Antarctic Islands

| Affected species group or community | Status | Extinct | Locally eliminated or reduced | Locally spreading or beneficiary | Island group | Animal responsible | Reason | Source |
|---|---|---|---|---|---|---|---|---|
| **Tussock grassland** | Native | | | | | | | |
| *Poa flabellata* | | | * | | South Georgia | Reindeer | Grazing | Leader-Williams *et al.* (1981) |
| *Poa foliosa* | | | * | | Macquarie | Rabbits | Grazing | Costin & Moore (1964) |
| *Poa cookii* | | | * | | Kerguelen | Rabbits | Grazing | Lesel & Derenne (1975) |
| **Heath** | Native | | | | | | | |
| *Acaena magellanica* | | | * | | South Georgia Kerguelen | Reindeer | Grazing | Leader-Williams *et al.* (1981) Lesel & Derenne (1975) |
| *Acaena magellanica* *Azorella selago* | | | * | * | Kerguelen | Rabbits | Dispersion of seeds; grazing | Lesel & Derenne (1975) |
| **Herbfield** | Native | | | | | | | |
| *Stilbocarpa polaris* | | | * | | Macquarie | Rabbits | Grazing | Costin & Moore (1964) |
| *Pringlea antiscorbutica* | | | * | | Kerguelen | Rabbits | | Lesel & Derenne (1975) |
| **Lichens** | Native | | * | | South Georgia Kerguelen | Reindeer | Grazing | Lindsay (1973) Lesel & Derenne (1975) |
| *Poa annua* | Alien | | | * | South Georgia Macquarie Kerguelen | Reindeer Rabbits Rabbits | Adapted to grazing | Walton (1975) |
| **Parakeet** *Cyanoramphus novae-zelandiae* | Endemic | 1880 – 1891 | | | Macquarie | Cats/Wekas | Predation pressure after arrival of rabbits | Taylor (1979) |
| **Rail** *Rallus philippensis* | Endemic | 1894 | | | Macquarie | Cats/Wekas | | |
| **Pipit** *Anthus antarcticus* | Endemic | | * | | South Georgia | Rats | Predation | Pye & Bonner (1980) |
| **Burrow-nesting prions and petrels** (Most species) | Native | | * | | South Georgia Macquarie Kerguelen Marion Crozet | Rats Cats | Predation | Pye & Bonner (1980) Jones (1977) Pascal (1982) Van Aarde (1980) Derenne & Mougin (1976) |
| **Skuas** *Stercorarius skua* | Native | | | * | Macquarie Kerguelen | Rabbits | Additional prey | Jones & Skira (1979) Lesel (1968) |
| **Wekas** *Gallirallus australis* | Alien | | | * | Macquarie | Rabbits | Prey | Jenkin *et al.* (1982) |

Fig. 6b.3. Feral cat on Marion Island (Photo: Kevin Hall).

no physical barriers to prevent dispersion (Macquarie Island). The second category is those which are either paired (Marion and Prince Edward Islands), or archipelagos (Iles Crozet, Iles Kerguelen), or dissected into 'islands' by large sea-entering glaciers (South Georgia), where single introductions have resulted only in more localized distributions. Thus, Prince Edward Island is still free of introduced mammals; South Georgian rats and Kerguelen rabbits are reasonably widespread only as a result of multiple introductions to different unglaciated areas.

## 6b.3. THE ESTABLISHMENT OF AN INTRODUCTION

Assuming that both climatic conditions and the range of food on an island are suitable for its requirements, an introduced mammal generally faces an ideal situation in which food is plentiful, competition from native species is lacking and predators are absent. Most introductions were made too long ago for their early history to be more than guesswork. Despite this, various aspects of their general biology have been recently examined.

### 6b.3.1. Feeding

The different herbivores appear to prefer various species of tussock grass, herbs and 'cabbage', where they are still available (Table 6b.2). Rodents appear to be primarily vegetarian, but may eat invertebrates, whilst rats also prey on native birds and scavenge upon carrion. The prey of cats depends to some extent upon whether rabbits and rodents have been introduced to the same island; burrow-nesting petrels and prions are the main native species preyed upon. With the passage of time since their introductions, it is likely that some herbivores have had to change their diet as several native species have been locally eliminated.

### 6b.3.2. Breeding

Reindeer, sheep and mouflon are natural seasonal breeders, but those introductions made directly from the Northern Hemisphere had to reverse their breeding seasons to produce young in the austral spring (Table 6b.3). The other introduced species all have a less distinct, or no, breeding season in their native habitats, but most have developed breeding seasons on sub-Antarctic islands. This probably arises from the pronounced change in photoperiod occurring at relatively low latitudes, and especially from marked seasonal changes in the availability of food and therefore in body condition.

### 6b.3.3. Population growth

The numbers of animals introduced originally were small, certainly in the case of deliberate introductions. Because many introductions were made before the era of animal census, reliable data are limited. However, it appears that numbers initially showed a rapid increase whilst food was abundant; later they peaked and finally stabilized at lower levels after reducing and then reaching a balance with their food supply (Fig. 6b.2). From these few available examples, (which agree with the expected pattern), it may be inferred that certain recent introductions (reindeer and mouflon on Iles Kerguelen) have yet to reach peak numbers. By contrast, the well established introductions will only show the more dampened fluctuations characteristic of natural populations: the peak will not appear again since the initially abundant food supply cannot be restored. For the established mammals, therefore, population numbers derived from recent censuses are unlikely to change dramatically under prevailing conditions (Table 6b.3).

## 6b.4. EFFECTS ON NATIVE FLORA AND FAUNA

The floras and faunas of islands with simple and species-poor communities are particularly prone to outside causes of disturbance, for they have little capacity for ecological re-adjustment. The sub-Antarctic islands are no exception and their biotas have altered dramatically since man's arrival.

The grazing of herbivores has led to the local elimination of the most palatable species amongst the native floras. Tussock grasslands and herbfields have been the most disturbed communities, and individual species have been heavily grazed in other habitats. In addition, alien grasses, notably *Poa annua*, have spread as a result of their suitability for grazing and are now dominant in certain communities. In areas where herbivores have been present for some while, accelerated soil erosion has occurred. Because of the importance of tussock grassland in stabilizing soils, and as a habitat for invertebrates and burrow- and ground-nesting birds, the consequences of overgrazing by herbivores are far-reaching in their effects (Table 6b.4).

Two endemic bird species on Macquarie Island, a banded rail (*Rallus philippensis*) and a parakeet (*Cyanorhampus novaezelandiae*), became extinct between 1880 and 1894. These birds survived the first 70 years after the arrival of cats, but declined to extinction after the introduction of rabbits. This latter introduction led to great increases in the numbers of feral cats and introduced wekas (*Gallirallus australis*), and in turn presumably to more intensified predation on the endemic birds.

Cats and rats have also been responsible for the local elimination of various native birds, notably the burrow-nesting petrels and prions. Cats, in particular, demonstrate the principle that the consequences of an introduction are usually unforeseen by the introducer. Originally cats were brought in to control rodent pests, but have usually found burrow-nesting birds (and rabbits) easier prey.

One native species has benefited from the presence of introduced mammals: Antarctic skuas (and also introduced wekas) have increased in number due to the presence of rabbits as prey on Macquarie Island and Iles Kerguelen.

The more subtle and less obvious effects of introduced animals upon the fauna still await investigation. They might include: a changed role for the native microfauna in nutrient cycling or in the decomposition of unaccustomed dung; the interaction of native invertebrates with intermediate stages of parasites brought in by herbivores; the effects of rodents upon populations of their invertebrate prey, and so on.

## 6b.5. CONTROL MEASURES

Calls by conservationists for the control and management of introduced species generally occur when undesirable results of the introduction become apparent and pressing. The sub-Antarctic islands lie outside the Antarctic Treaty area and are under the sovereign control of several different nations whose aims in their stewardship of the islands differ, as does the legislation affecting them. Because of their isolation and few inhabitants, widespread control measures have been relatively few. Such attempts as have been made have been limited to shoot-outs or the introduction of specific diseases which induce high rates of mortality.

The rabbit flea was established on Macquarie Island over a period of 10 years from 1968 as a vector for myxomatosis, before the myxoma virus was introduced in 1978. Local eradications have been impressive, but total eradication seems unlikely (Jenkin et al., 1982). On Iles Kerguelen, the earlier introduction of myxomatosis in 1955 failed because a vector was not previously or simultaneously introduced (Pascal, 1982).

Cats were shot on Iles Kerguelen in the late 1950s and again in the early 1970s, but those remaining were still sufficient to permit rapid population growth and dispersion (Fig. 6b.2). Feline panleucopaenia virus was introduced to Marion Island in 1977, and has had some effect in reducing cat numbers, which it is as yet too early to evaluate (Van Aarde and Skinner, 1982).

Control measures against rodents would be quite ineffective and inappropriate due to the difficult terrain of most islands. Most concern should be directed at those islands where introduced mammals are widespread, and refugia for the native floras and faunas of a group are limited, for example, Macquarie Island. Whether control is actually worthwhile is open to question, for most attempts so far have met with only limited success. The problems of a build-up of resistance to introduced diseases and of achieving total extermination should not be underestimated. Furthermore, where some success has been achieved, the introductions were so well established that native floras and faunas, by then irrevocably altered, could never be restored. New communities have developed in which alien species have become integrated. Some comfort can be derived from the fact that undisturbed and representative examples of at least six island groups' floras and faunas still exist, even if only in localized areas. Further extinctions of endemic species seem unlikely in future as breeding sites are available in areas where introduced mammals have not penetrated.

## 6b.6. INTEREST FOR MAN

Man can also be included amongst the introduced mammals on sub-Antarctic islands; however, in spite of many years of occupancy, there are as yet no established populations. The original aim in importing other mammals for food and companionship has to some extent been fulfilled, but these mammals have not been exploited on a rational or commercial basis.

The main interest of these introduced mammals lies elsewhere. The small numbers originally introduced, the climatic severity and relative simplicity of the terrestrial habitats, together with the spirit for research which pervades north from the Antarctic Treaty area to the sub-Antarctic islands, combine to make them good subjects for applied research and more fundamental studies of adaptation.

In the words of Charles Darwin 'Nothing would more aid Natural History, than careful collecting and investigating of all the productions of the most isolated islands, especially those of the Southern Hemisphere'.

At one end of the species range, Berry's studies of mice and Van Aarde's of cats living close to one limit of their range have been of fundamental interest to geneticists. At the other extreme, comparisons between reindeer on Arctic and sub-Antarctic islands have led to greater understanding of processes involved in irruptions and population crashes (Klein, 1968; Leader-Williams, 1980b). Many topics await further investigation and an approach which compares the same species on different sub-Antarctic islands might prove especially fruitful. In future it is to be hoped that the mistakes of making further introductions, made as recently as the 1950s, will not be repeated; however, the interest of natural experiments created by these introductions have, in recent years, been increasingly recognized.

## ACKNOWLEDGEMENTS

My interest in this topic arose whilst I studied reindeer on South Georgia for the British Antarctic Survey, and I am especially grateful to R.M. Laws and J.P. Croxall for their encouragement.

## REFERENCES

Berry, R.J., Bonner, W.N. and Peters, J. (1979) Natural selection in house mice (*Mus musculus*) from South Georgia (South Atlantic Ocean). *J. Zool., Lond.* 189, 385 – 98.

Berry, R.J. and Peters, J. (1975) Macquarie Island mice: a genetical isolate on a Subantarctic island. *J. Zool., Lond.* 176, 375 – 89.

Berry, R.J., Peters, J. and Van Aarde, R.J. (1978) Subantarctic house mice: colorisation, survival and selection. *J. Zool., Lond.* 184, 127 – 41.

Costin, A.B. and Moore, D.M. (1960) The effects of rabbit grazing on the grasslands of Macquarie Island. *J. Ecol.* 48, 729 – 39.

Derenne, P. (1976) Notes sur la biologie du chat haret de Kerguelen. *Mammalia* 40, 531 – 95.

Derenne, P. and Mougin, J.L. (1976) Données ecologiques sur les mammiferes introduits de l'Ile aux Cochons, Archipel Crozet (46°06'S, 50°14'E). *Mammalia* 40, 21 – 53.

Holdgate, M.W. and Wace, N.M. (1961) The influence of man on the floras and faunas of southern islands. *Polar Rec.* 10, 475 – 94.

Jenkin, J.F., Johnstone, G.W. and Copson, G.R. (1982) Introduced animal and plant species on Macquarie Island. *CNFRA* 51, 301-14.

Jones, E. (1977) Ecology of the feral cat, *Felis catus* (L.), (Carnivora:Felidae) on Macquarie Island. *Aust. Wildl.Res.* 4, 249 – 62.

Jones, E. and Skira, I.J. (1979) Breeding distribution of the great skua at Macquarie Island in relation to numbers of rabbits. *Emu* 79, 19 – 23.

Klein, D.R. (1968) The introduction, increase, and crash of reindeer on St. Matthew Island. *J. Wildl. Manage.* 32, 350 – 67.

Leader-Williams, N. (1980b) Population ecology of reindeer on South Georgia. In *Proc. 2nd Int. Reindeer/Caribou Symposium*, Eds E. Reimers, E. Gaase and S. Skjenneberg, pp.664 – 76 Røoros, Norway, Trondheim. Direktoratet for Viltog Ferskvannskfisk.

Leader-Williams, N., Scott, T.A. and Pratt, R.M. (1981) Forage selection by introduced reindeer on South Georgia, and its consequences for the flora. *J. appl.Ecol.* 18, 83 – 206.

Lesel, R. (1968) Essai d'estimation de la population d'*Oryctolagus cuniculus* L. sur la Peninsule Courbet (Ile de Kerguelen). *Mammalia* 32, 612 – 20.

Lesel, R. and Derenne, P. (1975) Introducing animals to Iles Kerguelen. *Polar Rec.* 17, 485 – 94.

Lindsay, D.C. (1973) Effects of reindeer on plant communities in the Royal Bay area of South Georgia. *Br. Antarct. Surv. Bull.* 35, 101 – 9.

Pascal, M. (1982) Les espéces mammaliennes introduits dan l'Archipel des Kerguelen (Territoire des T.A.A.F.). *CNFRA* 51, 269 – 80.

Pye, T. and Bonner, W.N. (1980) Feral brown rats, *Rattus norvegicus*, in South Georgia, South Atlantic. *J. Zool., Lond.* 192, 237 – 55.

Skinner, J.D., Condy, P.R., Van Aarde, R.J., Bester, M.N. and Robinson, T.J. (1978) The mammals of Marion Island: a review. *S. Afr. J. Antarct. Res.* 8, 35-38.

Skira, I.J. (1978) Reproduction of the rabbit, *Oryctolagus cuniculus* (L.), on Macquarie Island, Subantarctic. *Aust. Wildl. Res.* 5, 317 – 26.

Taylor, R.H. (1979) How the Macquarie Island parakeet became extinct. *N.Z.J. Ecol.* 2, 42 – 5.

Van Aarde, R.J. (1979) Distribution and density of the feral house cat, *Felis catus*, at Marion Island. *S. Afr. J. Antarct. Res.* 9, 14 – 19.

Van Aarde, R.J. (1980) The diet and feeding behaviour of feral cats, *Felis catus*, at Marion Island. *S. Afr. J. Wildl. Res.* 10, 123 – 8.

Van Aarde, R.J. and Skinner, J.D. (1982) The feral cat population at Marion Island: characteristics, colonisation and control. *CNFRA* 51, 281-8.

Walton, D.W.H. (1975) European weeds and other alien species in the Subantarctic. *Weed Res.* 15, 271 – 82.

# CHAPTER 7

# Food Webs and Interactions: an Overview of the Antarctic Ecosystem

ANDREW CLARKE

British Antarctic Survey, Natural Environment Research Council, High Cross, Madingley Road, Cambridge CB3 0ET, U.K.

## CONTENTS

## 7.1. INTRODUCTION

It is in the nature of human beings to classify and name. Thus the title of this book is Antarctica, the implication being that the Antarctic forms a discrete whole, an environment or ecosystem which can be delineated and described. To a certain extent this is true. The Antarctic as a whole may be defined as everything south of the Antarctic Convergence, and the terrestrial Antarctic obviously starts and stops where land meets sea. And yet, as must be obvious from the preceding chapters, the Antarctic environment consists of several very different components, each of which is usually treated independently and is not always so easily defined. For example the gradual change of the marine fauna

329

from north to south makes any division into sub-Antarctic and truly Antarctic communities somewhat arbitrary. A similar problem, though not so acute, exists for terrestrial plant communities, and freshwater lakes can vary so much that each almost needs a classification to itself. It is the purpose of this chapter to try and view the Antarctic as a whole; a sort of compare and contrast essay of the kind beloved of examiners.

Several of the broad generalizations which may be made about the Antarctic ecosystem were clear to Holdgate when he made the first (1967) attempt to discuss the biology of Antarctica as a whole. Since then a significant amount of research has been done, and the inevitable specialization has occurred. Such specialization has the advantage of producing increased understanding (for example, a more detailed knowledge of the levels of production in different habitats and energy flow between habitats, both emphasized by Holdgate as areas where more data were required); the disadvantage is the increasing difficulty of seeing the parallels and links between the various compartments we choose to study.

The first section of this chapter examines separately the three most usual subdivisions of the Antarctic (the marine, freshwater and terrestrial environments), looking in particular at primary production and how this is cycled within each habitat. The following section compares the effects of temperature and seasonality in the three environments; the final section looks at the possible reasons for the observed differences and the impact of man.

## 7.2. SEASONAL PATTERNS OF ENERGY AND NUTRIENT FLOW

### 7.2.1. The marine environment

It is convenient to start discussion of the marine environment with winter. This is because almost all biological activity occurs in the summer, and winter is a time when the environment stabilizes.

In winter sea ice forms, and this can reduce the area of open water in the Southern Ocean by as much as half, a reduction which has dramatic and far reaching effects. It alters the mean albedo of the polar regions (thus itself altering patterns of heat flow), it all but excludes atmosphere – seawater gas exchange (of fundamental importance to the bio-geochemical cycling of carbon dioxide), and it limits the availability of light for photosynthesis.

Primary production in the open ocean increases markedly as surface ice melts, but production is usually limited by water stability. Many studies have shown that phytoplankton production increases dramatically when water conditions remain stable; normally, however, the effect of wind-induced turbulence is sufficient to keep production well below the level where limitation by nutrients will occur. Indeed there is little direct evidence that marine primary productivity on the open ocean is ever limited by nutrients (see the chapter by El-Sayed). Nevertheless the observation that productivity is greater closer inshore and in areas of upwelling has been taken by some authors as indirect evidence of a nutrient limitation of Southern Ocean primary production (for example Everson, 1984). It is probable, however, that the effects of light, nutrient limitation, stability and grazing are complex and interactive, varying from area to area and making it unlikely that any single generalization will have widespread validity.

Despite these complexities, it is essentially true that once solar radiation increases in summer, sea ice melts and the summer pulse of primary productivity begins (see Fig. 2 in the chapter by El-Sayed). Two features of this pulse of carbon fixation are of fundamental significance to the marine ecosystem, its magnitude and its predictability. Both of these are well illustrated by the annual pattern of the standing crop of phytoplankton at an inshore site at Signy Island (Fig. 7.1). In three consecutive years the sharp rise in the amount of phytoplankton in the water column occurred within a period of only 10 days. For short periods of time the bloom can be so intense as to colour the sea, and a rain of particulate matter

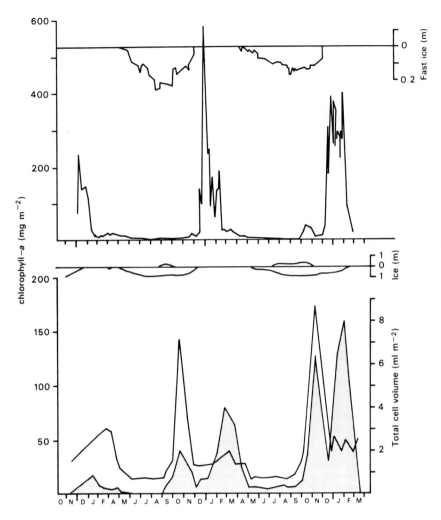

Fig. 7.1. The seasonal pattern of the standing stock of chlorophyll-*a* in two Antarctic aquatic habitats:
a) Marine; at Small Rock, Borge Bay, Signy Island during 1972 − 74. Water depth 16.5 m. Also shown is the duration and thickness of fast ice during the two winter periods. Reproduced, with permission, from Whitaker (1982).
b) Freshwater; Heywood Lake, Signy Island. Solid line is total water column chlorophyll-*a*, shaded area is total cell volume in same water column. Also shown is duration and thickness of surface ice, and superficial snow. Reproduced, with permission, from Light *et al.* (1981).

settles to the sea bed. Since all food webs in the sea lead back ultimately to primary production, many secondary production processes such as growth, reproduction and even feeding are also limited to summer, though the degree of seasonality lessens with the number of intervening links in the food web.

It is now clear that, as in other areas of the world, primary production in the Southern Ocean is very high only in inshore (neritic) areas. Whitaker (1982) measured annual carbon production as 86 g m$^{-2}$ a$^{-1}$ and 289 g m$^{-2}$ a$^{-1}$ in 1972/73 and 1973/74 respectively at an inshore site at Signy Island. These were estimated by $^{14}$C incorporation, and such year-to-year variation is typical of a neritic site. Over the Southern Ocean as a whole, mean production is probably much less, and a generally accepted figure at the moment is about 0.134 g m$^{-2}$ d$^{-1}$ (see the chapter by El-Sayed). Assuming the growing season to be 150 days this gives an annual carbon production of about 20 g m$^{-2}$ a$^{-1}$. If the area of Southern

Ocean is taken to be 32 million km$^2$, then annual production (as carbon) is 643.2 million tonnes (Holm-Hansen *et al.*, 1977). Phytoplankton are usually assumed to contain 10% of their fresh weight as carbon and so this is equivalent to a biomass production of about 6400 million tonnes per year. In Antarctic seas the plants involved are known to include diatoms, dinoflagellates and silicoflagellates, as well as several other groups. Until fairly recently the larger diatoms have attracted most attention, but recent studies from temperate and tropical seas have shown that very small photosynthetic organisms are considerably more important than had previously been thought (at this size it is not always easy to make clear-cut distinctions between animals and plants, since many organisms can utilize both heterotrophic and autotrophic nutrition). Whether this is also true in the Southern Ocean is as yet unknown, but in 24 stations in the Bellingshausen and Scotia Seas, von Bröckel (1981) showed that organisms passing through a 20 micron filter (nanoplankton) were responsible for about 70% of the total plant biomass and over 90% of the production (as measured by $^{14}$C incorporation).

One unusual habitat for phytoplankton is within the sea ice itself. In selected areas of infiltration ice blooms of the diatom *Navicula glaciei* can reach densities of 5.5 g m$^{-2}$, with an average annual carbon production in all suitable areas at Signy Island of $1-2$ g m$^{-2}$ a$^{-1}$ (Whitaker, 1977). The importance of sea ice phytoplankton in seeding the summer open water bloom has been discussed in the chapter by El-Sayed.

The Southern Ocean food web has often been described as simple, particularly in comparison with marine ecosystems elsewhere, as is typified by the often quoted food chain *Fragilariopsis antarctica* → *Euphausia superba* → baleen whales. In reality, the picture is more complex than this, and a representative food web for the Southern Ocean is shown in Fig. 7.2.

Some of the links in this web are conjectural, and in only a very few cases do we have any idea of either standing stocks or consumption. A value for annual primary production has been discussed above, and it has often been assumed that a major part of this is consumed by one organism alone, the Antarctic krill, *Euphausia superba*. It is now becoming clear, however, that krill probably do not feed exclusively on diatoms. Rather, they appear to have a feeding apparatus which allows the possibility of feeding on the much smaller nanoplankton and possibly even smaller organisms, as well as preying directly upon other herbivores such as copepods. In Fig. 7.2, therefore, several speculative arrows have been added describing these interactions.

A further complication is that the primary production is grazed by a variety of zooplankton other than krill (these have been grouped together in Fig. 7.2 as other zooplankton herbivores), and also by ciliate protozoans (especially tintinnids). These very small herbivores are major consumers of small phytoplankton. In Long Island Sound, for example, micro-zooplankton (predominantly tintinnids) removed up to 41% of the chlorophyll-*a* standing crop each day and at times consumed as much phytoplankton as copepods (Caprulio and Carpenter, 1980). Tintinnids were reported by the 'Discovery' Investigations but the quantitative importance of protozoa and bacteria in the Southern Ocean is largely unknown, although von Bröckel (1981) reported a mean protozoan biomass of 109 mg m$^{-2}$ (as carbon), and open ocean heterotrophic bacterial activity appears to be comparable with other areas of the world (Bölter and Dawson, 1982; Morita *et al.*, 1977). Ciliate protozoans are also active consumers of bacteria, as are the small early larval stages of crustaceans. At present, however, we have no quantitative idea of the importance of these relationships, although such an understanding is obviously vital to any complete picture of the interaction between krill and phytoplankton.

Despite so many uncertainties in the quantitative relationships between the various herbivores, it is true to say that the Southern Ocean food web differs strikingly from those described from more northern seas in the relative balance of the major types of primary consumer. In cold-temperate regions much of the primary production is grazed by copepods whereas in the Southern Ocean, although copepods are important, by far the single most important macro-zooplankton herbivore appears to be *Euphausia superba*. Unfortunately, despite much work neither the standing stock nor the annual

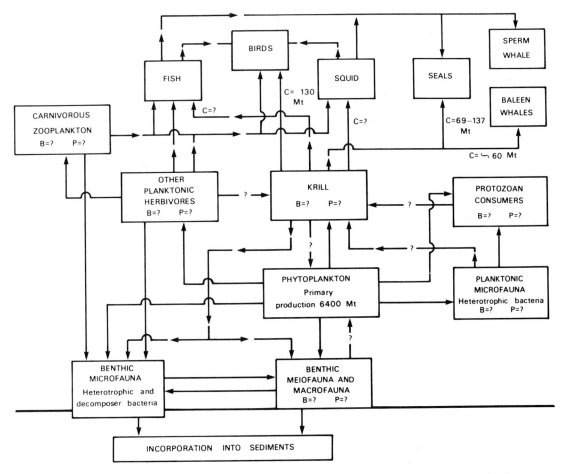

Fig. 7.2. Diagrammatic food web for the Southern Ocean. B, biomass; P, production; C, consumption. All units are in millions of tonnes (Mt) wet weight per year. Where figures are given for consumption of krill by seals the first assumes the seal biomass as in 1972, the second assumes a 7.5% annual increase in crabeater and leopard seal biomass to 1982 (Laws, 1984).

production of krill are accurately known; indeed we do not as yet have an agreed value for the duration of the life cycle. Some estimates of the food consumption of an individual krill have, however, recently been made. Clarke and Morris (1983) estimated that during the 190 day summer at South Georgia a mature male krill increases in fresh weight from about 0.4 g to about 1.6 g (based on the curve in Mauchline, 1980). Over this period the average energy intake calculated from growth, moulting and respiration data was about 5% body weight per day, a reasonable figure for a primary consumer.

The only data available for consumption by whole populations of organisms are for the higher predators of krill, and these are marked in Fig. 7.2. Everson (1984) has calculated that the currently accepted values for biomass and food consumption of seals and whales suggest conversion efficiencies from krill to seals, and from krill to whales, of about 0.6% and 1% respectively over the whole lifetime. These are lower values than might be expected, possibly because of the long life-span of these higher predators. In the younger stages conversion ratios are likely to be much higher, possibly as high as 30 − 40%.

The annual consumption values in Fig. 7.2 are for the whole of the Southern Ocean. The values for whales, seals and birds are believed to be reasonably accurate, the value for fish less reliable, and the value for squid little more than a guess. The best estimates of krill consumption by seabirds are available for dense colonies of seabirds at Bird Island, South Georgia. Croxall and Prince (1981) have estimated that breeding seabirds at South Georgia consume a minimum of 1.7 million tonnes of food during the summer, of which 88% is krill, 6% copepods, and about 3% each fish and squid. Over three quarters of this consumption is due to one species alone, the macaroni penguin, *Eudyptes chrysolophus*. Including estimates of the food consumption of non-breeders and making an allowance for the deposition of fat reserves for the post-breeding moult increases the total summer food consumption to 2.7 million tonnes. The summer predation of krill alone by birds at South Georgia is 10 − 20 million tonnes per summer and the peak removal rate of krill by all predators combined may be as high as 4 million tonnes per month. There is little information about the production of krill at South Georgia (indeed it is currently unclear whether the stock at South Georgia is self maintaining, or is continually replenished from further south), but an acoustic survey during December 1981 revealed a standing stock of only 0.5 million tonnes of krill around South Georgia. This discrepancy highlights one of the major difficulties in interpreting energy flow calculations, namely the varying nature of the interaction. The demand for food by the seabird population will vary throughout the summer, notably in relation to the age and food requirements of the chicks. The population of krill being fished will also vary, with krill arriving from further south and being removed by predators. The nature of the interaction between predator and prey is not fully understood, but it is clear that an unreplenished standing stock of 0.5 million tonnes cannot long sustain the removal of 4 million tonnes per month. Interestingly, Everson (1984) has noted that a suitable rate of replenishment of krill would be provided by a belt of water 100 m deep, 50 miles wide, moving at half a knot and containing an average density of only 1 krill per cubic metre.

The dependance of the breeding population of seabirds and fur seals, *Arctocephalus gazella*, at South Georgia on krill was illustrated dramatically in the austral summer 1977/78. In this season there was a natural shortage of krill around South Georgia, possibly because of a slight shift in the water currents. As a result the growth rates of fur seal pups were markedly reduced, and there was widespread chick failure in those birds which depend on krill to feed their growing chicks (for example, Black-browed albatross, *Diomedea melanophrys*, and gentoo penguin, *Pygoscelis papua*).

One conspicuous absence from Fig. 7.2 is any indication of consumption by man. This aspect of the Southern Ocean marine food web is very important, and is considered at the end of the chapter.

### 7.2.2. Freshwater environments

There is probably no such thing as a representative Antarctic lake or pool, since conditions vary so much from area to area. For example, lakes on sub-Antarctic islands are generally more complex than those further south, in some the surface freezes for only 3 − 4 months each winter, and the vegetation may include a few submerged vascular plants. At the other extreme meromictic lakes in the Dry Valleys of Southern Victoria Land and other continental Antarctic sites remain frozen to a considerable depth (several metres) all year round. These unusual lakes can contain quite warm bottom water, they may have saline layers at depth, and they generally show very low levels of primary productivity.

Between these two extremes are what may be regarded as the more typical maritime Antarctic lakes. These lakes are small (usually from about 1000 to 50,000 m$^2$ in area), often associated with areas of exposed rock, frequently close to the sea, and whose surface melts for at least part of the year. Even these lakes, however, show a greater range of nutrient status than most terrestrial or marine habitats, and no complete energy flow model has yet been put forward for an Antarctic lake. Nevertheless, it is possible to make some general comments about energy and nutrient flow in polar lakes, based mainly on work performed at Signy Island by the British Antarctic Survey.

When surface ice forms in autumn the water column stabilizes and most of the phytoplankton cells, deprived of adequate light, settle out to the bottom. Much of this material is decomposed by bacteria, both aerobic and anaerobic. The aerobic activity lowers the oxygen content of the water under the ice, whilst anaerobic activity (for example fermentation, nitrate and sulphate reduction) produces reduced inorganic and organic compounds. The resulting redox conditions permit a release of inorganic nutrients, notably phosphate, into the water column. These nutrients are of vital importance to the following spring phytoplankton bloom.

The annual pattern of primary production is shown for a representative lake of intermediate nutrient status (mesotrophic) in Fig. 7.1. As in the sea this pattern is both sharp and predictable. There is however, a notable difference in that there are two distinct peaks of chlorophyll-$a$. The first, and often major, peak in water column chlorophyll-$a$ occurs after superficial snow has disappeared, but well before the ice melts. This under-ice chlorophyll peak is a well described phenomenon from Arctic lakes, and is caused by an increase in the population of small chlorophyte and chrysophyte algae with a very high chlorophyll content per cell (which permits maximal use of the available radiation). This increase is fuelled by the products of mineralization during the winter, and hence does not occur in the more nutrient-poor (oligotrophic) lakes. In the latter lakes levels of inorganic nutrients in the water are low, and phosphorus limitation occurs very rapidly.

Although light is obviously a limiting factor in that growth of algae under ice cannot proceed until superficial snow has melted, the overall extent of the under ice peak is probably limited by available nutrients. Recent research has indicated that the first inorganic nutrient to be limiting (until the ice melts) is probably carbon dioxide. The decrease in the population of chlorophytes and chrysophytes is the result of grazing, loss of cells in the outflow (which is by now running), and nutrient limitation.

During the summer open water period incident light levels are much higher, wind induced turbulence destroys the water column stability and the phytoplankton population becomes dominated by cryptophyte algae. These have a much lower chlorophyll-$a$ content per cell than chlorophytes or chrysophytes, and as a result the open water period is characterized by a larger total cell volume and greater carbon fixation (as measured by $^{14}C$ incorporation) than the previous under-ice peak (Fig. 7.1). It is likely that the chlorophytes and chrysophytes are unable to compete with the more robust cryptophytes once water column stability has been disturbed.

The extent to which this primary production is grazed by herbivores is unknown, but grazing does appear to be important at times in the more productive eutrophic lakes. In oligotrophic lakes most of the small amount of primary production is utilized directly by the bacterial population (for example, Moss Lake at Signy Island: Ellis-Evans, 1982). Neither do we have any clear idea of the importance of sedimenting material as a food source for the benthic consumers, although wind-induced turbulence will keep much of the material in suspension in summer.

In some nutrient-poor upland lakes there are extensive growths of benthic plants. These are usually algal felts (filamentous blue-green algae and diatoms), but in some lakes there are also rich stands of epiphyte covered mosses. The productivity of these benthic communities can exceed that of the phytoplankton, but the productivity of the various individual components is as yet uncertain. Bacterial chemosynthetic processes occur only in the lake sediments and form a small proportion of the total autotrophic production; in Moss Lake they represent 2% of total production, and in the eutrophic Heywood Lake about 1% (Ellis-Evans, 1982). In Heywood Lake the total bacterial plankton increases sharply in October, before the ice cover melts, and are most productive during the summer open water months. This indicates a response by the bacteria to phytoplankton development, and recent work has demonstrated extensive nutrient exchange between these two groups during the summer.

Freshwater lakes contain a relatively small number of species of most typical small invertebrate groups, including protozoans, rotifers, tardigrades, nematodes, gastrotrichs, turbellarians, enchytraeids, annelids and ostracods. Although the details of these communities remain to be elucidated, they appear to be essentially opportunist herbivores which browse the benthic plant communities, especially the epiphytic algae (Priddle and Dartnall, 1978). One copepod and some larval stages of the anostracan *Branchinecta gainnii* migrate into the water column to graze the phytoplankton during summer, and there is one carnivorous copepod, *Parabroteas sarsi*.

The major features of these lakes have been combined into a generalized flow diagram for an ideal maritime Antarctic lake (Fig. 7.3). No figures have been applied since these will vary enormously from lake to lake. This diagram does, however, emphasize the importance of the bacterial populations, particularly those associated with the sediment, and also the importance of gaseous exchange. Certain continental Antarctic lakes, and those on sub-Antarctic islands will not be well described by the sort of scheme in Fig. 7.3, but both are discussed in the chapter by Priddle (Chapter 3c).

### 7.2.3. The terrestrial environment

The intensity of primary production in the terrestrial Antarctic environments varies from area to area. In particular there is a steady decrease towards the pole, and this is associated with a change in the relative importance of different plant types. In the lowlands of sub-Antarctic islands such as South Georgia (54°S) there are extensive swards of grass and other herbs, whereas at Signy island (61°S) in the maritime Antarctic almost all the primary production (leaving aside unicellular algae) is in the form of bryophytes and lichens, and further south on the Antarctic continent lichens become predominant.

Although the bulk of photosynthesis occurs once snow and ice have cleared and plants are exposed to the light, some activity does occur under winter snow and ice cover. In particular respiration continues, decreasing oxygen and increasing carbon dioxide levels beneath the ice. Once incident radiation increases with the onset of summer, up to 20% of the photosynthetically active radiation may penetrate thin snow cover, and photosynthesis can begin well before the snow melts. Indeed in the conditions of low oxygen and high carbon dioxide, photosynthesis can be stimulated to quite high rates. Eventually, photosynthesis may become limited by lack of carbon dioxide until the superficial ice melts and allows direct gaseous exchange with the atmosphere. Once free exchange between the plants and the atmosphere is possible, then it seems likely that primary production in the terrestrial environments is limited by the availability of nutrients and free water.

The fate of primary production in the terrestrial environment is dramatically different from that in both the marine and freshwater environments, as can be seen from the simplified food web for a Signy Island moss community in Fig. 7.4. The precise details of energy flow will differ between sites, but several broad generalizations can be made. The most important of these is that in comparison with the aquatic systems grazing of the primary production by herbivores is negligible, and most of the organic matter in the system is cycled by the decomposer community. The activity of this community, however, appears to be severely limited by temperature. The ground is frozen for 8 or 9 months of the year, and even when it has thawed, decomposer activity is frequently below the optimal rate which can be measured in the laboratory (water availability is frequently the limiting factor). As a result some areas of the Antarctic are characterized by accumulations of peat which may be as old as 4800 years (and in some sub-Antarctic areas as old as 11,000 years). The lack of herbivores is in striking contrast to the marine and freshwater environments, and it is of interest that in Arctic tundra ecosystems grazing by invertebrate herbivores is also minimal although the consumption by lemmings and other vertebrate herbivores is very important. In Antarctica there are no native vertebrate herbivores, and in areas where these have been introduced the result has been a drastic alteration in the structure of the plant communities (for example, the effects of reindeer, *Rangifer tarandus*, on the vegetation of South Georgia, as discussed in the chapter by Leader-Williams). It would seem that mosses make very poor food for invertebrates (Block, 1980).

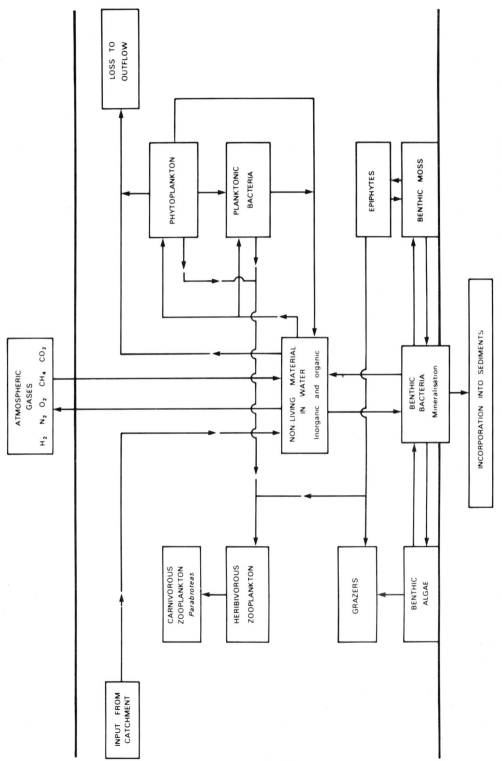

Fig. 7.3. Representative food web for a generalized Antarctic freshwater lake. From information supplied by Ellis-Evans, Hawes and Priddle (British Antarctic Survey).

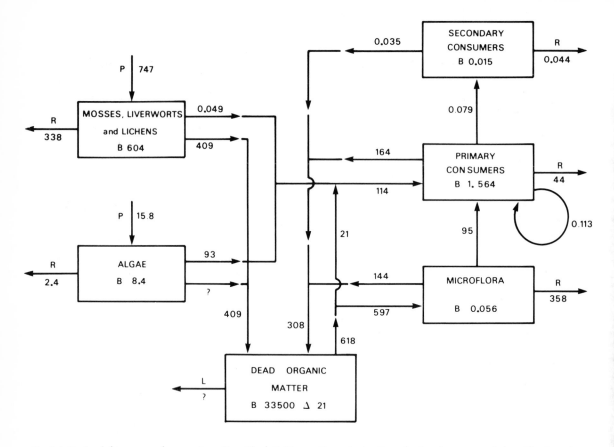

Fig. 7.4. Food web for a moss turf community at Signy Island. B, biomass; R, respiration; P, production; A, net accumulation. All units are in grams (dry wt) per m² per year. Redrawn, with permission, from Davis (1981).

The spring thaw also releases nutrients which are utilized by the microflora of bacteria, yeasts and fungi. Freeze — thaw damage to the older plant tissues releases small organic molecules such as sugars and amino acids, and with water now freely available there is a burst of respiratory activity in the microbial populations. The population then changes and community respiration decreases as the populations of testate amoebae and nematodes, creatures which eat microbes, increase (Wynn-Williams, 1983). Should a further freeze — thaw cycle occur during the summer, there is often no increase in microbial respiration similar.to that in spring. It is likely that this is due not to nutrient limitation but to a shortage of available free water. Once again the annual cycle of productivity is linked to the annual cycle of temperature, but in a subtly different manner.

Among the primary consumers protozoans are by far the most important organisms in terms of biomass and consumption of the microflora (Fig. 7.4). Their biomass is $1-2$ g dry wt m$^{-2}$, a typical value for terrestrial ecosystems, and their respiration exceeds that of any of the other invertebrate groups. Sarcodina were present in both communities examined (moss turf and moss carpet), but Mastigophora only in the moss turf. The other primary consumers of the microflora are rotifers, tardigrades, nematodes, mites and springtails. The secondary consumers are three carnivores, the nematode *Coomansus gerlachei*, the tardigrade *Macrobiotus furciger*, and the mite *Gamasellus racovitsai*.

## 7.2.4. Transfer of energy and nutrients between environments

Perhaps the most spectacular, certainly the most noticeable, evidence for the transfer of energy and nutrients between the various environments in polar regions is the fertilization of terrestrial habitats by nesting seabirds. The stimulation of plant growth in the area of seabird colonies was noted by the earliest travellers in polar regions, and can enable these colonies to be seen at a great distance.

Although seabirds have to spend the breeding season ashore they are essentially marine organisms. Most adults die at sea and are recycled within the marine ecosystem, either by decomposition or as food for other marine predators. However, while they are ashore they are contributing to the terrestrial environment through the excretion of waste products, loss of eggs, moulting of feathers, and the death of many chicks. The input of phosphorus and, particularly, nitrogen boosts levels of nutrients that are otherwise limiting, and promotes a luxuriant growth of plants. Such plant growth can give valuable clues to the existence of colonies of nocturnal burrowing petrels that might otherwise be missed.

So far this input has been quantified only on the sub-Antarctic Marion Island, where it has been estimated that about 90% of all nitrogen introduced annually into the vegetated areas comes from seabirds. Many of the nutrients deposited on land by seabirds are washed straight back to sea again, where they are important in stimulating primary production in shelf areas. The increased growth of vegetation also provides extra habitat for terrestrial invertebrate communities. These subjects are covered in detail in the chapter by Siegfried.

This example illustrates an important point about the nature of relationships between different environments. Although these are often expressed in terms of either biomass (for example Figs. 7.2 – 7.4) or energy, it is often not energy *per se* that is limiting in a particular environment, but some particular nutrient such as phosphorus, nitrogen or calcium. Expressions of inter-relationships merely as kJ flowing from one compartment to another may obscure important aspects of the relationship. For example at Marion Island there is evidence that in some areas it may be calcium that is limiting plant production; calcium may also be a limiting nutrient in the growth of the chicks of some seabirds.

Not only grasses and mosses but also lichens respond to nutrient input from seabirds. For example the distributions of the lichens *Caloplaca regalis* and *Xanthoria candelaria* at Signy island are clearly related to the presence of penguin colonies or cliff nesting petrels (Hooker, 1977). Prominent rocks used by skuas as look-out posts are also often characterized by these lichens.

Nutrients are also brought to land by seals, mainly as excreta and hair cast at moulting. Elephant seal wallows themselves support little life apart from some bacteria and dermatophagous fungi, although thallose algae such as *Prasiola* may grow around the edge. The outflow from these seal wallows and also bird colonies is, however, very important in governing the nutrient status of some freshwater lakes.

This last point highlights the intimate link between the terrestrial and freshwater habitats. In both cases the development of life depends on the inputs of both solar energy and nutrients (for the terrestrial environment nutrients will include water). No organism can survive in the absence of a range of macro- and micronutrients, and these nutrients have essentially two major sources, the underlying rocky substrate and the sea. Wind driven sea spray will carry marine salts long distances inland. For example the concentration of sodium in snow on the Ross Ice Shelf does not fall to an approximately stable value until about 400 km from the sea (Herron and Langway, 1979) and Holdgate *et al.* (1967) considered that all the soils of Signy Island were influenced by the proximity of the sea, for nowhere on the island is more than 1.5 km from the sea. Together with nutrients supplied by atmospheric fallout (such as sulphate and nitrate), even snow contains sufficient nutrients to support the growth of a community of microbial organisms. This community includes the snow algae *Chlamydomonas nivalis* which colours extensive areas of snow either red or green.

Meltwater from snow will contain a limited number of nutrients to supply freshwater lakes, but this nutrient input increases markedly once the run-off is leaching nutrients from the rock surface, and increases further still when the flowing water passes over vegetation. The nutrient status of a freshwater pool thus evolves with the catchment area of its freshwater supply (for a fuller discussion of this see the chapter by Priddle).

The extent to which the terrestrial environments supply nutrients to the sea is unclear, but the higher levels of primary production in inshore areas compared with the open ocean may reflect input of nutrients from the land.

## 7.3. THE EFFECTS OF TEMPERATURE

The two features more than any other serve to distinguish terrestrial environments from the marine and freshwater habitats in the Antarctic, namely fluctuations in temperature and variations in the availability of water.

The annual variation in seawater temperature is only about 6 degrees in sub-Antarctic waters, perhaps 3 or 4 degrees in the Scotia or Bellingshausen Seas, and at McMurdo Sound in the Ross Sea the temperature is perpetually within a single degree of $-1.8°C$. Freshwater lakes also tend to have stable temperature patterns. Water temperature is usually constant at about 0°C when the surface is frozen, and rises only to 5 or 6°C during the summer open-water period. Shallow shelf areas of the lakes obviously warm up more than the main water bodies, and in some continental lakes geothermal heating coupled with the insulating effect of a huge layer of permanent surface ice can result in quite warm bottom waters. In both marine and freshwater habitats the daily temperature variation is also usually small.

In contrast to the rather small annual variations in temperature experienced by organisms in the marine and freshwater environments, terrestrial habitats are subject to wide fluctuations. For example, the temperature recorded at the surface of a moss turf community at Signy Island (61°S) rose to over 30°C in summer, and fell below $-20°C$ in winter (Walton, 1982). Daily temperature variations can also be extensive, and in spring and autumn there may be rapid cycles of freezing and thawing (for example at two sites on Signy Island Walton recorded between 13 and 64 freeze/thaw cycles each year between 1972 and 1974). These cycles occur almost exclusively during the summer, for in winter the habitat is permanently below zero. These extremes of temperature are, however, not necessarily experienced by all organisms; many live within protective microclimates and inside the moss community the annual temperature cycle experienced by an arthropod such as the oribatid mite *Alaskozetes* may only be from about 5°C in summer to perhaps $-20°C$ in winter (Block, 1980).

The differing patterns of temperature variation between terrestrial and aquatic environments have two major physiological consequences. These are the need to avoid (or withstand) freezing, and the effects of temperature on metabolic rate.

### 7.3.1. Freezing

For the majority of organisms in the marine environment avoidance of freezing poses no real physiological problem. Most invertebrates are isosmotic or hyperosmotic to seawater, and so if the sea remains unfrozen then so do they. The only marine organisms which do run the risk of freezing are intertidal organisms exposed to very low air temperatures at low tide, such as the limpet *Nacella* (see chapter by Picken), those invertebrates encased within growing anchor ice, and fish. Sessile invertebrates frozen into anchor ice die, but limpets are able to withstand freezing temperatures, and even being trapped within ice. Tissue damage is avoided by allowing only extracellular water to freeze (withdrawal of water from the cells lowers their freezing point), and by encasing the body in a covering of mucus which possibly acts to prevent any contact with ice crystals (Kanwisher, 1966).

Fish differ from marine invertebrates in having tissue fluids that are less concentrated than seawater (hyposmotic). The blood serum freezing point of a typical temperate marine fish is about $-0.7°C$, significantly above the temperature of much of the Southern Ocean. Antarctic fish avoid freezing by the production of a series of protein or glycoprotein antifreeze molecules which lower the freezing point of the blood serum and other tissue fluids to about $-2°C$. In general, the more likely any species of fish is to come into contact with ice, the more antifreeze it contains. Thus *Pagothenia borchgrevinki* which feeds around the ice platelets under fast ice contains the highest concentration of serum antifreeze, whereas the liparid *Paraliparus devriesii* which lives in deep water in McMurdo Sound and probably never comes into contact with a seeding ice crystal, has no antifreeze at all and lives permanently supercooled. If one of these liparid fish is brought to the surface, contact with ice crystals in the surface waters can cause it to freeze (DeVries, 1980).

Invertebrates in freshwater lakes also have few problems with freezing as long as the lake itself does not freeze solid. Usually this does not occur, although ice may form along areas of shallow shelf. The overwintering of the anostracan *Branchinecta* as an egg, and the encysted overwintering stages of many of the freshwater organisms are more likely a response to the increasingly anoxic conditions and the low food availability, than a strategy for avoiding freezing.

By contrast, the need to avoid freezing is of paramount importance in the terrestrial environment. This is true not only in arthropods (see Chapter 3b by Sømme), but also for the many microscopic organisms which live in the moss communities. Thus the algae, fungi, bacteria, protozoa, tardigrades, and nematodes must all be able to withstand freezing temperatures, as indeed must the mosses and higher plants themselves. The mechanisms by which higher plants withstand freezing are not well understood (despite the commercial importance of frost damage), but this together with the ability of some unicellular algae and protozoa to survive freezing temperature are under active experimental investigation. It is fair to say, however, that at present there is no general theory of how organisms avoid freezing injury.

It is possible that many of the small organisms utilize resistant stages such as spores to survive the winter period, and these stages often involve some initial dehydration of the tissue. Encystment is important in unicellular algae, protozoa, rotifers and tardigrades, and in nematodes survival at very low temperatures depends upon extensive dehydration of the body although there is no encystment. Some field measurements have suggested that even at quite low temperatures ($-10°C$ for example) a significant proportion of soil water may remain unfrozen. The precise location of this water is unknown, but it is possible that if it is organized into layers around soil particles some bacteria may be able to survive in this undercooled water without encystment.

The avoidance of freezing by polar organisms has been most intensively studied in arthropods. In the majority of polar micro-arthropods freezing is lethal, and nucleation of the body fluids is avoided by a combination of the production of polyhydroxy compounds such as glycerol, and ridding the gut of any nucleating agents which might initiate freezing. The high levels of polyols in the tissue fluids depress their freezing point, and also decrease the temperature at which spontaneous (usually heterogenous) nucleation occurs. Furthermore in several polar insects proteins have now been detected which appear to interfere with the process of nucleation in a manner analogous to that of antifreeze proteins in fish. Only one Antarctic arthropod, the larva of the wingless midge *Belgica antarctica*, appears to be tolerant of freezing.

Tolerance of freezing appears to be related to an ability to withstand high osmotic gradients across cellular membranes and a low body water content. Interestingly, there is evidence that in micro-arthropods a low environmental relative humidity acts as a stimulus for the production of those polyols active in promoting the ability to supercool (despite the prevalence of snow and ice, at near freezing temperatures the relative humidity of terrestrial micro-habitats can be quite low). Some polar mosses have also been shown to be tolerant of extensive tissue desiccation (see chapter by Longton, 3b), but surprisingly little is known about how polar plants avoid freezing.

## 7.3.2. Temperature and physiology

Related to the differences in annual and diurnal temperature fluctuations between terrestrial and aquatic habitats are differences in the physiology of the organisms which live in these environments.

Most aquatic organisms in the Antarctic are subject to only small variations in temperature, and this has allowed the evolution of a physiology that is finely tuned to the temperature. In particular, the basal rates of metabolism (that is the energy cost of merely staying alive, separate from any costs of growth, reproduction, feeding or activity) are very low. As a result marine organisms can make much more efficient use of the food they capture than related organisms from warmer water, since for any given amount of food less goes in maintenance, and more into useful areas such as growth or reproduction.

With the exception of a few species of introduced mammals, no terrestrial Antarctic organisms are endothermic. All, therefore, have physiologies which are subject to major changes in temperature, both diurnally and seasonally. If a reasonable approximation to the annual temperature range experienced by a typical terrestrial micro-arthropod such as a mite is taken to be 10°C to − 20°C, then assuming there is no compensation for temperature in the mite's physiology, many of the energetic processes will operate about 15 times faster in summer than in winter (the $Q_{10}$ for many biological processes in unstressed organisms is usually between 2 and 3). It might be expected that this enormous difference would be offset by a seasonal adjustment of the organism's physiology. Such seasonal acclimatization has been widely reported from temperate organisms, but has not yet been investigated in Antarctic species. Even more of a problem is that the daily variation in temperature may be as much as 20 degrees. How Antarctic terrestrial organisms cope with this variation is not understood. However it seems that unlike marine invertebrates, whose basal metabolic rate is very low (see above), terrestrial invertebrates have fairly high basal metabolic rates (Block, 1980). The reason for this is not clear, and indeed it would seem to be energetically disadvantageous to increase basal rate, for any such increase must be paid for with energy that cannot then be used for useful processes such as growth. Nevertheless the data do suggest an elevated metabolic rate, and it must be concluded that in some manner as yet unknown the ability to tolerate wide diurnal fluctuations in temperature superimposed on a generally low mean temperature necessitates a relatively high basal metabolic rate.

Despite much research on the effects of temperature on enzyme activity, we still have no general picture of how organisms cope at the cellular level with alterations in temperature. Increases in the concentration of enzymes, synthesis of new enzyme variants and the alteration of the lipid environment of membrane-bound enzymes have all been demonstrated to be of importance in temperature adaptation in one organism or another, but perhaps the best candidate for a general mechanism which will at least maintain enzyme activity in the face of the effects of a decreased cell temperature is regulation of the tissue fluid pH (Clarke, 1983).

## 7.3.3. Water availability

In order to survive, all living creatures must exchange water and salts with their environment. The only exceptions are some resting stages and those organisms such as nematodes able to withstand anhydrobiosis; in these cases, however, metabolism has all but ceased. For aquatic organisms such exchange may pose energetic problems but there is obviously no problem of water availability. This is not so for terrestrial organisms where, perhaps surprisingly in view of the prevalence of snow and ice, water availability can be a major environmental limitation.

The smaller organisms of the microflora and microfauna live permanently in water layers associated with the vegetation or soil particles, and probably encyst when temperatures drop and the water freezes. For larger organisms, however, loss of water from the tissues by evaporation or during respiration can be a major hazard at certain times of the year. Thus the outer surface structures of many plants and animals show features characteristic of desert organisms, including waxy coatings and impervious sculptured cuticles. In the Dry Valleys of Victoria Land humidity is so low that virtually no organisms can survive, despite the lack of ice cover and the continuous daylight in summer. The most notable exceptions are the communities of endolithic organisms (algae, lichens and cyanobacteria) found growing within rocks. Here they can derive salts from the slow weathering of the rocks, and utilize what little water there is trapped between the mineral crystals.

For an organism living in polar regions there is nothing special about the temperature of 0°C, other than that below this temperature water starts to be locked up in the form of ice. As long as tissues do not freeze and gas exchange is possible, then metabolism will continue well below zero. In winter it is more the lack of free water than the low temperature *per se* which forces organisms to enter resting stages. This is well illustrated by the effect of freeze/thaw cycles on microbial respiration already discussed. When liquid water is present in abundance (spring) nutrients released by freeze/thaw damage can be rapidly utilized; in summer the lack of liquid water severely limits such activity. Details of the interaction of water availability with temperature and other environmental variables and their importance in the ecology of plant communities are given in the chapter by Longton.

## 7.4. SEASONALITY

It is probably fair to say that in Antarctica the most important seasonable variable, and ultimately the driving force for all seasonality in polar regions, is the annual variation in solar radiation. At high latitudes this variation is extreme in that incident radiation is zero (or almost so) for long periods of the year, when the land and sea are in complete darkness. Further north this variation is much less extreme, and there is some solar radiation at all times of the year. This variation in radiation has two important consequences. Firstly it governs the maximum amount of energy available for photosynthesis, and secondly it regulates ambient temperature.

The primary production in all three major ecosystems in the Antarctic is intensely seasonal, and this seasonality presents consumer organisms with a variety of problems. Since food is available for only limited periods, they must confine all tissue growth and reproductive material to that period when food is plentiful, and they must also survive the (sometimes long) periods when food is scarce or almost non-existent.

In the marine environment most herbivores can probably feed only during the summer phytoplankton bloom. In order to survive the winter they must either synthesize a store of reserves during the summer, switch to another source of food in winter, or reduce their winter metabolic demands to survive despite the lack of both food and reserve of energy. The herbivorous euphausiid *Euphausia crystallorophias* synthesizes a lipid store during the summer phytoplankton bloom at McMurdo Sound. During the winter under the fast ice the euphausiids cannot feed, and they use the lipid reserves to fuel both maintenance and the synthesis of eggs. Total lipid content thus decreases steadily during the winter. As soon as summer returns and phytoplankton food is once more available, the eggs are released into the plankton, and the adults can feed again and replenish their lipid reserves. The utilization of stored lipid

AN-W

for production of eggs during the winter is critical, since it allows the adult to release these eggs in time for the newly-hatched larvae themselves to make use of the phytoplankton bloom for food.

A different pattern is shown by the Antarctic krill *Euphausia superba*. In this species most of the production of lipid for eggs takes place during the early part of the summer, and most eggs are released in January/February though still in time for the larvae to feed on the late summer phytoplankton bloom. *E. superba*, however, does not appear to synthesize a lipid store for winter, and so this species must either switch to a more omnivorous diet in winter (perhaps predation of smaller herbivores such as copepods, or perhaps feeding on very small organisms such as bacteria), or survive by utilizing body material, as has recently been suggested (Ikeda and Dixon, 1982).

Benthic particulate feeders depend for their food on the primary production in the water column above them settling out. How these organisms survive the winter is not yet known. What evidence there is suggests that they do not synthesize a lipid reserve for winter, and it must be supposed that they therefore starve. Since benthic marine invertebrates have very low basal metabolic rates, in the absence of any growth or reproductive activity energetic requirements will be very low. Indeed Russian workers have reported that the metabolic rates of these benthic organisms in winter are so low that they have been described as overwintering in a state of suspended animation or winter dormancy.

For organisms higher up the food web and eating animal rather than plant food, energy intake will be less markedly seasonal. Nevertheless the benthic scavenging shrimp *Chorismus antarcticus* does show a clear seasonal pattern of growth, and most of the production of egg yolk occurs in summer. The eggs are large, and development rates therefore slow (Clarke, 1982). In consequence, female shrimps must feed to produce the eggs in one summer, spawn in autumn and then brood the eggs over the winter; the following spring the larvae hatch in an advanced state of development in time to utilize the spring increase in food availability. This pattern of large eggs hatching advanced larvae is very common in the polar benthos (see the chapter by Picken, 4b). In the isopod *Serolis cornuta* brooding of eggs to maturity takes almost 18 months and so successive broods can be produced only every two years. Other scavenging forms are less tied to the annual cycle of production and breed year round (for example the large isopod *Glyptonotus antarcticus*, although here again the eggs are large and take a long time to develop).

A parallel may be drawn here with the long fledging periods of many Antarctic seabirds. In at least one species, the Wandering Albatross *Diomedea exulans*, chick development takes over a year so that females breed only every other year. The Grey-headed Albatross, *D. chysostoma* also breeds every two years, and king penguins, *Aptenodytes patagonica*, at South Georgia in only two out of every three years.

It has been argued (for example Southwood, 1977) that when an environment is either stable, or variations in the major environmental parameters are predictable, then selection will favour the development of communities of organisms with very efficient life-styles. In the benthic marine environment the major seasonable variable, primary production, is very predictable. Coupled with the efficient use of food made possible by the low basal metabolic rate (itself a consequence of the low temperature), the result has been the widespread evolution of typically slow-growing very efficient life-styles (for a fuller discussion of this see Clarke, 1983).

Less is known about the physiological ecology of the zooplankton. Growth rates are seasonal in those species so far studied, and among herbivorous species large lipid stores are frequent. This may be because the metabolic cost of staying in the water column is sufficiently great that when feeding is not possible, some form of energy reserve is necessary.

In many planktonic species the lipid store is wax ester rather than the more usual triglyceride fat; this is particularly so in many copepods and euphausiids. Thus many herbivorous copepods living in high latitudes where their phytoplankton food is plentiful only for a short period of the year, synthesize an enormous wax ester reserve as a late (stage IV and V) copepodid. This wax store can form as much as 80% of the total organic matter in the animal, and is used up during maturation to the adult stage and production of eggs. Benthic invertebrates which are either subject to a less strictly seasonal supply of food, or which can overwinter successfully by starvation and ceasing all but maintenance metabolism, rarely synthesize wax esters (Clarke, 1983).

Primary production is also strongly seasonal in the freshwater lakes, and only during the summer months is there sufficient material in the water column to act as food for particulate feeding organisms (Fig. 7.1). Most of the algae pass the winter in an encysted form, inactive. Organisms which feed on the algae must therefore also themselves spend the winter either as a non-feeding resting stage, or switch to alternative foods. The anostracan *Branchinecta* overwinters as an egg (no adults survive over winter), and the newly hatched larvae the next spring move up into the water column to feed on the phytoplankton bloom. The herbivorous copepod *Pseudoboeckella* however, has a less synchronous breeding cycle (ovigerous females may be found at all times of year) and a variety of different stages of this copepod therefore pass the winter, probably feeding on the benthic microflora and microfauna.

The water column under the winter ice can become severely anoxic in winter, due to the activity of bacteria. The lack of oxygen together with the low availability of food means that many of the microfauna also probably overwinter as resting cysts or spores. The effects of seasonality in the freshwater lakes is modified to an extent by the nutrient status of the lake, but the overall pattern remains similar from lake to lake.

Although terrestrial plants may have the option of overwintering as an inactive spore or seed, they must also attempt to survive in a vegetative form. In winter photosynthesis is reduced and, like animals, the plants must rely on stored reserves to survive. In many sub-Antarctic plants an important fraction of the products of summer photosynthesis are stored immediately, often in shoot bases, roots or rhizomes, rather than converted to new green tissue. These reserves allow the plant to overwinter, and even more important to produce new photosynthetic tissue early in the spring to make maximal use of the early summer radiation. Some plants will photosynthesize actively under ice as long as there is sufficient light and carbon dioxide. This tendency to spread the products of photosynthesis in time is reflected in the fact that there are no annual species on sub-Antarctic islands, many of the grasses are wintergreen, and preformation of flowers occurs.

For example, in the tussock grass *Poa flabellata* at South Georgia the major storage product is fructan (or fructosan, a polymer of the sugar fructose). In leaves the fructan content ranges from 2% to 40% of the leaf dry weight, increasing with leaf age. The shoot contains very high levels (58 – 73%), and the rhizome 11 to 32%. These are not inactive stores, but are continually being used and replaced, with the size of the fructan molecule (from 3 to 12 fructose units linked together) varying from tissue to tissue and also with the season. The total size of the fructan store enables the plant to survive two consecutive poor growing seasons, and is coupled with a fairly low seed output (Gunn and Walton, in press).

A low seed output and a slow growth rate, with much of the new plant growth being produced vegetatively, are widespread strategies in Antarctic plants. Variations in the balance of sexual to vegetative production also occur within the same species growing at different sites. Thus at South Georgia *Phleum alpinum* (one of very few truly bipolar species) reproduces predominantly vegetatively, and the amount of seed produced varies with the severity of the habitat, being highest in enriched lowland sites and lowest on exposed fell-field sites. Overall, sexual reproduction is low in comparison with Arctic and alpine regions (Callaghan and Lewis, 1971).

In many species the seeds are produced at the end of the growing season, and overwinter before germinating the following summer. Seeds of the grass *Festuca contracta* at South Georgia have a requirement for a post-drop ripening period. This ensures that the seeds are unlikely to germinate during transient mild periods during the winter, when the resultant young seedlings would be killed either by the return of winter frost conditions, or by desiccation as a result of frost heaving. In many areas soil instability resulting from the frequent freezing and thawing of the ground is a major environmental factor. Not only is the substrate in which plants are rooted often mobile, but frost heaving renders the chances of new seedlings becoming established very low. Coupled with a low seed output this makes the problem of dispersal for the plant difficult, and many sub-Antarctic plant seeds are specially adapted for long distance dispersal (on the plumage of birds, for example). A parallel may be drawn here with the tendency for many benthic marine invertebrates to produce small numbers of large eggs which are often brooded by the female. The absence of the free swimming planktonic larvae which are so characteristic of northern temperate waters means that dispersal is difficult, and as a result there is a high degree of endemism (see the chapter by Picken, 4b).

Although there is no evidence of a greatly-increased seed size in polar plants (compare the egg size of many marine invertebrates), there are strong analogies in the life history strategies of these plants and aquatic animals. In both cases selection has produced a community of organisms that grow slowly and efficiently, and which utilize sometimes large reserves of stored material for overcoming periods of low energy intake. This would suggest that, as has been argued for marine invertebrates and fish, the slow growth of polar plants is an adaptation or strategy, and not due to direct limitation by the low temperature. It would therefore be expected that the enzymes of these polar plants would be found to operate efficiently when examined in the laboratory, and not to be temperature limited.

In the maritime Antarctic the flora is dominated by cryptogams rather than flowering plants and grasses. Unfortunately, little appears to be known about the chemical composition or storage products of Antarctic mosses. Radiotracer studies do, however, show that as in higher plants a large proportion of the products of summer photosynthesis stored in the older plant tissues above the peat layer. As with flowering plants the incidence of sexual reproduction (fruiting bodies) is generally low, and decreases in higher latitudes. Often fruiting bodies are produced only in good summers. The growth form of many mosses tends to trap water between the stems, and this forms a valuable microhabitat for smaller organisms. When this water freezes in winter the microflora (unicellular algae, bacteria and yeasts) probably overwinter as spores or cysts, as do most of the microfauna. The microhabitats in moss carpets under the winter snow may thaw at intervals throughout the winter, and when this occurs the larger fauna (mites and springtails) may even be seen feeding. Food particles in the gut would seem likely to act as nucleating agents for ice formation when temperatures drop once more, and this would be fatal, but this period of the life history and ecology of the larger invertebrate fauna is little studied.

## 7.5. PAST HISTORY AND FUTURE PROSPECTS

Several differences between the three major habitats in Antarctica have already been described, but one major distinction which has not been discussed is the degree of richness of the flora and fauna. Contrary to many early reports it is now clear that the Antarctic marine fauna can be every bit as rich in species and dense in biomass as marine communities elsewhere. In one area of McMurdo Sound, for example, infaunal biomass can reach densities as great as anywhere in the world (Dayton and Oliver, 1977). This richness is in stark contrast to Antarctic terrestrial and freshwater habitats where species diversity is low.

The likely reason for this difference is not the harshness of the climate, but the length of time these communities have had to evolve. The mean temperature of the seas has decreased by about 11°C since the beginning of the Tertiary (roughly 50 million years ago). The rate of cooling can be determined from oxygen isotope studies of the foraminifera in ocean sediment cores and has been fairly gradual apart from three periods of rapid cooling, the most recent being about 4/5 million years ago. Despite these occasional rapid drops in temperature marine organisms have probably been able to evolve gradually to keep pace with the changing temperature (Clarke, 1983). Most of the marine fauna has thus probably been able to evolve *in situ*, and a nice example is provided by fossil echinoids from Australia and nearby Antarctica. Here it is possible to see the switch from species which release their eggs directly into the sea, to species which brood their eggs and young. This is an ecological change, and one that is usually taken to indicate the appearance of cool water. It most likely marks the onset of the circum-Antarctic current (Foster, 1974).

On land, however, both the strictly terrestrial habitats and the freshwater pools have been in existence for only a relatively short period of time. At South Georgia, for example, the ice sheet has probably left terrestrial habitats open for colonization for only about 12,000 years (Sugden and Clapperton, 1977). During the periods of maximum ice cover there would have been very few pockets of surviving flora and fauna to act as later centres of dispersal and most of those organisms now established must therefore have arrived from outside. The scattered distribution and limited extent of Antarctic terrestrial and freshwater habitats means that colonization rates will inevitably be slow, and they are likely to be even lower than in a comparable warmer area, for two reasons. The first is that those organisms whose life-style makes them liable to succeed in a newly-exposed Antarctic habitat have low dispersal rates (see previous section). Secondly, and probably more importantly, the physical nature of the newly-exposed ground is not conducive to the easy establishment of new plants. In particular, the mobile surface and the low water availability makes for a particularly hostile environment for colonizing plants and under natural circumstances terrestrial and freshwater habitats are thus likely to remain poor in species for some time. However, a new agent of dispersal is now active, namely man. Walton (1975) for example lists 32 species of alien vascular plant introduced to sub-Antarctic islands by the activities of man, all of which have persisted or become naturalized.

## 7.5.1. Pollution

Man has had a decisive impact on the Antarctic ecosystem since it was first explored. Seals, fur seals and whales have all been heavily exploited, fish and krill are being harvested now, and many alien organisms have been introduced. (Previous exploitation and introduced organisms are discussed in the chapters by Sage and Leader-Williams respectively, 8 and 6b). Although the deliberate introduction of alien organisms to the Antarctic Treaty area is now carefully controlled, some inadvertent introductions are still likely. Increasing numbers of tourists and scientists visiting Antarctica will inevitably bring alien species with them, particularly micro-organisms. Such introductions are unlikely to have the drastic effects of measles or smallpox introduced to island native peoples, but alien micro-organisms will pollute terrestrial habitats and are a potential hazard to colonies of mammals or birds.

Increased numbers of people will also damage many delicate habitats. Footsteps across a moss bank may take years to disappear, and the unthinking collection of a lichen (by scientist as well as tourist) can remove 200 years of slow growth at one fell swoop. Tourist pressure at bird colonies can also have damaging effects by upsetting breeding behaviour, but it is chemical pollution which probably poses the greatest threat.

Although early studies detected pesticide residues in some Antarctic vertebrates, what few results there are to date suggest that the Antarctic is still relatively free of contamination by pollutants (see the chapter by Sage). Should exploitation of minerals start, then the situation will likely change. Unfortunately cold water slows the rate of microbial degradation of crude oil, and the slow turnover of many of the invertebrate communities means that these will not easily recover from a major spillage of oil. It seems probable that exploitation of Antarctica's mineral resources will occur eventually. It would make good sense to know the background levels of, for example, petroleum hydrocarbons before exploitation starts, rather than after. Knowing what levels of such compounds are present naturally makes it easier to detect chronic pollution when it occurs.

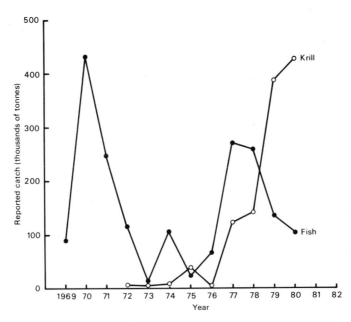

Fig. 7.5. Reported catches of krill and fish (in thousands of tonnes) in the Southern Ocean.
● = total fish caught in FAO areas 41 and 51.
o = total krill caught in waters south of the Antarctic Convergence.
The year 1969 refers to the austral summer 1969/70, and so on. All data from Everson (1978) and FAO Yearbook of Fisheries Statistics (FAO, 1981).

## 7.5.2. Fishing

Although exploitation of seals and seabirds has now ceased, and the taking of whales almost so, commercial harvesting of fish and krill is still important. In 1970/71 eastern block fleets removed over 400,000 tonnes of fish in one season, mostly from the waters around South Georgia. There are clear indications that this rate of removal was more than the stock could sustain, and in subsequent years catches were smaller, and were mostly taken from around Kerguelen. Since 1975/76, fish catches have increased again and South Georgia has once more become an important fishing ground (Fig. 7.5). The fleets are now also turning their attention to krill, and in 1980 over 400,000 tonnes were removed. Current estimates of krill abundance, and of sustainable yields, vary widely, and despite the research into krill biology currently being undertaken by many countries it looks as though once again exploitation may outstrip knowledge. The Southern Ocean food web is complex and meaningful figures for population parameters can only be guessed at until many of the links in the food web are better understood (Fig. 7.2). For example, a currently-accepted general figure for Southern Ocean primary productivity is 6400 million tonnes per year (Holm-Hansen *et al.*, 1977). This figure has been used to estimate how many krill might be sustained by the annual production of phytoplankton, but until we know whether this figure includes all the production by very small photosynthetic organisms, what proportion of this production passes through the micro-zooplankton, and how much is consumed by the larger zooplankton other than krill (salps, for example, are probably very important), such a figure has only limited usefulness. The modelling of complex ecosystems is a difficult procedure, and even in well-studied systems (such as the Scotian shelf in the western North Atlantic: Mills, 1980) not all the figures can be persuaded to balance.

There is an obvious need for more research, and it is to be hoped that the world political and economic climates will allow such research to proceed before one of the world's last relatively-unexploited areas is damaged beyond repair.

## ACKNOWLEDGEMENTS

Writing a chapter such as this would not be possible without the help of many people. I would particularly like to acknowledge the help provided by Cynan Ellis-Evans, Ian Hawes and Julian Priddle with the section on freshwater environments, and useful discussions with Bill Block and David Walton (terrestrial environments), and Inigo Everson (marine environments and fisheries statistics).

## REFERENCES

Block, W. (1980) Survival strategies in polar terrestrial arthropods. *Biol J. Linn. Soc.* 14, 29 – 38.

Bölter, M. and Dawson, R. (1982) Heterotrophic utilization of biochemical compounds in Antarctic waters. *Neth. J. Sea Res.* 16, 315 – 32.

von Bröckel, K. (1981) The importance of nanoplankton within the pelagic Antarctic ecosystem. *Kieler Meeresforsch.* 5, 61 – 7.

Callaghan, T.V. and Lewis, M.C. (1971) The growth of *Phleum alpinum* L. in contrasting habitats at a sub-Antarctic station. *New Phytol.* 70, 1143 – 54.

Caprulio, G.M. and Carpenter, E.J. (1980) Grazing by 35 to 202 μm micro-zooplankton in Long Island Sound. *Mar. Biol. (Berl.)* 56, 319 – 26.

Clarke, A. (1982) Temperature and embryonic development in polar marine invertebrates. *Int. J. Invert. Repr.* 5, 71 – 82.

Clarke, A. (1983) Life in cold water: the physiological ecology of polar marine ectotherms. *Oceanogr. Mar. Biol. Ann. Rev.* 21, 341 – 453.

Clarke, A. and Morris, D.J. (1983) Towards an energy budget for krill: the biochemistry and physiology of *Euphausia superba* Dana. *Polar Biol.* 2, 69 – 86.

Croxall, J.P. and Prince, P.A. (1981) A preliminary assessment of the impact of seabirds on marine resources at South Georgia. *Colloque sur les Ecosystémes SubAntarctiques. 1981, Paimpont, C.N.F.R.A.*, no. 51, 501 – 9.

Davis, R.C. (1981) Structure and function of two Antarctic terrestrial moss communities. *Ecol. Monogr.* 51, 125 – 43.

Dayton, P.K. and Oliver, J.S. (1977) Antarctic soft-bottom benthos in oligotrophic and entrophic environments. *Science* 197, 55 – 8.

De Vries, A.L. (1980) Biological antifreezes and survival in freezing environments. In *Animals and Environmental Fitness*, Ed. R. Gilles, pp. 583 – 607. Pergamon, Oxford.

Ellis-Evans, J.C. (1982) Seasonal microbial activity in Antarctic freshwater lake sediments. *Polar Biol.* 1, 129 – 40.

Everson, I. (1978) Antarctic fisheries. *Polar Rec.* 19, 233 – 57.

Everson, I. (1984) Zooplankton. In *Antarctic Ecology*, Ed. R.M. Laws, pp. 463 – 90. Academic Press, London.

FAO (1981) Yearbook of Fisheries Statistics, 1980. 50. Food and Agriculture Organisation of the United Nations, Rome.

Gunn, T.C. and Walton, D.W.H. (in press) Storage carbohydrate production and overwintering strategy in a winter-green tussock grass on South Georgia (sub-Antarctic). *Polar Biol.*

Herron, M.M. and Langway, C.C. (1979) Dating of Ross Ice shelf coves by chemical analysis. *J. Glaciol.* 24, 345 – 57.

Holdgate, M.W. (1967) The Antarctic ecosystem. *Phil. Trans. Roy. Soc., Lond, B* 252, 363 – 83.

Holdgate, M.W., Allen, S.E. and Chambers, M.J.G. (1967) A preliminary investigation of the soils of Signy Island, South Orkney Islands. *Bull. Br. Antarct. Surv.* 12, 53 – 71.

Holm-Hansen, O., El-Sayed, S.Z., Franceschini, G.A. and Cuhel, R.L. (1977) Primary production and the factors controlling phytoplankton growth in the Southern Ocean. In *Adaptations within Antarctic Ecosystems*, Ed. E.A. Llano, pp. 11 – 50. The Smithsonian Institution, Washington D.C.

Hooker, T.N. (1977) *The growth and physiology of Antarctic lichens*. Ph.D. thesis, University of Bristol.

Ikeda, T. and Dixon, P. (1982) Body shrinkage as a possible over-wintering mechanism of the Antarctic krill, *Euphansia superba* Dana. *J. exp. mar. Biol. Ecol.* 62, 143 – 151.

Kanwisher, J.W. (1966) Freezing in intertidal animals. In *Cryobiology*, Ed. H.T. Meryman, pp. 487 – 94. Academic Press, London.

Laws, R.M. (1984) Seals. In *Antarctic Ecology*, Ed. R.M. Laws, pp. 621 – 715. Academic Press, London.

Light, J.J., Ellis-Evans, J.C. and Priddle, J. (1981) Phytoplankton ecology in an Antarctic lake. *Freshwat. Biol.* 11, 11 – 26.

Mauchline, J. (1980) The biology of mysids and euphausiids. *Adv. Mar. Biol.* 18, 1 – 681.

Mills, E.L. (1980) The structure and dynamics of shelf and slope ecosystems off the north east coast of North America. In *Marine Benthic Dynamics*, Eds K.R. Tenore and B.C. Coull, pp. 25 – 47. Belle W. Baruch Library in Marine Science, University of South Carolina Press.

Morita, R.Y., Griffiths, R.P. and Hayasaka, S.S. (1977) Heterotrophic activity of micro-organisms in Antarctic waters. In *Adaptations within Antarctic Ecosystems*, Ed. E.A. Llano, pp. 99 – 113. The Smithsonian Institution, Washington D.C.

Priddle, J. and Dartnall, H.J.E. (1978) The biology of an Antarctic aquatic moss community. *Freshwat. Biol.* 8, 469 – 80.

Southwood, T.R.E. (1977) Habitat, the templet for ecological strategies? *J. anim. Ecol.* 46, 337 – 65.

Sugden, D.E. and Clapperton, C.M. (1977) The maximum ice extent on island groups in the Scotia Sea, Antarctica. *Quat. Res.* 7, 268 – 82.

Walton, D.W.H. (1975) European weeds and other alien species in the sub-Antarctic. *Weed Res.* 15, 271 – 82.

Walton, D.W.H. (1982) The Signy Island terrestrial reference sites: XV, Micro-climate monitoring. *Bull. Br. Antarct. Surv.* 55, 111 – 26.

Whitaker, T.M. (1977) Sea ice habitats of Signy island (South Orkneys) and their primary productivity. In *Adaptations within Antarctic Ecosystems*, Ed. G.A. Llano, pp. 75 – 82. The Smithsonian Institution, Washington D.C.

Whitaker, T.M. (1982) Primary production of phytoplankton off Signy Island, South Orkneys, the Antarctic. *Proc. Roy. Soc. Lond., B* 214, 169 – 89.

Wynn-Williams, D.D. (1983) Microflora-microfauna interactions in Antarctic moss peat decomposition processes. In *New Trends in Soil Biology (Proc. VIII Intl. Colloq. Soil Zool.)*, Eds P. Lebrun, H.M. André, A. de Medts, C. Grégoire-wibo and G. Wanthy, pp. 237 – 45. Dieu-brichart, Ottignies-Louvain-la-Neuve.

CHAPTER 8

# Conservation and Exploitation

BRYAN SAGE

Waveney House, Waveney Close, Wells-next-the-Sea, Norfolk NR23 1HU, U.K.

## CONTENTS

## 8.1. INTRODUCTION

The Antarctic is unique in several respects and its scientific importance is self-evident. Its living resources include more than 40 species of breeding seabirds (of which some, such as the Adélie penguin *Pygoscelis adeliae* and emperor penguin *Aptenodytes forsteri*, are endemic to the area), and six species each of seals and whales. Antarctic seas (the Southern Ocean) are locally and seasonally among the most biologically productive in the world, and their fishery resources are considerable. The Antarctic Continent itself has unique terrestrial and freshwater ecosystems which, by virtue of the extreme environmental conditions, are very susceptible to damage.

Largely as a result of the coming into force of the Antarctic Treaty in 1961, there have been more than two decades of peaceful scientific co-operation in the area, and today more than a dozen nations mount year-round research programmes in the area covered by the Treaty, that is to say south of latitude 60°S. In the legal and political context Antarctica is unique in that it is the only place in the world where there is no general agreement on either the extent or existence of national sovereignty claims. However, the time has come when some major political and diplomatic issues will have to be resolved, and just how this is done may have far reaching effects in terms of conservation prospects in the future. Given the absence of generally recognized sovereignty over most of the region and conflicting claims for jurisdiction, it is obvious that the administrative and legal problems are enormous and complex. Despite the Antarctic Treaty, no claimed jurisdiction by any nation has been surrendered.

## 8.2. THE EXPLOITATION BACKGROUND

When a new resource is discovered there is usually an immediate rush to exploit it, and the Antarctic has not been exempt from this general rule. It was the reports prepared by Captain James Cook following his circumnavigation of the Antarctic some 200 years ago that precipitated the first era of major resource exploitation in the area.

### 8.2.1. Marine resources

Attention was directed to the Antarctic fur seal *Artocephalus gazella* and then to the southern elephant seal *Mirounga leonina*. Within a short space of time both were severely reduced in numbers, and the former locally exterminated. The first sealing expedition arrived at South Georgia in 1778, and by 1791 there were at least 102 ships exploiting the stocks of fur and elephant seals. James Weddell (1825), himself a sealer, estimated that by 1822 no fewer that 1,200,000 skins had been taken, and that the fur seal was to all intents and purposes extinct on the island. Sealing commenced in the South Shetland Islands in 1819, the year that they were discovered, and in the peak season of 1820 – 1821 sealers from 47 American and British ships took about 250,000 fur seals; the following season brought the species to the verge of extinction in the islands. By the time this first extensive period of sealing came to an end about 1825, the Scotia Arc populations of fur seals were virtually exterminated. The present populations are the result of a slow build-up from the remnants that survived this period of slaughter.

Wherever fur seals were found, the Indian Ocean islands, Macquarie Island, the islands off New Zealand, or those around the tip of South America, the story was the same. Within a few seasons of discovery the seal population was exterminated or reduced to a tiny fraction of its original size.

Elephant seals were hunted at the same time as the fur seals but their exploitation (for oil) was less severe. Even so, many island populations were greatly reduced in numbers. Ship-based hunting of this species died out with the disappearance of the South Sea whalers. Elephant seal stocks recovered more rapidly than those of the fur seal, and by 1910 had increased on South Georgia to the point where a land-based industry was possible (Fig. 8.1). This was a controlled operation, in contrast to the earlier events, and adult males were exploited from 1910 – 1964, mainly at near the maximum sustainable yield for the population (Bonner, 1982). Recent changes in the structure of the South Georgia population as described by McCann (1980), can be attributed to the cessation of exploitation. The whaling company which cropped these seals ceased operations in 1964 due to a lack of whales to catch.

The second phase of exploitation was whaling which began in 1904 and was initially shore-based. By 1911 eight whaling station leases had been granted at South Georgia (Fig. 8.2) and nine shore-based factory ships and a shore station were operating in the South Shetland Islands. British governmental control of the industry came to an end in 1925 with the advent of the first stern-slip floating whale

Fig. 8.1. Sealers at South Georgia flense a bull elephant seal. (Photo: W.N. Bonner)

factory. This allowed all stages of the whaling operation to be carried out on the high seas beyond the reach of any restrictive legislation. No species or race of whale has been exterminated, but there is no clearly-proved evidence of any stock recovering despite the conservation measures that have been applied, so that now only the minke whale *Balaenoptera rostrata* may be taken (Bonner, 1981). The biomass of baleen whales has been reduced to about 16% of its initial value, or about 35% of their initial numbers (Laws, 1977). Currently only Japan and USSR catch whales in the Southern Ocean.

There was only one exploitative industry based on birds, and this concerned the boiling down for oil of king penguins *Aptenodytes patagonicus* and royal penguins *Eudyptes schlegeli* on Macquarie Island during a period of some 25 years from 1891. Approximately 150,000 birds were processed each year, but

Fig. 8.2. A fin whale on the flensing plan at Grytviken, where the Antarctic whaling industry started in 1904. (Photo: W.N. Bonner)

TABLE 8.1. Catches of Antarctic Krill, Metric Tonnes, as Reported for Three Fishing Areas in the FAO Yearbooks of Fisheries Statistics

| Season | 1973/74 | 1974/75 | 1975/76 | 1976/77 | 1977/78 | 1978/79 | 1979/80 | 1980/81 |
|---|---|---|---|---|---|---|---|---|
| Area |  |  |  |  |  |  |  |  |
| C 48 (Atlantic) | 21,700 | 38,900 | 521 | 106,794 | 89,923 | 266,534 | 356,978 | 285,117 |
| C 58 (India) | 643 | 1,081 | 2,266 | 12,383 | 52,844 | 65,431 | 120,047 | 160,069 |
| C 88 (Pacific) | — | — | — | 3,355 | 36 | 600 | — | 3,080 |
| Total | 22,343 | 39,981 | 2,787 | 119,177 | 142,767 | 332,565 | 477,025 | 448,266 |

purely fortuitously this had little effect on their populations since the numbers taken did not exceed the annual increment. According to Mitchell and Tinker (1980), proposals have been made to revive this trade.

The third and most recent phase of exploitation concerned fish, and the 1976 – 77 Antarctic fin fish catch reported to the Food and Agriculture Organization (FAO) was 272,000 tonnes, taken by the USSR, East Germany and Poland. There have been signs of over-fishing, and the area around South Georgia almost certainly suffered from this in the early 1970s; the USSR caught 411,000 tonnes of fish there in 1969 – 70. There are indications that these three communist countries are rapidly increasing their Antarctic fin-fish catch (Mitchell and Tinker, 1980).

Conventional fin-fish in the Southern Ocean are a less important component of the total biomass than in the northern seas, and the trawlers have therefore turned their attention to the super-abundant Antarctic krill, *Euphausia superba*, which has the potential to become the world's major fisheries product. Krill fishing on an experimental basis began in the early 1960s, primarily by the USSR, but it is only in the last five years that the emergence of a large-scale commercial fishery has become apparent.

The reported catch of krill in 1976 – 77 was just over 119,000 tonnes and it is believed that the 1980 – 81 catch was around 448,000 tonnes (Table 8.1). The major krill fishing nations at the moment are Japan, Poland, and the USSR, but five other nations have been involved in experimental fishing. The problem is that at the present time the scientific data on which krill catch quotas should be calculated simply does not exist. Estimates of the standing stock of krill in the Southern Ocean have varied greatly, the largest, of 1,350,000 tonnes, has been put forward by the USSR, significantly one of the nations most heavily committed to harvesting this resource. The size of the annual harvest of krill is clearly a vital issue from the conservation viewpoint. Krill is the principal link in the food web of the Southern Ocean (Chapter 7), being the primary food supply of five species of whale, three species of seal, 20 species of fish, three species of squid, and numerous species of seabird. Establishment of a major fishery at such a low level in the food chain would be unprecedented. If the precedent of whaling is anything to go by, it is quite possible that some nations might be willing to sacrifice the Antarctic ecosystem in order to maximize profits and returns.

### 8.2.2. Mineral resources

There has long been a considerable amount of speculation concerning the extent of the mineral resources of the Antarctic Continent, but even today any discussion about these potential resources has to be based to a large degree on analogy. Substantial deposits of coal (the largest field in the world) and iron have been found in Antarctica (Chapter 2b) and it is known that other minerals occur there, as for example copper, gold, molybdenum, nickel, titantium and uranium. So far no deposits of any mineral

Fig. 8.3. A Russian stern trawler lies alongside her mother ship off South Georgia. These vessels can catch and process either conventional fin fish or krill, depending on availability. (Photo: D.W.H. Walton)

have yet been found which could be economically exploited on the basis of current prices and technology, and it will be some considerable time before it becomes clear whether or not this situation will change. What is certain is that given the extreme environmental conditions and high costs involved, any proven deposits would have to be exceptionally rich to be economically exploitable.

Whilst there is certainly no immediate prospect of any development of those minerals mentioned above, the same may not necessarily be true in the case of oil. So far only traces of gaseous hydrocarbons have been discovered in Antarctica (Chapter 2b), but there is strong circumstantial evidence that there is oil there. Any reserves which may in due course be located will almost certainly be offshore and a map showing the location of potential areas was published in Zumberge (1979). The most likely areas are the Ross Sea basin, Weddell Sea basin, Amundsen Sea continental margin, and the Bellingshausen Sea continental margin. A recent United States government report, based entirely on analogy with sedimentary structures elsewhere in the world, stated that recoverable reserves in promising offshore areas could be in the order of tens of billions of barrels, and in February 1979 a Gulf Oil executive stated that the oil potential of the two most promising areas in the Ross and Weddell Seas was in the range of 50 billion barrels, but could be much more. Five years earlier a Soviet geologist had predicted that the oil resources of Antarctica would exceed those of Alaska (Mitchell and Tinker, 1979). All of these predictions are at present based on minimal geological data.

Several nations have made no secret of their interest in Antarctica's potential mineral resources. Soviet interest has been constantly reiterated, and Druzhnaya Station was established in the Weddell Sea sector in 1975 as the first base overtly dedicated to mineral prospecting. In 1979 Japan announced plans for a three year survey for oil and gas off the Antarctica Continent to begin in 1980. It is also significant that West Germany, Norway, the USA and USSR, have recently been carrying out (purportedly scientific) seismic surveys in the Ross and Weddell Sea areas.

There was little serious speculation about the utilization of Antarctic minerals until the 1970s for several reasons, one of which was that since the area was administered by international agreement for

TABLE 8.2. Antarctic Resources and their Possible Exploitation Dates

| Resource | Location | Potential Yield | When Exploitable |
|---|---|---|---|
| Krill | Antarctic seas, mainly South Atlantic, Indian Ocean and Ross Sea | ? 110 – 150 million tonnes per annum | Now |
| Fish | Continental shelf areas of Antarctic Continent and islands | Substantial, 1977 fin-fish catch 279,000 tonnes | Now |
| Seals | Antarctic Continent and islands | Considerable | Now, but no interest at present |
| Whales | Southern Ocean | Potential high, but stocks low and apparently not increasing | ? |
| Oil | Continental shelf areas of Antarctic Continent, particularly Amundsen, Bellingshausen, Ross and Weddell Seas | ? Tens of billions of barrels | Early 1990s onwards? |
| Coal | Transantarctic Mountains | Largest field in the world with approximately 11% of global reserves | ? |
| Iron | Prince Charles Mountains, Greater Antarctica | Substantial, but low grade | ? |
| Other minerals | Greater and Lesser Antarctica | Not known | ? |

scientific purposes, it was generally considered to lie outside the commercial sphere. However, attitudes are changing and there is no doubt that several nations are poised ready to move rapidly into oil exploration as soon as this becomes feasible. The legal and political complexities attached to the minerals issue are discussed in detail by Auburn (1982). Technology should not be regarded as a limiting factor. The oil industry now has considerable experience of drilling in Arctic waters, and the technology for deep sea drilling is advancing rapidly. Other technological developments already in mind for Antarctic conditions could probably be made operational within five years or so.

The situation with regard to mineral resources and development in the Antarctic is summarized in Table 8.2.

In July 1981 the Antarctic Treaty nations adopted a recommendation calling for the urgent negotiation of a régime on Antarctic mineral resources, and laid down certain principles on which such a régime should be based. The first meeting to discuss this recommendation was held in Wellington, New Zealand, in June 1982, and other meetings have been held since.

## 8.3. PRESENT ENVIRONMENTAL IMPACTS IN THE ANTARCTIC

Whilst Antarctica is not pristine, it has been substantially less affected by man's activities than any other continent. Many of the sub-Antarctic islands on and near the Antarctic Convergence have, however, been influenced to a greater or lesser degree by human activities.

Every year thousands of tonnes of cargo and millions of gallons of fuel are brought in to Antarctica (90% of it by ship), and very little of this is ever removed. Most remain as buildings, machinery, solid wastes, particulate matter, and gases. The main products of burning millions of gallons of oil and gasoline each year are heat, carbon dioxide, nitrogen oxides, sulphur oxides, hydrocarbons and particles. Wherever these products are emitted, they increase background concentrations in the air and become incorporated into the seas, ice, snow, and soils. However, because concentrations of combustion products are low the environmental effects are generally localized.

The general criteria for selection of a suitable site for a base are reasonably level ground, freedom from permanent snow and ice, and good access to the sea in summer. These same factors also make such sites likely to support permanent plant communities, or breeding colonies of sea birds, or both. It is therefore inevitable that such bases must have some impact, such as the disturbance of nesting birds. It has been shown, for example, by Muller-Schwarze and Berlander (1978), that close approach to penguins may cause them to desert the nest and cease breeding for that year. At the Cape Royds Adélie penguin *Pygoscelis adeliae* rookery the number of nests started to decline soon after the establishment of the research station at McMurdo Sound, and the situation became worse when tourist parties started visiting the area (Fig. 8.5). This downward trend was reversed when visits by scientists and tourists were restricted (Thomson, 1977).

A major result of human occupation of bases is the accumulation of waste and the contamination of the environment by it. A conspicuous feature of many Antarctic bases is their associated dumps of worn out or abandoned equipment and packaging (Fig. 8.4). Whilst aesthetically displeasing, such dumps have little biological effect. Quantities of waste have been disposed of by dumping offshore (often on the ice) and the seabed at McMurdo Sound, for example, is now littered with barrels, fuel lines, tractors, and beer cans. The Antarctic Peninsula has been the scene of a great deal of human activity over a long period. Man's impact on this area, both historical and recent has been extensively reviewed by Lipps (1978).

A code of conduct was drawn up by the Antarctic Treaty nations in 1975, and covers the disposal of wastes by expeditions and bases. It requires that for coastal bases, with certain exceptions, liquid waste should be macerated and flushed to sea, non-combustibles dumped in the sea, combustibles burnt, but batteries, all plastics, rubber and radio-isotopes should be removed from the Treaty area. However, it appears that little effort is made at many research stations to observe this code.

## 8.4. EXTERNAL INFLUENCES ON THE ANTARCTIC ENVIRONMENT

The first announcements (George and Frear, 1966; Sladen *et al.*, 1966) that DDT compounds had been detected in Antarctic wildlife received wide publicity. There has been continuing research into the extent to which the Antarctic ecosystem has been penetrated by pesticides, hydrocarbons and heavy metals, but many questions remain unanswered. So far as atmospheric pollution is concerned, since more than 90% of man's emissions into the atmosphere occur in the northern hemisphere, and as the equator is a barrier for tropospheric interhemispheric exchanges, the inference is that the influence of global pollution in Antarctica must be very weak. A recent paper by Boutron and Delmas (1980) does in fact confirm that man's impact on the heavy metals and sulphur compounds content of the atmosphere in the remote areas of the southern hemisphere is currently negligible.

The Antarctic marine ecosystem is remote from sources of pollutant contamination. Nevertheless, organochlorine pollutants of agricultural and industrial origin have been detected in wildlife in both the Antarctic and sub-Antarctic. The species concerned include the Weddell seal *Leptonychotes weddellii*, two species of penguin, three species of petrel, two species of skua, and the Antarctic cormorant

Fig. 8.4. Derelict store boxes outside the United States Antarctic Expedition's East Base on Stonington Island, Marguerite Bay. (Photo: W.N. Bonner)

*Phalacrocorax atriceps.* The highest residues recorded in Antarctic wildlife have been DDE and PCB concentrations of 46ppm and 185ppm respectively, in Wilson's petrels *Oceanites oceanicus* taken at Palmer Station (Risebrough, 1977).

The available evidence suggests that the scientific bases are not major sources of organochlorine contamination. Transfer processes that may be responsible for the occurrence of these pollutants in the Antarctic include the migration of seabirds and marine mammals from the north to feeding or breeding grounds in the Antarctic, transport by ocean currents, or transportation in the atmosphere. In the case of resident wildlife, the latter seems the most likely route, but the case is not yet proven.

There seems to be very little agreement concerning the origin of hydrocarbon pollution in the Antarctic and clearly much more research is needed. It was argued by Platt and Mackie (1979) that a world-wide dissemination of pyrolysis products was partly responsible for the presence of some of the hydrocarbons found in surface sediments in King Edward Cove, South Georgia. This hypothesis was refuted by Clarke and Law (1981) on the basis of studies on aliphatic and aromatic hydrocarbons in marine benthos collected in Antarctica. They attributed the presence of these hydrocarbons to a local

Fig. 8.5. Tourists from *Lindblad Explorer* visiting a king penguin rookery at the Bay of Isles, South Georgia. This well-controlled party is staying at the periphery of the rookery, minimizing disturbance to the penguins. (Photo: W.N. Bonner)

source, and concluded that the Antarctic generally remains unpolluted by the pyrolysis products of fossil fuels. The discussion has been continued by Platt and Mackie (1981) and the basic disagreement remains. However, a local source was presumably responsible for the oiled Adélie penguins *Pygoscelis adeliae* reported from Cape Bird by Wilson (1979).

## 8.5. ENVIRONMENTAL EFFECTS OF FUTURE DEVELOPMENTS

Terrestrial habitat destruction or modification in the Antarctic (excluding the effects of introduced species) has so far been confined primarily to the vicinity of scientific bases, and to some of the earlier whaling stations. This type of impact which may well be highly significant in the local context, has very little effect on the ecosystem as a whole. However, future large-scale exploitation of natural resources would certainly create an entirely different set of circumstances. One point that can be made is that in general terms, the marine ecosystem of the Southern Ocean would be far more resistant to impact than the terrestrial or freshwater systems of the Antarctic Continent and its associated islands. Environmental conditions in the great expanse of the Southern Ocean are less extreme, more uniform and more continuous than on land, all of which confers a fairly high degree of stability. The ocean also has great buffering capacity, and most environmental impacts would probably be localized, at least initially.

From what has been said earlier it is apparent that there is no immediate threat of mineral exploitation on land, and the exploration activities which would precede exploitation would have minimal environmental impact. Nevertheless, it must be borne in mind that should circumstances in the future change to the extent that hard rock mineral exploitation on land becomes a reality, then some of the resulting impacts would be severe, even though they might be localized.

Assuming that an economically viable, highly concentrated ore were found and that it would not require processing before being taken out of the Antarctic, the associated facilities required to exploit it would be substantial. The necessary infrastructure would, for example, include accommodation for workers, probably runways for large aircraft, waste disposal facilities, storage facilities, power stations, water supply, tailings dumps, and a shipping terminal with access to the sea in summer. The latter would have to be sited on the very limited ice-free coastal areas so often already occupied by scientific bases, and/or seabird and seal colonies; the possible adverse impacts are obvious. The key impacts of on-land mineral development have been clearly stated by Holdgate and Tinker (1979), and they include the physical alteration of the land surface over large areas; the deposit of rock waste and tailings on land; the spread of dust over adjacent ice, altering its albedo (reflectivity) and perhaps causing melting; the washing of pollutants such as fine tailings and leachites into freshwater lakes and coastal seas, with considerable potential ecological effects; some air pollution from energy generation; the accumulation of wastes, including sewage; and pressures from recreational activities and transport of personnel.

As a result of increasing economic pressures related to the world energy situation, it is the possible crude oil resources of Antarctica that are likely to attract commercial interest in the relatively near future and, as stated earlier, the necessary technology for offshore oil exploitation could be developed fairly quickly. Despite experience gained by the oil industry in Arctic waters, the fact remains that no attempts have so far been made to exploit offshore oil resources in an environment where heavy sea-ice and huge icebergs (extending as much as 250 metres below the surface) are found in combination. It can reasonably be assumed that the preliminary stages of oil exploration that do not involve drilling would have minimal environmental impact. It can also be assumed that those oil companies interested in Antarctic oil resources already have a good idea of the general location of the most promising areas, and these were mentioned earlier.

The unique problems associated with oil production and transportation in the Antarctic are considered in some detail by Holdgate and Tinker (1979) and Zumberge (1979), the former with particular reference to the Ross Sea. In Antarctica the sedimentary deposits in which oil might be found are primarily confined to the narrow continental shelves, and it is these same areas that have high biological productivity. In the event that oil exploration and production required extensive land-based facilities, then the environmental impacts would be similar to those already discussed above. It is possible that the operations might be largely self-contained offshore, and any resulting impacts would be on the marine environment. There are three main stages during which accidents could release crude oil into the sea or on to the ice, and these are exploratory drilling, production (from submarine well heads), and transportation by surface tankers or (much further in the future), by submarine tankers.

Predicting the precise effects of oil spills in Antarctic waters is very much in the realm of speculation due to a lack of data, and this includes not knowing the chemical composition of any crude oil that might be found. The many questions to which answers would need to be found include:

What would be the effects of oil spilt *on* the pack-ice or trapped beneath it?

What would be the rate of natural decomposition of Antarctic crude?

What would be the likelihood of spilt oil damaging scientifically important coastal habitats?

Would the oil have significant effects on populations of vertebrates such as seals and seabirds, particularly penguins?

However, in view of the central role of krill in the Antarctic marine ecosystem, and its future economic importance, the most vital question of all is: what impacts would crude oil have on the krill itself, and as a result on populations higher up the food chain? At present it is impossible to answer any of these questions. In addition to the oil itself, there are other potential environmental impacts arising for example from the flaring of natural gas, disposal of wastes from drilling platforms and other installations, the effects of drilling muds, and so on.

It would seem that there are two certainties. The first is that at least in the foreseeable future, only very large oilfields would be worth developing in the Antarctic, and that to do so would involve major industrial activity at sea and possibly also on land. Secondly, if oil is produced in the Antarctic then accidents will be inevitable as a result of mechanical failure or human error.

## 8.6. THE LEGAL BACKGROUND

The legal framework within which the problems of conservation in the Antarctic have to be considered consists primarily of the Antarctic Treaty, and various other subsequent pieces of legislation.

### 8.6.1. The Antarctic Treaty

This Treaty was signed in Washington D.C. (USA) on 1st December 1959, by 12 nations (the consultative parties) — Argentina, Australia, Belgium, Chile, France, Japan, New Zealand, Norway, South Africa, United Kingdom, USA and USSR, and it came into force in 1961. Poland became the 13th signatory in 1977 and the Federal Republic of Germany the 14th in 1981. Since then Brazil and India have also become consultative parties. There are also 15 acceding states, so the Treaty has 30 members in all. The Treaty, the full text of which is published by each of the member countries, contains 14 Articles.

Article 1 states that the Antarctic shall be used for peaceful purposes only. No military bases are to be established and no military manoeuvres or weapon testing are to be carried out. The second Article establishes freedom of scientific investigation, and co-operation towards that end. Article 4 states that nothing contained in the Treaty may be interpreted as a renunciation, denial or support of any claim to territorial sovereignty so long as the Treaty is in force. This Article is important because it effectively freezes the claims issue. The fifth Article states that Antarctica may not be used for nuclear explosions or disposal of nuclear waste, but does not prevent the peaceful use of nuclear energy in the Antarctic. The sixth Article provides that the Treaty shall apply to the area south of 60° South latitude, including all ice shelves. Article 9 requires the contracting parties to settle their disputes by peaceful means. The twelfth Article provides for amendments to the Treaty. Also, after 30 years have elapsed since the Treaty entered into force (i.e. on 23rd June 1991), the Treaty may be reviewed if any of the contracting parties so desires. It is important to note that on this date the Treaty does not 'end'; it is merely available for review.

For some 20 years the Antarctic Treaty has provided the framework for peaceful scientific co-operation in Antarctica, and environmental protection has been a major focus of its activities. Nevertheless, signs of strain have recently emerged. Although the Treaty has proved to be something of a model in terms of international co-operation, it is even more significant for what it omits. The question of territorial claims, for example, was simply frozen and no attempt made to solve the issue, so the disputes (particularly between Argentina, Chile and Britain) remain. The Treaty makes no mention of Antarctic resources since it was based on the assumption (reasonable enough in 1960) that the area did not contain resources likely to be exploited in the near future, but recent developments have clearly shown that this is not the case. The area to which the Treaty applies is all land south of the 60th parallel, including all ice shelves, but it does not affect the rights of any state under international law with regard to the high seas within that area. There are two particular factors which should be borne in mind in connection with this exclusion of the high seas. The first is that the high seas boundary has never been clearly defined and, secondly, the recent widespread introduction of 200 mile exclusive economic zones (EEZs) off coastal territory is significant in that seven of the Treaty nations have territorial claims in the Antarctic. The legal basis for the Antarctic Treaty's jurisdiction over Southern Ocean resources is questionable.

Officially, non-claimant signatories to the Treaty recognize no ownership of land, sea or resources, so in theory there is freedom of access to the resources for all nations. In practice, there are indications that the consultative parties of the Treaty regard themselves as having special rights to the resources.

## 8.6.2. The 'Agreed Measures'

The Antarctic Treaty specifically refers to the need for the preservation and conservation of living resources, and in 1964 the Treaty nations adopted the 'Agreed Measures for the Conservation of Antarctic Fauna and Flora'. In the preamble to these measures the Antarctic is recognized as a Special Conservation area. There are a total of 14 Articles in the Agreed Measures and these are quoted in full by Auburn (1982). The four most important of these are Article VI which deals with the protection of native fauna and includes special protection for species listed in an Annex. These currently include all species of fur seals of the genus *Arctocephalus*, and the Ross Seal *Ommatophoca rossii*. Article VII requires each participating government to take appropriate measures to minimize harmful interference with the normal living conditions of native mammals and birds. Article VIII allows for the declaration of areas of outstanding scientific interest as Specially Protected Areas (SPAs). Article IX regulates the introduction of non-indigenous species.

The Agreed Measures have been reviewed periodically at meetings of the Treaty nations and some reinterpretation has taken place, particularly with regard to SPAs. At the seventh Consultative Meeting in 1972 it was decided to allow marine SPAs to be considered, and provision was made for the designation of Sites of Special Scientific Interest (SSSIs), which would be sites having a lesser degree of protection than SPAs, thus facilitating scientific research. The current SPAs and SSSIs are listed in Tables 8.3 and 8.4, and an annotated guide to conservation areas is in preparation by SCAR.

There have been various problems associated with the practical application of the Agreed Measures, particularly in the marine context, and these are reviewed in detail by Auburn (1982). Despite whatever drawbacks they may have, the Agreed Measures constitute the first internationally agreed system for monitoring man's direct impact on the Antarctic ecosystem.

One growth area which has been taken increasingly seriously by the Consultative Parties is tourism. This began over 25 years ago and with the employment of ever larger cruise ships and aircraft may now be approaching a phase of rapid expansion. Recent studies of visitor patterns and numbers (Codling, 1982; Reiche, 1980) have shown that over most of this period the numbers of visitors landing annually from both ships and aircraft was around 840. About three quarters of their visits were concentrated on nine places, all of which were occupied by scientific stations.

Anxiety over tourism arises for a number of reasons. Firstly, the landing of large parties of tourists who cannot be properly supervised is likely to have important and lasting effects on the local flora and fauna. Secondly, the visits to scientific stations, although often welcomed by the scientists, cause disruption of work and social stresses. And thirdly, accidents associated with the cruise ships or aircraft often require the use of scarce personnel, time or fuel from one or more of the Treaty countries. Although there have been no fatalities associated with the cruise ships yet, there have been several groundings and cases of passengers left stranded. There has also been the serious air crash of the Air New Zealand DC10 with the death of 257 people. Rescue and enquiry work for this accident disrupted all the scientific programmes at Scott Base and McMurdo Sound for most of the 1979 – 80 season.

Treaty discussions, begun in 1966 and continued at nearly all subsequent meetings, have reached the point at which a category called 'Areas of Special Tourist Interest' has been agreed on (Recommendation VIII – 9 Annex B) but so far there has been no decision on the designation of these areas. SCAR has undertaken the production of a brochure 'A visitor's guide to the Antarctic and its environment' to aid tourists in understanding the rules they should abide by but, despite policy statements from countries such as New Zealand and Argentina on how tour operators should organize the visits, there is little evidence of these being enforced. The registration of two of the major cruise ships, *World Discoverer* and *Linblad Explorer*, is in non-Treaty countries so that any suggestions of legal enforcement must be put aside.

TABLE 8.3. Specially Protected Areas (SPAs) Recognized Under the Agreed Measures of the Antarctic Treaty

| | | |
|---|---|---|
| 1. | Taylor Rookery, MacRobertson Land<br>67°26′S 60°50′E | (Emperor penguin rookery) |
| 2. | Rookery Islands, Holm Bay (Mawson)<br>67°37′S 62°33′E | (Bird colonies) |
| 3. | Ardery Island & Odbert Island, Budd Coast<br>66°22′S 110°28′E, 66°22′S 110°33′E | (Bird colonies and habitat) |
| 4. | Sabrina Island, Balleny Islands<br>66°54′S 163°20′E | (Representative area) |
| 5. | Beaufort Island, Ross Sea<br>76°58′S 167°03′E | (Bird colonies, ecosystem) |
| 6. | Cape Crozier (Now SSSI #4) | |
| 7. | Cape Hallet, Victoria Land<br>72°18′S 170°19′E | (Vegetation) |
| 8. | Dion Islands, Marguerite Bay<br>67°52′S 68°43′W | (Emperor penguins) |
| 9. | Green Island, Berthelot Islands<br>65°19′S 64°10′W | (Vegetation) |
| 10. | Byers Peninsula (Now SSSI #6) | |
| 11. | Cape Shirreff, Livingstone Island<br>62°28′S 60°48′W | (Diversity and seals) |
| 12. | Fildes Peninsula (Now SSSI #5) | |
| 13. | Moe Island, South Orkney Islands<br>60°45′S 45°41′W | (Representative area) |
| 14. | Lynch Island, South Orkney Islands<br>60°40′S 45°38′W | (Antarctic hair grass) |
| 15. | Southern Powell and adjacent Islands,<br>South Orkney Islands<br>60°45′S 45°02′W | (Representative area<br>and fur seals) |
| 16. | Coppermine Peninsula, Robert Islands<br>62°23′S 59°42′W | (Diversity and birds) |
| 17. | Litchfield Island, Arthur Harbour<br>66°16′S 64°06′W | (Diversity and birds) |

It is possible that the present situation is less serious than generally supposed. An analysis of a recent *World Discoverer* cruise to the Antarctic Peninsula showed that only 18 hours was spent on shore in the Antarctic in a 23 day cruise. The size of the party appears to be critical for adequate control purposes and 130 has been suggested as the maximum. This is unfortunate in view of the plans to build 300+ berth cruise ships. An important comment made by Rosamunde Codling was that the passengers were in many cases much better behaved and more conservation minded than many of the scientific station

TABLE 8.4. Sites of Special Scientific Interest Designated by Recommendations of Antarctic Treaty Consultation Meeting

| | |
|---|---|
| 1. | Cape Royds, Ross Island. |
| 2. | Arrival Heights, Hut Point Peninsula, Ross Island. |
| 3. | Barwick Valley, Victoria Land. |
| 4. | Cape Crozier, Ross Island. |
| 5. | Fildes Peninsula, King George Island, South Shetland Islands. |
| 6. | Byers Peninsula, Livingston Island, South Shetland Islands. |
| 7. | Haswell Island. 61°31′S, 90°00′E. |
| 8. | Western Shores of Admiralty Bay, King George Island, South Shetland Islands. |

personnel encountered. This has also been shown very clearly in the operations undertaken in recent years by several Governments in extending and rebuilding their Antarctic bases. If the scientists cannot agree to uphold the necessary standards of conservation themselves there seems little chance of persuading the public of the necessity for them. Agreement is needed between the Consultative Parties both on a code of practice for conservation and tourism and on the will to enforce it.

## 8.6.3. The Convention for the Conservation of Antarctic Seals

The Antarctic Treaty failed to address itself in any adequate way to the problem of the marine resources of the Southern Ocean, and when the prospect of a resumption of sealing (which did not in fact materialize) in the Antarctic was raised the need for further legislation became apparent. The next international achievement (after the Agreed Measures) was the signing in London in 1972 of the Convention for the Conservation of Antarctic Seals, which is unique in that it is the first international agreement to make detailed provisions for protecting species *before* an industry to exploit them has actually developed. The convention, which came into force on 11th March, 1978, has been ratified by ten nations (Argentina, Belgium, Chile, France, Japan, Norway, South Africa, UK, USA and USSR). Poland has acceded to the Convention.

The convention was signed by the 12 original signatories of the Antarctic Treaty, but is open for accession by any country 'which may be invited to accede with the consent of all the contracting parties'. It applies to the sea areas south of latitude 60°S, though provision is made for reporting catches in the area of floating sea-ice north of 60°S. The convention covers all six species of seal that occur in the specified area, and catch limits are set for the crabeater *Lobodon carcinophagus*, leopard *Hydrurga leptonyx* and Weddell *Leptonychotes weddellii* seals. The taking of Ross seals *Omatophoca rossii*, elephant seals *Mirounga* species, and fur seals *Arctocephalus* species is completely banned, and the adult stock of Weddell seals is protected during the period when it is concentrated for breeding on fast-ice. The Southern Ocean is divided into six zones and in the event of sealing taking place one would be closed each year in rotation, and there are three sealing reserves — around the South Orkney Islands; the southwestern Ross Sea; and at Edisto Inlet. Scientific advice will be provided as required by the Scientific Committee on Antarctic Research (SCAR). Should a sealing industry develop again at any time in the future, then this convention provides the means for its regulation and monitoring.

## 8.6.4. The Convention for the Conservation of Antarctic Marine Living Resources

The most immediate targets for exploitation in the living resources of the Antarctic are fish and krill, and large-scale exploitation of the latter could very well have serious consequences for the Antarctic marine ecosystem as a whole. If krill biomass is markedly reduced then this could lead to the depletion of the populations of dependent species such as seals, fish, and seabirds, and (perhaps more importantly) could cause a slowing in the rate of recovery of whale stocks. Factors tending to encourage the development of a major krill fishing industry include the increasing pressures on conventional fisheries worldwide, and the establishment of 200 mile Exclusive Economic Zones by coastal states, which has effectively excluded many nations from traditional fishing grounds.

The conservation of the Southern Ocean represents a major challenge, and the Antarctic Treaty nations have been conscious of the problem of its living resources for some time. In 1975 they noted 'the need to promote and achieve within the framework of the Antarctic Treaty the objectives of protection, scientific study and rational use of Southern Ocean marine living resources'. By 1977 this subject had become the major issue on the Antarctic Treaty agenda and a detailed recommendation was agreed upon in September of that year.

The outlines of the living resources convention agreed in 1977 were further discussed in Canberra in 1978 when a revised draft was agreed. After further protracted meetings in Buenos Aires, Washington and Berne, the Convention on the Conservation of Antarctic Marine Living Resources was signed in Canberra in May 1980, and it finally came into force on 7th April 1982. The Convention provides the essential legal framework for management of the living resources of the Southern Ocean. It adopts an ecosystem approach and, unlike the Antarctic Treaty, embraces the whole of the Southern Ocean north to the biological boundary of the Antarctic Convergence. The objective of the Convention is conservation which, in this case, specifically includes the concept of rational use of the resources. The Convention is open to accession by any state interested in researching or harvesting in relation to marine living resources in the Southern Ocean.

The primary objectives of the Convention are embodied in three conservation principles. The first of these concerns the prevention of decrease in the size of any harvested population to levels below those which ensure its stable recruitment. For this purpose its size should not be allowed to fall below a level close to that which ensures the greatest net annual increment. The second deals with the ecological relationships between harvested, dependent and related populations of Antarctic marine living resources. These are to be maintained and depleted populations should be managed so as to be restored to the levels defined in the first principle. The third principle sets out to prevent change or minimize the risk of changes in the marine ecosystem which are not potentially reversible over two or three decades, taking into account the state of available knowledge of the direct and indirect impact of harvesting, the effects of the introduction of alien species, the effects of associated activities on the marine ecosystem and of the effects of environmental changes, with the aim of making possible the sustained conservation of Antarctic marine living resources.

In order to achieve its aims the Convention provided for the setting up of an international body to facilitate research, collect and analyse data, and adopt various conservation measures. The Commission for the Antarctic Marine Living Resources Convention (CAMLR) was duly established and held its first meeting in Hobart in May – June 1982.

Fundamental to the conservation principles is the availability of adequate scientific knowledge for the stated purposes, and the Convention itself in its preamble states that it is essential to increase knowledge of the Antarctic marine ecosystem so as to be able to base harvesting decisions on sound scientific information. Such information was not available, but fortunately this need had been anticipated by biologists working in the Antarctic. A proposal for an international co-operative research programme was put forward by SCAR (the Scientific Committee on Antarctic Research) and SCOR (the Scientific Committee on Oceanic Research) and was adopted in 1976. This programme was given the name BIOMASS (Biological Investigations of Marine Antarctic Systems and Stocks). BIOMASS has involved most of the nations who research in the Antarctic and successfully staged the largest multi-ship co-ordinated biological experiment, FIBEX, in 1980 – 81. BIOMASS plans to culminate in a major synthesis of data at a symposium which is scheduled to take place in about 1986.

The full text of the Articles of the Convention, and a discussion of its legal complexities, will be found in Auburn (1982). The whole problem of the management of the Southern Ocean, including historical, biological and legal aspects, is dealt with in detail by Mitchell and Sandbrook (1980).

## 8.7. DISCUSSION

The whole question of conservation in the Antarctic is likely to be resolved one way or the other within the next few years. The Antarctic Treaty worked extremely well as long as Antarctica apparently contained nothing of commercial value, but now that its possible commercial potential is realized the signs of strain are all too evident. It may be that the Treaty will not even survive in its present form until 1991 when it comes up for review. But even if it does, it seems inevitable that it will then be the subject of extensive review. The greatest danger lies in the fact that the issue no longer lies purely within the realm of science, but that large industrial corporations will be prompting individual governments to act in their own (and industry's) best interests. It would seem that there are two alternatives insofar as the Antarctic Treaty is concerned. It can either evolve into a régime with much wider international support and participation, or fall victim to the claims and counter-claims of rival nationalisms.

It is necessary in this discussion to make a distinction between the Antarctic Continent and its associated islands, and the Southern Ocean. The living resources of the latter are going to be exploited, and indeed are already being exploited by those nations economically and technologically equipped to do so. The negotiation of the Convention on Antarctic Marine Living Resources is an encouraging step and if it receives international support, then effective conservation may well become a reality. However, like all international agreements, this one is a compromise and not entirely satisfactory to all parties, and it has its deficiencies. For example, for management purposes, the marine areas south of the Antarctic Convergence should be treated as an entity. Nevertheless, Article IV,2 states (in part):

> Nothing in this Convention and no acts or activities taking place while the present Convention is in force shall:
> (a) constitute a basis for asserting, supporting or denying a claim to territorial sovereignty in the Antarctic Treaty area or create any rights of sovereignty in the Antarctic Treaty area;
> (b) be interpreted as a renunciation or diminution by any Contracting Party of, or as prejudicing, any right or claim or basis of claim to exercise coastal state jurisdiction under international law within the area to which this Convention applies:
> (c) be interpreted as prejudicing the position of any Contracting Party as regards its recognition or non-recognition of any such right, claim or basis of claim.

The text therefore establishes a fairly firm foundation for the exemption of extensive areas south of the Convergence from any conservation regulations that may be proposed by CCAMLR. Additionally, it also reveals a certain ambiguity over the precise geographical scope of CCAMLR itself, thereby sowing the seeds of future conflict.

The International Union for the Conservation of Nature and Natural Resources (IUCN), under the auspices of the United Nations Environment Programme (UNEP) and the World Wildlife Fund (WWF), in 1981 launched its *World Conservation Strategy* simultaneously in 32 countries. The question of the resources of the Southern Ocean was discussed. The Strategy considers that the Antarctic Treaty powers and nations fishing or intending to fish in the Southern Ocean should exercize extreme restraint on catch levels until the understanding of this uniquely productive ecosystem improves. All harvesting should be on an experimental basis as part of a scientific research programme to improve knowledge of krill and the Southern Ocean as a whole. There should be base-line areas set aside and given complete protection where no krill or other living or non-living resources may be taken, so that impacts outside can be monitored and evaluated correctly.

Whether or not extreme restraint will be exercized is questionable. The major pressure for good and effective conservation has come from the USA, while opposition has been from Japan, Poland and the USSR, the three nations currently most involved in extensive krill fishing. It is worth noting that at the first CCAMLR meeting in Hobart in 1982 the USSR adopted an intractable position regarding the reporting of fin fish and krill catches. The inference is that Russia is now focusing its attention on the living resources of the Southern Ocean as an alternative to her exclusion from 200 mile fishing zones, and a succession of poor domestic wheat harvests. While the Convention on the Conservation of Antarctic Marine Living Resources is a major achievement, whether or not it is going to work effectively is by no means certain and only time will tell.

Whatever measure of agreement may have been reached in respect of the marine resources, no agreement is at present in sight for oil or on-land mineral resources. The minerals question was taken up in earnest by the Treaty countries in 1982 and looks like being a dominant item in future meetings. There are important divisions between the Treaty nations on if and how the exploitation of oil and land-based minerals should take place and these are to a considerable extent bound up with sovereignty implications, which was the reason resources were excluded from the provisions of the original Treaty. Despite the suggestion in some quarters that the 15 signatory nations are attempting to exclude all outside initiatives and carve up Antarctic resources for themselves, this is not a practical possibility in legal terms. Whatever arrangements are agreed within the framework of the Treaty, the fact remains that the real problems of exploiting Antarctica's resources are practical rather than legal. Amongst the Treaty countries are not only the world's richest nations, able to support the massive investment necessary, but virtually all the expertise necessary for organizing the exploitation efficiently within such a restrictive environment.

Fig. 8.6. A gentoo penguin broods its chicks at Port Lockroy, Wiencke Island. The huge pile of whale bones in the background bears witness to the activities of the whalers who used this harbour in the early years of this century. (Photo: W.N. Bonner)

Indeed, it might prove difficult to enforce any international agreement arising from the Treaty if non-Treaty countries decide to mount their own exploitation exercizes. If, however, the present negotiations result in the acceptance by the Treaty countries of a global perspective, including elements of the 'common heritage of mankind' proposals put forward by Third World countries it may be possible to gain general acceptance for a Minerals Convention. It is significant that the proposals for the designation of Antarctic resources as a 'global common', made with some force at the United Nations by countries such as Sri Lanka, have nothing to do with conservation and are designed only to ensure participation in any profits that might result from exploitation. It is perhaps pertinent that even IUCN in its World Conservation strategy seems to accept that there will be exploitation since it says: '...the feasibility of oil exploration and exploitation in particular should be approached with the utmost caution'. This philosophy of accepting that exploitation is inevitable has not been widely endorsed by other organizations concerned with the conservation of environment and wildlife.

Failure to reach agreement on the issue of oil and land-based minerals could be disastrous. At best it could threaten the continued existence of the Antarctic Treaty, and at worst could lead to economic or military confrontation. An example of the attitudes likely to prevail in the event of non-agreement came as long ago as 1975 when it was suggested that in the absence of a shared understanding, those countries which did not recognize claims to sovereignty might have to assert the right to commence mineral resource activities at their will. By 1979, USA determination to secure some access to Antarctic minerals, especially oil, had strengthened still further, and in April 1979, 20 environmental and conservation organizations in the USA wrote to President Carter expressing their serious concern with the emerging government position on Antarctic minerals.

There is no clear and obvious answer that is likely to meet all the political, economic and scientific objectives. One possibility, put forward strongly by several major conservation bodies, is not to exploit the oil and mineral resources at all, (assuming that commercially viable reserves are in fact found). This is a strong case on both scientific and aesthetic grounds but it ignores the political forces ranged against it. The conservation argument that any exploitation could not take place without causing major damage and disruption to the terrestrial and marine ecosystems is not supportable on those terms. True, there is likely to be fairly severe localized damage, especially to some of the terrestrial communities, but it is by no means clear that this would greatly exceed the damage currently being caused by the establishment and use of an increasing number of scientific stations on most accessible areas of the Continent. Damage to the marine ecosystem is much less likely to be significant since the area under consideration is over 10% of the world's oceans with an enormous buffering capacity. Equally it is not certain at the moment when the mineral resources are likely to become economically useful and if this does not occur for another twenty or thirty years technological advances may make extraction a much more acceptable process than is currently envisaged.

It would be exciting and unusually altruistic if international agreement was reached to preserve Antarctica as an area dedicated to scientific research. The Second World Conference on National Parks, held at Grand Teton in the USA in 1972, unanimously recommended that the Antarctic Continent and its surrounding seas be established as the first World Park under the auspices of the United Nations. New Zealand took this proposal to the 1975 Antarctic Treaty meeting. These biennial meetings are held behind closed doors, but since nothing further happened it is safe to assume that the proposal met with considerable resistance. In 1980 non-governmental conservation groups in Australia called for Antarctica to be made a World Preserve. There is no doubt that Antarctica falls within the first and third categories of Natural Heritage in the Convention for the Protection of the World Cultural and Natural Heritage (the World Heritage List), adopted in 1972 by UNESCO.

The problems associated with Antarctica are potentially explosive and the continent too important to the world, and too unique, to be carved up for commercial gain. However, given the examples of past history, the chances of Antarctica becoming the first World Park are probably remote. It will doubtless be argued that conservation must be combined with measures to meet short-term economic needs, particularly those of the developing countries. If this is the framework adopted conservationists and scientists should strive to reduce the impact of exploitation on one of the world's most exciting wilderness areas.

# REFERENCES

Auburn, F.M. (1982) *Antarctic Law and Politics*. C. Hurst & Co., London.

Bonner, W.N. (1981) The krill problem in Antarctica. *Oryx* 16, 31 – 7.

Bonner, W.N. (1982) *Seals and Man: a Study of Interactions*. University of Washington Press, Seattle.

Boutron, C. and Delmas, R. (1980) Historical record of global atmospheric pollution revealed in polar ice sheets. *Ambio* 9, 210 – 15.

Clarke, A. and Law, R.J. (1981) Aliphatic and aromatic hydrocarbons in benthic invertebrates from two sites in Antarctica. *Mar. Pollut. Bull.* 12, 10 – 14.

Codling, R.J. (1982) Seaborne tourism in the Antarctic: an evaluation. *Polar Rec.* 21, 3-10.

George, J.L. and Frear, D.E.H. (1966) Pesticides in the Antarctic. *J. appl. Ecol.* 3 (Supplement), 155 – 67.

Holdgate, M.W. and Tinker, J. (1979) *Oil and Other Minerals in the Antarctic*. Scientific Committee on Antarctic Research, Cambridge.

Laws, R.M. (1977) Seals and whales of the Southern Ocean. *Phil. Trans. R. Soc. Ser. B* 279, 81 – 96.

Lipps, J.H. (1978) Man's impact along the Antarctic Peninsula. In *Environmental Impact in Antarctica*, Eds B.C. Parker and M.C. Holliman, pp. 333 – 71. Virginia Polytechnic Institute and State University, Blacksburg.

McCann, T.S. (1980) Population structure and social organization of southern elephant seals, *Mirounga leonina* (L.). *Biol. J. Linn. Soc.* 14, 133 – 50.

Mitchell, B. and Sandbrook, R. (1980) *The Management of the Southern Ocean*. International Institute for Environment and Development, London.

Mitchell, B. and Tinker, J. (1980) *Antarctica and its Resources*. International Institute for Environment and Development, London.

Muller-Schwarze, D. and Belanger, P. (1978) Man's impact on Antarctic birds. In *Environmental Impact in Antarctica*, Eds B.C. Parker and M.C. Holliman, pp. 373 – 83. Virginia Polytechnic Institute and State University, Blacksburg.

Platt, H.M. and Mackie, P.R. (1979) Analysis of aliphatic and aromatic hydrocarbons in Antarctic marine sediment layers. *Nature, Lond.* 250, 567 – 78.

Platt, H.M. and Mackie, P.R. (1981) Sources of Antarctic Hydrocarbons. *Mar. Poll. Bull.* 12, 407 – 9.

Reich, R.J. (1980) The development of Antarctic tourism. *Polar Rec.* 20, 203 – 14.

Risebrough, R.W. (1977) Transfer of organochlorine pollutants to Antarctica. In *Adaptations Within Antarctic Ecosystems*, Ed. G.A. Llano, pp. 1203 – 10, Smithsonian Institution, Washington, D.C.

Sladen, W.J.L., Menzie, C.M. and Reichle, W.L. (1966) DDT residues in Adélie penguins and a crabeater seal from Antarctica. *Nature, Lond.* 210, 670 – 3.

Thomson, R.B. (1977) Effects of human disturbance on an Adélie penguin rookery and measures of control. In *Adaptations Within Antarctic Ecosystems*, Ed. G.A. Llano, pp. 1177 – 80. Smithsonian Institution, Washington D.C.

Weddell, J. (1825) *A Voyage Towards the South Pole, Performed in the Years 1822 – 24*. Longman, Hurst, Reese, Orme, Brown and Green, London.

Wilson, G.J. (1979) *Possible Environmental Effects of Mineral Exploration and Exploitation in Antarctica*. Scientific Committee on Antarctic Research, Cambridge.

Zumberge, J.H. (1979) Mineral resources and geopolitics in Antarctica. *Am. Sci.* 67, 68 – 77.

# Index

# KEY ENVIRONMENTS

*Other Titles in the Series*

---